III

IV

V

VI

I Haus des Rundfunks, plan for the competition scheme. Source: TU Berlin Architekturmuseum, Inv. Nr. 4577.

II Funkhaus Berlin-Nalepastrasse, view of Block A. Source: Bundesarchiv. Bild 183-54922-0004.

III Berlin Staatsbibliothek (State Library) designed by Hans Scharoun as part of the Kulturforum. Photograph by the author.

IV An angel flies over the Funkturm in West Berlin and inside the Berlin State Library. Source: *Der Himmel über Berlin / Wings of Desire*, directed by Wim Wenders. West Germany: Road Movies Filmproduktion; Argos Films; Westdeutscher Rundfunk, 1987.

V Funkhaus Berlin-Nalepastrasse, view of the elevated circulation connecting Block A with Block B. Photographs by the author and Pablo Thiermann.

VI Interior view of Saal I inside Block B of the Funkhaus Berlin-Nalepastrasse showing a recording session with the Berlin Radio Symphony Orchestra conducted by Rolf Kleinert. In this room, there is no space for a live audience. Source: Bundesarchiv. DM 3/27362.

INTRODUCTION

III → P. 4

IV → P. 5

In the opening sequence of the Wim Wenders film *Der Himmel über Berlin* (*Wings of Desire*), angels appear to fly freely between the two sides of a divided Berlin, soaring over buildings, entering them through windows, passing through walls and doors.[1] When the angel played by Bruno Ganz flies into Hans Scharoun's Staatsbibliothek (State Library) we see, through his eyes, a serene space that is frictionless, boundary-free, and continuous.[2] (Fig. III, IV) However, when the same angel encounters the Berlin radio tower his ears are assailed by a cacophony of voices—of overlapping radio signals. Although the angels could easily transgress the formidable Berlin Wall, they were unable to circumvent the radio waves that inundated the ethereal space of the city, as if the space defined by those electromagnetic signals as they crisscrossed the Wall was even more fragmented, divided, and to a certain extent more differentiated than the urban fabric of the city below. The angels perceived the presence of certain objects that made the seemingly stable notion of architecture tremble. By virtue of their size, form, materiality, or technical complexity, or simply because of their unique mode of existence, certain objects push the conceptual plasticity of architecture to its limit.

I

On August 13, 1961, the German Democratic Republic (GDR) began construction of the Berlin Wall. Built within East German territory, but encircling the Western sectors, it turned West Berlin into an island situated within the Eastern zone. Both symbolically and materially, the Wall cut through everything that had previously connected the city and made it a singular entity. It broke the continuity of roads and streets, telephone cables, train tracks, and subway lines, and even cut straight through buildings. At the same time, it separated families and friends and severed social ties as much as economic and commercial exchange. As an architectural element, the Wall was faultlessly averse to all things material. Only the invisible transmission of music, sounds, and information disseminated via electromagnetic waves

1 *Der Himmel über Berlin*, dir. Wim Wenders (West Germany: Road Movies Filmproduktion; Argos Films; Westdeutscher Rundfunk, 1987).

2 The Staatsbibliothek is located in the Kulturforum in former West Berlin. Scharoun won the competition for its design in 1964, and the building was finally completed in 1978.

lay outside its control. In other words, the Berlin Wall broke the continuity of everything except the agency of a system with no apparent design, and no cohesion in and of itself. Built up over a span of nearly one hundred years, this system operated through numerous technological objects and cultural practices, all housed in a series of institutions and facilities dispersed throughout the city. Taken together, they constituted a materially grounded infrastructure with an invisible but not immaterial agency that effectively challenged the divisive power of the notorious Wall.

II → P. 2

V → P. 6

The GDR broadcasting house, the Funkhaus Berlin-Nalepastrasse, was part of this system—and one of those objects that stretch the conceptual plasticity of architecture to its extreme limit, by virtue of its capacity to give shape to the immaterial. (Fig. II) Even today, long after the fall of the Wall, it retains something of a forbidding air. Located in the industrial area of Köpenick, the complex is accessed through a steel gate. The first thing one sees, to the left, is a massive block made of dark red brick, six stories high, 150 m long, and terminated at one end by a tower. The rigorous pattern of the fenestration of the main facade makes the heavy structure look as if it is made of delicate wickerwork. This feature—common in *Backsteinexpressionismus* (brick expressionism)—is achieved here with a very simple tectonic trope: pairs of white-framed windows are placed between concrete lesenes, which accentuate their verticality. While this tectonic expression has something of a civic character that does not entirely mesh with the building's actual purpose, it nevertheless exudes a certain boldness that makes it fit in quite naturally with the industrial box-like structures in the immediate surroundings.

(Fig. V) But crossing this block and turning left, parallel to the River Spree, it becomes apparent that the Funkhaus is neither a single building nor an industrial complex. Rather, it has the appearance of a multiplicity of fragments that are formally and programmatically independent from each other, yet at the same time articulated in a sophisticated manner:

VI → P. 7

there are bridges connecting one block to another, and enough material continuity to convey a sense of cohesion; the windows, though their numbers and proportions differ significantly, still share a common vocabulary. Similar to Emil Kaufmann's understanding of Claude-Nicolas Ledoux's Ideal City at the Royal Saltworks in Arc-et-Senans, each fragment seems to express an autonomous function.[3] The next building, several hundred meters further on, but connected to the first one by a bridge, is a curvilinear block. The entrance facade has hardly any windows and appears as an almost monolithic brick box. Go inside, however, and things change dramatically. The entrance hall has a solemn authority, with its floor of polished black stone and black concrete columns that guide the way up the stairs to a foyer with spectacular views of the river. All the walls and columns here are clad with 3 cm-wide semicircular wooden strips, and a perforated white dropped ceiling with a vibrant circular pattern extends throughout the space. From the foyer, you enter a large recording hall, Saal I, to find a space that is even harder to describe. The material and geometric composition of the walls defining this room, and many others, shaped the city's acoustic political space—precisely the sounds and voices that those filmic angels perceived when flying over Berlin. (Fig. VI)

The Funkhaus Berlin-Nalepastrasse was not merely another structure built in the context of the political division of Berlin in 1949, but part of a much larger *ensemble of technical forms* defined by institutions, cultural practices, political decisions, and electronic technical media, as well as other related buildings.[4] What follows amounts to an environmental history of radio performed through the media-archaeological reconstruction of this ensemble. This approach aims to apprehend the conditions of its existence, its historical context, its spatial and, thus, political consequences, which reached far beyond the limits of Berlin.[5] In the context of this project, I will define this technical ensemble as the *architecture of radio*. In analyzing its existence—in delving into the indissoluble marriage between buildings, electronic technical media, cultural

3 Emil Kaufmann, *Von Ledoux bis Le Corbusier: Ursprung und Entwicklung der autonomen Architektur* (Stuttgart: G. Hatje, 1985).

4 Gilbert Simondon, *On the Mode of Existence of Technical Objects*, trans. Cecile Malaspina and John Rogove (Minneapolis: Univocal, 2017), 63–83.

5 The idea of an environmental history of radio echoes an idea briefly sketched by British architectural historian Robin Evans, in an essay published in 1971. To read more, consult Robin Evans, "The Rights of Retreat and the Rites of Exclusion: Notes towards the Definition of Wall," *Architectural Design* 41, no. 6 (1971): 335–339. The notion of media archaeology is associated with the reception of the work of German media theoretician Wolfgang Ernst. See Wolfgang Ernst, *Digital Memory and the Archive* (Minneapolis: University of Minnesota Press, 2013), 55–73. For more on the reception of the term media archaeology, see Jussi Parikka, *What is Media Archaeology?* (Cambridge: Polity Press, 2012), 62–89. Also relevant to this discussion is the work of Cornelia Vismann. See Cornelia Vismann, *Files: Law and Media Technology* (Stanford: Stanford University Press, 2008).

and political practices, as well as urban and territorial organization—I propose to test the limits of the term *architecture* by reconsidering its role in the age of radio and the advent of the remote transmission of information. In an era in which information has become ubiquitous—extending even beyond the limits of the planet while also penetrating deep into our bodies and saturating almost every aspect of our daily lives—it seems urgent to historically reposition the material ethos of information as it intersects with architecture. The politics of information circulation can only be properly apprehended once it has been materially grounded. By looking at Berlin from this perspective, we confront a different modernity, one that prompts us to reconsider the role and agency of buildings within a much larger network composed of artifacts and practices dedicated to the production, processing, transmission, and reception of information.

Without a doubt, we live in the age of radio. We hardly talk to each other in person anymore, being more absorbed in sending and receiving radio electromagnetic signals that are decoded into countless images, sounds, and texts. Radio-emitting and radio-receiving devices surround us, and we are so accustomed to broadcasts—even to broadcasting ourselves—that we barely register this everyday practice. Radio has been within and around our techno-natural environment for more than a century, redefining our modes of existence, and yet its environmental impact has been widely overlooked. Avoiding the narratives of invention or innovation that so often populate histories of architecture and technology, the focus of this book is on the environmental impact of radio in Cold War Berlin. Chronologically, it centers on a late and mature state of radio technology, a period in which radio had already become a set of *radio techniques*.[6] It is precisely in this later stage of its transformation that we can see—and clearly hear—the architectural impact of the entangled development of technological inventions, political disputes concerning these inventions, and their role in shaping the environment. Given that architecture is

6 For more on the relationship between technique and technology as it relates to the notion of techniques of the body, see Marcel Mauss, *Techniques, Technology and Civilisation* (Oxford: Berghahn Books, 2006), 77–93. The notion of *Kulturtechnik* or cultural techniques associated with the reception of the work of German media historian Bernhard Siegert has also been relevant for this project. See Bernhard Siegert, *Cultural Techniques: Grids, Filters, Doors, and Other Articulations of the Real,* trans. Geoffrey Winthrop-Young (New York: Fordham University Press, 2014). See also Bernhard Siegert and Geoffrey Winthrop-Young, “Cacography or Communication? Cultural Techniques in German Media Studies,” *Grey Room* 29 (Fall 2007): 26–47.

a relatively slow medium, it is often more fruitful to look beyond the periods of invention and to consider instead those other moments when there was a sudden investment of technological efforts and resources and recalibration of bureaucratic procedures, aesthetic classifications, and related cultural practices—moments when political and ideological constructions took form in a set of both visible and invisible artifacts. This ensemble of techniques and artifacts is what I refer to here as the *architecture of radio*.

The architecture of radio is understood in connection with, and as an extension of, what Michel Foucault defined as apparatus or *dispositif*. According to Foucault, an apparatus is "a thoroughly heterogeneous set consisting of discourses, institutions, architectural forms, regulatory decisions, laws, administrative measures, scientific statements, philosophical, moral, and philanthropic positions—in short, the said as much as the unsaid. The apparatus itself is the system of relations that can be established between them."[7] Beyond discourse analysis, this book entails a careful examination of the "system of relations" behind highly material and often overlooked aspects of information technologies. Due to its ethereal nature and ability to penetrate and basically disregard the existence of walls, doors, and windows, radio has rendered its intrinsically material grounds invisible. By framing radio as a particular form of architecture in its own right, as a system of relations strongly dependent on built artifacts, I also aim in this project to rearticulate the relationship between the virtual and the actual. Architecture is always grounded in matter, but it is never about matter alone.

A material understanding of Berlin's architecture of radio during the Cold War period reveals global and local under-researched continuities between electronic technical media and architecture, between discrete walls and political strategies, and between buildings and large territorial organizations. This book unravels these trajectories and networks in an effort to articulate a retroactive theory of design. In other words, it focuses on the role of the built artifact during a period in

7 Michel Foucault, *Power/Knowledge: Selected Interviews and Other Writings, 1972–1977,* ed. C. Gordon (New York: Pantheon Books, 1980), 194–196, here 194. For a further interpretation of the term apparatus or *dispositif*, see Giorgio Agamben, *What is an Apparatus and Other Essays* (Stanford: Stanford University Press, 2009), 1–23. The term *Apparat* was used by Bertolt Brecht when he speculated about a form of radio capable of sending as well as receiving. See Bertolt Brecht, "Der Rundfunk als Kommunikationsapparat," *Blätter des Hessischen Landestheaters*, no. 16 (July 1932).

which the historic solidity of architecture was radically challenged by the technology and techniques of radio. In doing so, this work transposes the same question to our present time to interrogate, with historical distance, the status and agency of buildings with regard to our increasingly hyperconnected, ubiquitous, and apparently invisible modes of co-existence.

II

The *technical medium* of radio became *mass media* in Berlin on October 29, 1923, at precisely 8:00 p.m., when the radio station Sendestelle Berlin made its first broadcast to a dispersed audience from its base in Vox-Haus, on the fourth and fifth floors of an existing building adjacent to Potsdamer Platz. Up until the end of World War I, radio had operated as a one-to-one communication channel; with this broadcast, it began its transformation into a one-to-many network for the dissemination of sonic-based content almost at the speed of light.[8] From this event onwards, the *medium* of radio would become ubiquitous, as electromagnetic waves began to cover vast swaths of territory and radio transmissions came within the reach of anyone with a radio receiver capable of transforming the electromagnetic waves back into the realm of the mechanical—into sounds that could be heard by listeners.

Beginning in the 1920s and 1930s, the impact of radio on German culture attracted the critical attention of politicians, artists, engineers, and intellectuals alike. The speed and scope of information dissemination through wire or wireless electronic technical media had already been transformed by telegraphy and telephony at the turn of the century.[9] Now radio, turned into mass media, was transforming not only the means but also the content of communication, reconfiguring the traditional artistic practices of music and theater in line with their renewed media-specific conditions.[10] In a lucid critique from 1932, Bertolt Brecht polemically defined radio as "one-sided when it should be two," "an apparatus for distribution, for

8 Hans Jürgen Koch and Hermann Glaser, *Ganz Ohr: eine Kulturgeschichte des Radios in Deutschland* (Cologne: Böhlau, 2005), 5–66.

9 Dwayne R. Winseck and Robert M. Pike, *Communication and Empire: Media, Markets, and Globalization, 1860–1930* (Durham: Duke University Press, 2007), and Richard R. John, *Network Nation: Inventing American Telecommunications* (Cambridge, MA: Harvard University Press, 2010).

10 One of the first to perceive the intertwining threads of electromagnetic remote telecommunications and culture was none other than Aby Warburg, just a few months before the first open radio transmission in Berlin in 1923. In his famous lecture on "serpent ritual," Warburg warned his listeners that collapse of physical distance imposed by telegraphy and telephony threatened to exterminate the space for contemplation (*Andachtsraum*), which was for him the necessary condition for the space of thinking (*Denkraum*). For further reading, see Aby Warburg, *Schlangenritual: Ein Reisebericht: Bilder aus dem Gebiet der Pueblo-Indianer in Nord-Amerika* [1923] (Berlin: Verlag Klaus Wagenbach, 1988). For an English translation: Aby Warburg, "A Lecture on Serpent Ritual," *Journal of the Warburg Institute* 2, no. 4 (April 1939): 277–292. I would like to thank Niklas Maak for pointing me to this reference.

mere sharing out." What he called for instead, in that politically turbulent last year of the Weimar Republic, was a two-way medium, switched from "distribution to communication."[11] Brecht saw the potential of radio to become "the finest possible communication apparatus in public life, a vast network of pipes."[12] However, of all the early theoreticians engaging with radio during the Weimar Republic, it was Rudolf Arnheim who developed the most consistent study of the emerging medium in his book *Rundfunk als Hörkunst* (*Radio: An Art of Sound*), published in 1936. Here, Arnheim aimed to articulate the artistic potential of radio, imagining it as "the great miracle of wireless. The omnipresence of what people are singing or saying anywhere, the overlapping of frontiers, the conquest of spatial isolation, the importation of culture in the waves of the ether, the same fare for all, sound in silence."[13] In turn, a young Theodor W. Adorno was a frequent contributor to the avant-garde journal *Musikblätter des Anbruch* from as early as 1925, developing the seeds of his later critical understanding of the relationship between technology, music, and popular culture.[14] After emigrating to the United States in 1938, Adorno collaborated with the Rockefeller-funded Princeton Radio Research Project, aiming to bridge critical social theory with the empirical research methods used by North American sociologists, most notably Paul F. Lazarsfeld. During both his life in the Weimar Republic and his time in the United States, Adorno developed a thorough critique of the political and ideological effects of radio-transmitted music on capitalist consumer societies. In "A Social Critique of Radio Music," a text tangential to his critique of the "culture industry" articulated in *Dialectic of Enlightenment* (1947), he engages with popular culture and portrays radio as just "an essentially new technique of musical reproduction," famously blaming it for "keeping listeners from criticizing social realities."[15]

At the intersection of these three early critical responses to radio's social and political effects, we can identify the locus for a much-needed historiographical

11 Bertolt Brecht, "The Radio as an Apparatus of Communication," in *The Weimar Republic Sourcebook*, ed. Anton Kaes, Martin Jay, and Edward Dimendberg (Berkeley: University of California Press, 1994), 615–617, here 616. For the original German, see Brecht, "Der Rundfunk als Kommunikationsapparat."

12 Brecht, "The Radio as an Apparatus of Communication," 616. For more on Brecht's understanding of media as it relates to the avant-garde in Berlin during the 1920s and 1930s, see Roswitha Müller, *Bertolt Brecht and the Theory of Media* (Lincoln: University of Nebraska Press, 1989), 1–21.

13 Rudolf Arnheim, *Radio: An Art of Sound* (London: Faber & Faber, 1936), 14. This is the first publication of the German manuscript that Arnheim wrote in 1933 with the title "Der Rundfunk sucht seine Form." The book was later published in Italian as Rudolf Arnheim, *La radio cerca la sua forma* (Milan: Ulrico Hoepli, 1937). A German edition of this book appeared for the first time in 1979 as Rudolf Arnheim, *Rundfunk als Hörkunst* (Munich; Vienna: Carl Hanser Verlag, 1979). For the last edition with a critical commentary by Helmut H. Diederichs, see Rudolf Arnheim, *Rundfunk als Hörkunst* (Frankfurt am Main: Suhrkamp, 2001).

14 Thomas Y. Levin, "Elements of a Radio Theory: Adorno and the Princeton Radio Research Project," *Musical Quarterly* 78, no. 2 (1994): 316–324.

15 Theodor W. Adorno, "A Social Critique of Radio Music," in *Current of Music: Elements of a Radio Theory,* ed. Robert Hullot-Kentor (Frankfurt am Main: Suhrkamp, 2006), 201–217, here 209–210. For more on Adorno's understanding of early electronic music, see Theodor W. Adorno, "Technique, Technology, and Music To-day," *Gravesaner Blätter* 11/12, no. 4 (1958): 51–61. The article is interesting since, at some level, it contradicts the argument he developed in "A Social Critique of Radio Music." There, Adorno strongly criticizes radio for not producing a musical form of its own, instead replicating previous forms unsuited to radio reproduction. What he fails to see, however, is that in his critique of electronic music published in 1958, the examples he discusses come exactly from the extended industry of radio, bearing in mind that the studio in which Karlheinz Stockhausen was developing his compositions was physically and institutionally bound to Nordwestdeutscher Rundfunk (NWDR) in Cologne.

correction—one that can only be done, I argue, by analyzing the role of architecture at the dawn of the fusion of cultural and industrial production. First, Brecht identified the social effects of radio at the exact moment when it was transitioning from a point-to-point conversational medium to a means of massive "indoctrination" and "manipulation."[16] Second, Arnheim offered an almost phenomenological account of the artistic potential of radio beyond the mere functional effects of communication. Third, over the years, Adorno constructed a critique of radio, and by extension, of the so-called "culture industry," that was based primarily on an analysis of the *content* of radio broadcasts. All three authors recognized the latent potential and danger behind the fusion of culture and industry, yet there was no analysis of the actual processes, institutions, and technical and bureaucratic mechanisms that made that fusion possible in the first place. In other words, they failed to consider what makes culture effectively, rather than metaphorically, an industry. One of the intentions of this book is to demonstrate that the transformation of radio from an intersubjective medium of communication into a one-directional mass media connected to the industry was made possible by the emergence of a new type of building that reified institutions and modes of production in a previously unthinkable way. To ask whether the institutional reforms that originated in the interwar period demanded a new kind of building, or if the conception of the built artifact allowed the cultural/industrial institutions to operate in such a way, is a chicken-and-egg question of no interest. What is clear, though, is that all three authors discussed above wrote around the time of the inauguration in 1931 of the Haus des Rundfunks, the building designed by Hans Poelzig for the Reichs-Rundfunk-Gesellschaft, or RRG (Reich Broadcasting Corporation), which could be considered the first ground-up construction specifically designed for radio broadcasting. The absence of critical scrutiny of this building is symptomatic of the longstanding inability of critical discourse to grasp the material backbone of apparently immaterial

16 In the 1970s, Hans Magnus Enzensberger offered the political left one of the most insightful internal critiques with his analysis of the relationship that socialist intellectuals have had with the emerging electronic media. The notions of "indoctrination" and "manipulation" are taken directly from this article. See Hans Magnus Enzensberger, "Constituents of a Theory of Media," *New Left Review*, no. 64 (1970): 13–36.

information and communication industries. Years passed, and the argument that media, and specifically radio, would overcome all material boundaries intensified. A young Marshall McLuhan theorized that mass media "cracked the very walls of the classroom";[17] that new technologies such as radio broke down "all city walls" like the trumpets in Jericho;[18] and, most famously, that media have always been nothing less than "extensions of man."[19] Radicalizing Marshall McLuhan's position—with an added critical stance towards the Frankfurt School's understanding of media—Friedrich Kittler claimed that far from being *just* "extensions of man," in the 1880s electronic media determined the point of emergence of a decentered human subject, the "so-called-man."[20] Building explicitly on Kittler, Wolfgang Hagen's book of 2005, *Das Radio,* presented arguably the first historical account of radio understood in its dual dimension as intersecting the material history of the *technical medium* with the operative principles of radio as *mass media.*[21] Alternatively, and using radio to demonstrate his theory of media archaeology, Wolfgang Ernst proposed a radical break between the media-archaeological analysis of radio as a *technical medium* and the cultural analysis of radio as *mass media.* Contrary to Hagen, Ernst thoroughly uncoupled the techno-epistemological ethos of radio from its historical narrative understood as mass media.[22]

In all these works, however—the material history offered by Hagen, the early media theories deployed by McLuhan, the media-archaeological approach articulated by Ernst, the ideological critique of Adorno, or the almost utopian impact envisioned by Brecht—there is no treatment of the built infrastructure of radio. Though the "hundred years of radio" in Germany are mirrored and echoed by almost one hundred years of radio history and theory, there has been hardly any critical analysis of the built infrastructure that allows for the existence of this apparently immaterial medium. In other words, the intangible, wireless nature of

17 Marshal McLuhan, "Classroom Without Walls," *Explorations: Studies in Culture and Communication*, no. 7 (March 1957): 22–29, here 23.

18 Marshall McLuhan, "Media Fit the Battle of Jericho," *Explorations: Studies in Culture and Communication*, no. 6 (July 1956): 15–19, here 16.

19 Marshall McLuhan, *Understanding Media: The Extensions of Man* (Toronto: McGraw-Hill, 1964).

20 Friedrich Kittler, *Gramophone, Film, Typewriter*, trans. Geoffrey Winthrop-Young (Stanford: Stanford University Press, 1999), 16. For the original in German, see Friedrich Kittler, *Grammophon Film Typewriter* (Berlin: Brinkmann & Bose, 1986).

21 Wolfgang Hagen, *Das Radio: Zur Geschichte und Theorie des Hörfunks—Deutschland/USA* (Munich: Wilhelm Fink, 2005).

22 Wolfgang Ernst, "Distory: One Hundred Years of Electron Tubes, Media-Archaeologically Interpreted, Vis-à-Vis One Hundred Years of Radio," in *Digital Memory and the Archive*, ed. Jussi Parikka (Minneapolis: University of Minnesota Press, 2012), 158–171. More recently, a compelling and thorough study by Carolyn Birdsall reflects on the role played by modern sound technologies (with a particular focus on radio) in developing the sonic experience during the Third Reich. Building upon the notion of "ear witnessing"—after conducting several interviews with "ear-witnesses" of National Socialism in Germany—Birdsall's study offers new insight into the "conditions in which listening and sound-making take place, and the spatial characteristics influencing sound as such." Carolyn Birdsall, *Nazi Soundscapes: Sound, Technology and Urban Space in Germany, 1933–1945* (Amsterdam: Amsterdam University Press, 2012), 30.

the technical medium, coupled with its ubiquity, seems to have rendered the material grounds of radio effectively invisible to historians as well.[23]

On the other hand, if the built condition of possibility of radio has become invisible to its own historiography, radio as a bundle of *techniques* and *technologies* has been absent from modern architecture's century-long obsession with technology. Ironically, radio embodied most of the aesthetic values celebrated in modernist art and architecture: it was simultaneously a technical medium, an industry, and a means of transport; it radically redefined the relationship between speed and space; it had an impact on the formation of subjectivity on a broad scale, all of which demanded the design and construction of new building types to operate effectively. And yet none of the "modern masters"—Ludwig Mies van der Rohe, Le Corbusier, or Walter Gropius, for instance—ever showed any real interest in radio. Both Alexander Rodchenko and László Moholy-Nagy famously photographed radio transmission towers as symbols and emblems of the modern world, but even those notable observers of technological change completely overlooked the massive complexes in which the transmitted content was actually produced. It is as if the modernist obsession with immateriality, coupled with the apparently immaterial nature of broadcasting, dematerialized buildings into the translucent and linear transmission towers, rendering the act and fact of information production and transmission completely invisible. Absorbed by the aesthetics of circulation, modernist observers paid no attention to the politics of production happening inside opaque and apparently much more conservative brick-and-mortar buildings. To use Reyner Banham's phrase, in their obsession with the "First Machine Age," architects and historians—consciously or not—overlooked the emergence of the first *Electromagnetic Age*, even if the likes of Hermann Muthesius and Hans Poelzig were actively shaping the architecture of radio.

"It may well be that what we have hitherto understood as architecture, and what we are beginning to

23 Within this genealogy of histories and theories of radio, it is worth including the book by Brian Larkin, *Signal and Noise: Media, Infrastructure, and Urban Culture in Nigeria* (Durham: Duke University Press, 2008). In a section dedicated to radio (pp. 48–72), Larkin introduces a compelling reading of collective radio listening during the early years of radio, analyzing, for example, the material infrastructure used under British colonial rule to broadcast, through wires, to the Nigerian population. For more on the role of radio in colonial and postcolonial Africa, see Frantz Fanon, *A Dying Colonialism*, trans. Haakon Chevalier (New York: Grove Press, 1967), 60–97; Marissa J. Moorman, *Powerful Frequencies: Radio, State Power, and Cold War in Angola, 1931–2002* (Athens, OH: Ohio University Press, 2019); Fred Moten, "Not in Between," in *Black and Blur*, ed. Fred Moten (Durham: Duke University Press, 2017), 1–27.

understand of technology are incompatible disciplines," wrote Banham in his *Theory of Design* (1960), urging architects to cast off the "cultural load" of the profession and embrace the challenges and opportunities of technological culture.[24] And he warned that those who failed to do so could well be left behind: "The architect who proposes to run with technology knows now that he will be in fast company." If he cannot keep up, "he may find that a technological culture has decided to go on without him."[25] Banham attributed a "business-like and ship-shape appearance" to Le Corbusier's Villa Savoye, and defined Mies's Barcelona Pavilion as "so purely symbolic in intention that the concept of Functionalism would need to be stretched to the point of unrecognizability before it could be made to fit it."[26] For him, the *real* modern architect was R. Buckminster Fuller, whose "entirely radical" Dymaxion House was "an adaptation of light-metal methods employed in aircraft construction."[27] Banham may have been right in identifying the academicism and symbolic appropriation of a "Machine Age style." But what he failed to identify was the revolution propelled by the production and dissemination of information—a revolution that had been unleashed during the First Electromagnetic Age and was taking on a more radical form at the very time he was writing those words.[28]

Both trajectories described above—the historical and theoretical understanding of electronic media technologies on the one hand, and the architectural and spatial understanding of those technologies on the other—merged for the first time in the multidisciplinary context of the Culture and Communications seminar held at the University of Toronto between 1954 and 1955, an event supported by the Ford Foundation. The idea of *acoustic space* was vigorously debated in the context of the seminar and accompanying publication, *Explorations in Communications*, serving literally as a platform for intellectual exchange between media theorist Marshall McLuhan, architectural historian Sigfried Giedion, and urban planner Jaqueline Tyrwhitt.[29] For McLuhan, the idea of acoustic space was important for describing the non-visual space

24 Reyner Banham, *Theory and Design in the First Machine Age* (New York: Praeger, 1967), 329. Relevant to this particular discussion, and to Banham's book in general, is Mark Wigley's book *Buckminster Fuller Inc: Architecture in the Age of Radio* (Zurich: Lars Müller, 2015). In this intellectual biography, Wigley discusses the general project of Fuller as being directly influenced by radio. He convincingly demonstrates that the Dymaxion House, used by Banham as the epitome of true modern architecture, has nothing to do with the "machine age" but rather is a direct outcome of Fuller's engagement with radio. The house could be detached from the ground—and thus liberated from land ownership and the burden of pipes and infrastructure—by being connected to the world through radio. Also relevant in this context is Shundana Yusaf's reconstruction of the history of the BBC's broadcasting of architectural ideas through radio, *Broadcasting Buildings: Architecture on the Wireless, 1927–1945* (Cambridge, MA: MIT Press, 2014). While both books have been historically and conceptually useful for this project, the question regarding radio's material infrastructure nevertheless remains widely unexplored.

25 Banham, *Theory and Design in the First Machine Age*, 329.

26 Banham, *Theory and Design in the First Machine Age*, 321.

27 Banham, *Theory and Design in the First Machine Age*, 326.

28 The argument of the invisibility of information processing as a vector of historical change as it relates to the work of Reyner Banham has previously been presented in Zeynep Çelik Alexander's introduction to *Design Technics: Archaeologies of Architectural Practice*, ed. Zeynep Çelik Alexander and John May (Minneapolis: University of Minnesota Press, 2020), ix–xviii, here xv, and in Zeynep Çelik Alexander, "The Larkin's Architectural Technologies of Trust," *Journal of the Society of Architectural Historians* 77, no. 3 (September 2018): 300–318.

29 I consider this particular debate as one of the origins of an intellectual dialogue between architectural theory and media studies that would have ramifications traceable up to the present. For more on the role played by Jaqueline Tyrwhitt in the Culture and Communications seminar, see Michael Darroch, "Bridging Urban and Media Studies: Jaqueline Tyrwhitt and the Explorations Group, 1951–1957," *Canadian Journal of Communication* 33, no. 2, (2008): 147–169. For more on the relationship between Marshall McLuhan and Sigfried Giedion as it relates to the idea of *acoustic space,* see Reto Geiser, *Giedion and America: Repositioning the History of Modern Architecture* (Zurich: gta Verlag, 2018), 372–404.

produced by the ubiquitous presence of electronic technical media. It provided a conceptual framework that allowed him to define the epistemological shift that had followed the advent of electronic media technologies, reconfiguring both the lineal space of the book and the flat space of the image. As he saw it, new electronic media technologies turned space into an overlap of different sensorial stimuli, but also brought dissimilar forms of expression into simultaneous coexistence. In this sense, radio was the epitome of acoustic space, being the ultimate non-visual, electronic technical medium that allowed different realities to exist concurrently by actually blurring the physical boundaries of space and making the world a unified and continuous field.[30]

Its invisibility in this century-long debate notwithstanding, radio as both a technical medium and mass media inevitably intersects with the built environment, either in its indifference to or reliance on the city's material infrastructure. In this sense, this book looks at the built structures that are intrinsically bound to radio as well as to the urban fabric of Berlin. Yet, far from constructing an architecturally deterministic narrative—or a building *a priori,* to put it in Kittlerian terms—this book aims instead to understand the specific role that built artifacts play in a new condition defined by the politics of information production and dissemination through a new technological regime. Buildings understood as historical constructs—and different from other storing devices—possess a double capacity. On the one hand, they are *documents*, acting as sensors on which political, economic, cultural, and technological forces leave their traces. Equipped with the right disciplinary, conceptual, and technical tools, buildings understood as *documents* can be interrogated. Through them, one can make matter speak.[31] On the other hand, buildings can also be considered *agents*. They not only passively register change, but actively produce and accelerate it as well. The angle of inclination of a wall, the size of a room, or the organization of a floor plan, for example, record a technological and cultural operation

30 The first documented use of *acoustic space* in this context was in fact in the title of an essay published in 1955 by psychologist D. Carl Williams in a journal created in 1953 by Marshall McLuhan and Edmund Carpenter. See D. C. Williams, "Acoustic Space," *Explorations: Studies in Culture and Communication* 4 (February 1955): 15–20. While the article was originally attributed to Williams, it would be credited to Carpenter and McLuhan when it was republished five years later in *Explorations in Communications: An Anthology*, ed. Edmund Carpenter and Marshall McLuhan (London: Beacon Press, 1960), 65–70. Regarding Giedion's later engagement with the term, see Sigfried Giedion, *The Eternal Present: The Beginnings of Art* (New York: Pantheon Books, 1962), 526–527. The idea of acoustic space was reconceptualized in Wolfgang Ernst, *Sonic Time Machines: Explicit Sound, Sirenic Voices, and Implicit Sonicity* (Amsterdam: Amsterdam University Press, 2016), 21–34. For a more recent study on the idea of acoustic space see Joseph L. Clarke, *Echo's Chambers: Architecture and the Idea of Acoustic Space* (Pittsburgh: University of Pittsburgh Press, 2021).

31 In discussing the relationship between documents and monuments as they relate to history, I am building on the theory articulated in Michel Foucault's *The Archaeology of Knowledge*. Here, Foucault discusses explicitly how his understanding of archaeology suggests the transformation of documents into monuments, reversing a procedure that normally operates in the exact opposite direction. Avoiding the idea of memory—which suggests that monuments commemorate events of the past—I also understand the analysis of buildings as documents transformed into monuments, that is, not only as containers of discourse, but rather as active producers of discourse. See Michel Foucault, *The Archaeology of Knowledge* (New York: Pantheon Books, 1972), 3–17. For a more recent study on the relationship between buildings, monuments, and documents, see Lucia Allais, *Design of Destruction: The Making of Monuments in the Twentieth Century* (Chicago: University of Chicago Press, 2018).

32 Kittler, *Gramophone, Film, Typewriter.* Here, Kittler interprets and adopts Jacques Lacan's conceptual trilogy of the *symbolic*, the *imaginary*, and the *real* as the basis of his media theory, thus establishing a parallel between each of these registers with a specific electronic media technology.

that otherwise remains invisible and inaudible. But at the same time, those very characteristics of a building are responsible for that operation in the first place.

To be more explicit, there is no recording device that can fully grasp all the effects of radio. In this context, the intrinsically relational ethos embedded in architecture becomes indispensable for grasping the political implications of this form of production. Written documents such as correspondence and meeting protocols might filter certain aspects through the grid of the *symbolic*. Photography might duplicate certain actions at the level of the *imaginary*. But there are other artifacts—namely, the technical media devices themselves—that, when combined with buildings, can capture from within the intrinsic material relationships that testify to specific historical conditions and their concrete political consequences. When properly interrogated, these artifacts speak of the realm of the *real* they have previously registered.[32] In this sense—by bridging the increasingly small scale of electronic technical media and the vast scale of their territorial impact—buildings mediate our simultaneous existence on drastically different scales. In the context of this book, I interrogate buildings in this double sense—as *documents* and as *agents*—seeking to illuminate a widely unexplored historical episode in which two confronting political ideologies met in a new electromagnetic terrain. That condition has deep resonances with the present, and it is through this double understanding of the built artifact, that is to say, through turning *documents* into *monuments,* that I aim to construct an environmental history of Berlin's architecture of radio.

III

This book is structured in three chapters that, rather than following a strict chronological order, are organized according to three scales. Beginning with the largest scale, I analyze the organization of the city of Berlin and how it relates to the larger global territorial disputes and reconfigurations that emerged in Germany after World War II. From there, the analysis scales down to the level of buildings, focusing specifically on the origin and development of the typology of the broadcasting house. Finally, it scales down even further to the operative elements themselves, analyzing the space and practices of the invention and development of electroacoustic technical media within the laboratory. The goal is to demonstrate specifically how these elements influenced and determined the other scales under investigation. Using three different scalar lenses to examine this phenomenon, each chapter addresses in a different form—and with different emphases—the three registers through which the techniques and technology of radio operate: production, transmission, and reception of information. While positioned *in* and focusing *on* the Cold War period, each scalar lens makes a cross-section through a broader timespan, tracing genealogies back to the Weimar era and, in some cases, to the decades preceding World War I. Spanning from Wilhelmine Germany, through the interwar period, the Third Reich, and World War II, to the postwar decades, each scalar lens is used to study the development and transformation of radio from a technical medium into a politically contested form of mass media.

Each section focuses on a particular object, and the extensive archival research is mobilized towards understanding the conditions of existence of those sampled artifacts. This tripartite scalar structure problematizes the limits of linear, chronological histories. Looking through one scalar lens, the book identifies continuities across a much longer historical period: for example, there was virtually no break in the careers of certain scientists and artists—or even the trajectories of certain institutions—between the interwar and the postwar periods. On the other hand, when that same period is viewed through a different scalar lens, previously overlooked epistemological breaks, ruptures, and discontinuities come to light. The architecture of radio analyzed in the book was neither a cohesive design project drafted by one individual or a single institution, nor, on the contrary, an unrelated series of buildings, practices, and events. It was a new form that called for different ways of thinking and acting in and through space. By adhering to a scalar logic—by identifying the tectonic links connecting laboratories with broadcasting stations, artists with engineers, politicians with listeners, buildings with territories—the book aims to unravel these forms of thinking and to retroactively construct this different mode of acting and intervening in, modifying, controlling, and, above all, designing space.

Chapter 1 addresses the scale of the territory of the architecture of radio by analyzing the role and the strategic position of Berlin in a dual sense. On the one hand, I discuss the location of the city in the context of the broader division of the country and the continent after World War II. Rather than focusing only on the limit demarcated by the Iron Curtain, I consider how that very boundary was porous and transgressed by radio. On the other hand, I analyze the inner structure of the city—seen from the same perspective—in order to understand how it was defined and shaped by the strategic location of built artifacts dedicated to the transmission and reception of electromagnetic waves. With its focus on Berlin, this project can only scratch the surface of the global history of the architecture of radio. But as a contribution to that history, important to this chapter are the institutional, technological, and infrastructural continuities I have identified between the emergence of radiotelegraphy as a global colonial project in Wilhelmine Germany and the contested existence of radio broadcasting that defined Berlin during the Cold War.

The central object of study of this chapter is a map of the city of Berlin developed by the Stasi after the erection of the Berlin Wall. Whereas the map itself depicts a divided city, the GDR's secret intelligence service noted with grave concern how electromagnetic waves crossed the Wall from the West, violating the GDR's space of information sovereignty. The map also shows the jamming transmitters that were meant to block those invasive radio waves—and that made the city a contested territory of rival signals and noise. This particular map is analyzed in combination with several other cartographic representations developed by political and cultural institutions on both sides of the Wall, to provide a broader understanding of how the territory was conceptualized and perceived through the perspective of radio. In analyzing the cartographic representations in this chapter, I seek to understand the different basis—a technological rationality—on which space was defined, occupied, protected, and contested.[33] When we look at the territory of Cold War Berlin from the perspective of radio, we do not see the usual spatial binary of a divided city with buildings and monuments encoded with signs representative of this or that political ideology. What we get, instead, is closer to what the French philosopher Paul Virilio referred to as a *habitable circulation*, which in Berlin manifested in a contested terrain for the exchange of information.[34] By considering the city in terms of the circulation of sounds and ideas rather than goods and people, I aim to redefine the elements that constitute both urban and domestic space in the era of remote transmission of information. Viewed from this new perspective, it becomes apparent that the fundamental element of the Berlin Wall was one among many media technologies defining a larger network. What divided the city was not *the* Wall understood as *an* architecture, as Rem Koolhaas once put it, but rather the architecture of radio in which the Wall was effectively immersed.[35]

Chapter 2 addresses the scale of the building in the architecture of radio, and traces the invention, development, and transformation of the typology of the broadcasting house. The central object of analysis is the Funkhaus Berlin-Nalepastrasse, the East German broadcasting house built in 1952. Conceived after the political division of the city in 1949—but before the erection of the Berlin Wall in 1961—the Funkhaus was the most radical iteration of the broadcasting house typology, concentrating in one building the media industry of the entire GDR. The building also emerged as a defensive vehicle for countering the signals emitted by RIAS and other Western stations, which strategically

33 For more on the relationship between reason and rationality in the Cold War period, see Lorraine Daston et al., eds., *How Reason Almost Lost Its Mind: The Strange Career of Cold War Rationality* (Chicago: University of Chicago Press, 2013).

34 For more on the notion of habitable circulation, see Paul Virilio, *Speed and Politics* (Los Angeles: Semiotext(e), 2006), 29–47.

35 Rem Koolhaas, "The Berlin Wall as Architecture," in *S,M,L,XL: Office for Metropolitan Architecture,* ed. Rem Koolhaas and Bruce Mau (New York: Monacelli Press, 1995), 217–231.

used radio to target and destabilize countries of the Eastern Bloc. Utilized to secure and eventually expand the impact of East Germany beyond its fixed borders, the question arises as to whether the Funkhaus—and its extended constellation of artifacts—acted as an invisible predecessor of the (concrete) Berlin Wall.[36]

As a building type, however, the broadcasting house emerged in Berlin during the Weimar Republic, with the inauguration in 1931 of the Haus des Rundfunks, designed by Hans Poelzig for the Reich Broadcasting Corporation. As a new typology, it emerged from the entangled development of scientific research and new institutions while also establishing a new relationship between culture and industry during the interwar period in Germany. In this sense, the typology of the broadcasting house brought together a series of cultural practices that emerged from radio's transformation into a mass media. Taking programmatic fragments from cultural institutions such as theaters and concert halls, adopting the scale and processes of mass industrial production, and incorporating technical knowledge and technological gadgets produced in laboratories, the broadcasting house gave form to a series of joint cultural, technological, and political transformations. The novel typology of the broadcasting house emerged as the medium for translating sounds belonging to the mechanical realm into electrical impulses ready to be broadcast. It constituted, moreover, the missing link between electronic technical media—such as microphones, tape recorders, and amplifiers—and vast territorial organizations. Mediating between musicians and politicians, between wires and violins, between culture and industry, between a concentrated production and a dispersed audience, the typology of the broadcasting house represents a missing piece in the history of modern architecture.[37]

By analyzing the role that the Haus des Rundfunks played in building culture and politics from the Weimar Republic all the way through the postwar decades, and by connecting that history with the even more

36 The "performativity" of objects and media beyond the scope of subjects' intentions is discussed in Cornelia Vismann, "Cultural Techniques and Sovereignty," *Theory, Culture & Society* 30, no. 6 (2013): 83–93.

37 For more on the absence of the typology of the broadcasting house in the historiography of modern architecture, see Nikolaus Pevsner, *A History of Building Types* (London: Thames & Hudson, 1976); Adolf Behne, *Der moderne Zweckbau* (Berlin: Drei Masken Verlag, 1926).

influential typological variation of the GDR Funkhaus in Nalepastrasse, we can see that this new way of conceiving buildings made them formative actors in shaping space and, with it, politics and culture at an unprecedented scale. That relevance forces us to analyze these buildings not as singular inventions but rather as *types* that are agents and products of much more profound transformations. By situating them within the debate concerning type, I recast the broadcasting house as the basis for rethinking the relationship between typology and modern architecture more broadly. Within its long and polysemic existence, which can be traced back to the eighteenth century, the discourse that links ideas of type and architecture has at least two identifiable continuities. On the one hand, it has been consistently used to measure and establish the "internal" limits of the discipline of architecture against the emergence of other historical forces perceived as "external." Symptomatic of that line of argument are, for example, the ideas on type that Quatremère de Quincy developed against theories of natural mimesis that emerged during the eighteenth century; the writings on type by Viollet-le-Duc at the rise of national identity and "race science" in the nineteenth century; and the debate on *Typisierung* that emerged within Deutscher Werkbund circles aiming to promote the aesthetic values of German industry through standardized consumer goods.[38] On the other hand, the second discursive continuity that can be identified in the long existence of the ideas of type involves balancing the importance of form and function and, thus, of continuity and change. Exemplary of that debate is, for instance, Gottfried Semper's emphasis on the techniques of construction as means to establish historical continuity; Moisei Ginzburg's attempts to explain the taming of social change through form, as proposed in his "transitional types" in the Soviet Union; or Aldo Rossi's ideas of returning to the pre-modern European city as a way to reorient architecture through form and overcome the restrictions imposed by modern planning.[39] This last iteration in the semantic adaptation of the idea of type is what Anthony Vidler famously referred to as "the third typology" in his

38 Antoine-Chrysostome Quatremère de Quincy, "Type," *in Encyclopedie Methodique, Architecture* (Paris, 1825), reprinted in English in *Oppositions*, no. 8 (Spring 1977): 148. On the connection between type and race see Eugène-Emmanuel Viollet-le-Duc, *The Habitations of Man in All Ages* (Boston: James R. Osgood, 1876). A recent critical study of this text appears in Irene Cheng, "Structural Racialism in Modern Architectural Theory," in *Race and Modern Architecture: A Critical History from the Enlightenment to the Present*, ed. Irene Cheng, Charles L. Davis II, and Mabel O. Wilson (Baltimore: Project Muse, 2020), 134–152. For more on the debate on type in the circles of the Deutsche Werkbund see Frederic J. Schwartz, "The Type," in *The Werkbund: Design Theory and Mass Culture before the First World War* (New Haven: Yale University Press, 1996), 121–146; Frederic J. Schwartz, "Commodity Signs: Peter Behrens, the AEG, and the Trademark," *Journal of Design History* 9, no. 3 (1996): 153–185; and Hermann Muthesius and Henry van de Velde, "Werkbund Theses and Antitheses" [1914], in *Programs and Manifestoes on Twentieth-Century Architecture*, ed. Ulrich Conrads (Cambridge, MA: MIT Press, 1971), 28–31. To read more on the genealogy of the idea of type, see Adrian Forty, *Words and Buildings: A Vocabulary of Modern Architecture* (New York: Thames and Hudson, 2000), 304–311.

39 Gottfried Semper, *Der Stil in den technischen und tektonischen Künsten, oder Praktische Aesthetik: Ein Handbuch für Techniker, Künstler und Kunstfreunde* (Frankfurt am Main, Verlag für Kunst und Wissenschaft, 1860). For a critical reception of Semper's idea on type, see Charles L. Davis II, *Building Character: The Racial Politics of Modern Architectural Style* (Pittsburgh: University of Pittsburgh Press, 2019), 71–126. For an interpretation of debate on type and standardization between Germany and the Soviet Union see Alla Vronskaya, "Process: Organicist Aesthetics of Soviet Standardization," in *Architecture of Life: Soviet Modernism and the Human Sciences* (Minneapolis: University of Minnesota Press, 2022), 107–139; and Moisei Ginzburg, "Transitional Types of Inhabitation: Work by the Standardization Department at Stroykom RSFSR, 1928–1929" [1934] and "The Commune House" [1934], in *Dwelling: Five Years' Work on the Problem of the Habitation* (London: Fontanka, 2017), 66–81; 138–147. For Rossi's ideas on type, read Aldo Rossi, *The Architecture of the City* (Cambridge, MA: MIT Press, 1984). See also Rafael Moneo, "On Typology," *Opposition 13* (1978): 22–45.

canonical essay of that name in the 1970s.[40] Trapped between notions of form and function, between disciplinary continuity and change, discussions of type have been mostly absent from the critical analysis of architecture produced during the first half of the twentieth century. With this study, I aim for a further resemanticization of the idea of type, suggesting that with the emergence of the broadcasting house we can speak about a *fourth* typology. As presented here, the fourth typology incorporates the impact of both political ideology and electronic technical media on the production of architectural form.

If radio was the technical medium that inaugurated the dissemination of information in a disembodied form through the electromagnetic spectrum, the broadcasting house, conceived as a building type, initiated the decisive interaction between buildings and electronic technical media. The philosopher Jean-François Lyotard analyzed the role of architecture in the context of what he called the third technological revolution when he discussed the conceptual framework of *Les Immatériaux*, the Pompidou Center exhibition he curated with design theorist Thierry Chaput in 1985. For Lyotard, modernity was a project aimed at conquering and governing space through bodies, but in the new condition he defined as the *immaterial*, the frontier to be conquered was no longer bodies, but time. Seen retrospectively in this light, the broadcasting house was the first building typology that technically finetuned the relationship between space and time through sound. Hearing is the human sense that is closer and more sensitive to the perception of time. Precisely in the definition of the interiors of the broadcasting house, building culture came closer to the emergent modern science of electroacoustics, which until then had been the preserve of the military and experiments performed in laboratories. Chapter 2 thus presents the broadcasting house as the first built artifact through which bodies began to be systematically dissolved into sounds—and therefore time-based signals—enabling them, by extension, to transgress boundaries and occupy territories at the speed of light.[41]

40 Anthony Vidler, "The Third Typology," *Oppositions* 7 (Winter 1977): 1–4. See also Anthony Vidler, "The Idea of Type: The Transformation of the Academic Ideal 1750–1830," *Oppositions* 8 (Spring 1977): 95–115.

41 For more on the relationship between materials and the *immaterial*, see Jean-François Lyotard, "After Six Months of Work... [1984]," in *30 Years after Les Immatériaux: Art, Science, and Theory*, ed. Yuk Hui and Andreas Broeckmann (Lüneburg: Meson Press, 2015), 27–66.

Chapter 3 addresses the scale of the technical object in the architecture of radio by taking the laboratory as the construction site for the electrification of space. Focusing now on sound archive material, the central object of study in this chapter is a music composition titled "Zoologischer Garten," recorded by North American composer Frederic Rzewski in East Berlin in 1965. While a fellow of the Ford Foundation in West Berlin, Rzewski accepted a commission from an East German radio station to produce an original composition with the newly developed electronic synthesizer, the Subharchord, located in the recently established Labor für akustisch-musikalische Grenzprobleme (Laboratory for Border Problems in Musical Acoustics). The resulting musical composition can be appraised as a precise commentary on the early effects of the Berlin Wall at the scale of the city. Based on the experience of being confronted with two radically different realities on either side of an enclosure, Rzewski recorded how the Wall divided the city and simultaneously created contrasting proximities. At the same time, the political, technical, and aesthetic conditions under which the music was conceived were, in reality, the symptoms of a new form of construction engendered within the laboratory for scientific acoustic research.

This interpretation of Rzewski's piece relates directly to the material covered earlier in the book. In chapter 1, I analyze how the city was delineated, defined, and occupied by the dissemination of sounds through electromagnetic signals. In chapter 2, I explain how a new building type emerged at the crossroads between mechanical and electronic technical media. Following the decreasing scalar progression, in chapter 3, I analyze the technical objects through which this relationship between the mechanical, electronic, and electromagnetic realms was constructed. In doing so, I examine the electroacoustic laboratory as a new form of construction site.[42] Situated as an output of the electroacoustic laboratory, the music piece composed by Rzewski could be understood as a condensation and continuation of a series of scientific, technological,

42 Bruno Latour has theorized the relationship between construction and the space of the laboratory. According to him, the laboratory constitutes the construction site of scientific reality in that it actively transforms *facts* into *artifacts*. For further reading, see Bruno Latour and Steve Woolgar, *Laboratory Life: The Construction of Scientific Facts* (Princeton: Princeton University Press, 1986), 235–261.

institutional, and political practices originating in Berlin during the 1920s and 1930s, all of which were directly related to radio. Therefore, in this section, I trace the institutional and technological genealogy of a specific kind of laboratory space that emerged in Berlin and was dedicated to understanding the entangled relationship between sounds, space, and electricity.

Rzewski's piece was composed in the Labor für akustisch-musikalische Grenzprobleme, created in East Berlin in 1956. The Labor was modeled after the Rundfunkversuchsstelle (Radio Research Laboratory), which was founded in Berlin in 1929 and depended on support from the Academy of Music and the Reich Broadcasting Corporation. Consolidating various cultural and scientific interests, the Rundfunkversuchsstelle was an embryonic version of an interdisciplinary institution that aimed to rethink the intersection of sound and music in view of emerging electroacoustic technologies and to bridge the gap between culture and technology that radio had opened up. The Heinrich-Hertz-Institut für Schwingungsforschung (Heinrich Hertz Institute for Research on Oscillations, or HHI), founded in 1928, was another pioneering interwar organization, committed to unlocking the commercial and technological potential latent in the technical medium of radio. With the reorganization of institutions that followed the division of Germany at the end of the war, the work of the HHI was continued by the Rundfunk- und fernsehtechnisches Zentralamt (RFZ), which became, in 1950, the GDR's central institution dedicated to all technical innovations related to radio, television, and electronic information technologies. The Labor für akustisch-musikalische Grenzprobleme operated within the institutional frame of the RFZ. It was due to the trans-scalar nature of these laboratories that space—understood as the medium for the production, transmission, and reception of information—eventually became electrified. Moreover, the very practices and technologies originating in those laboratories served as the epistemological grounds for Berlin's electrified—and highly contested—

acoustic space after World War II. In sum, this book undertakes an environmental investigation of radio and its impact on space, time, and energy as analyzed at the scales of the city territory, buildings, and technical objects, respectively.[43]

43 Lefebvre theorized the relationship between space, time, and energy as it relates to rhythms. See Henri Lefebvre, *Rhythmanalysis: Space, Time and Everyday Life* (London: Bloomsbury, 2004).

For millennia, the experience of hearing and listening was intrinsically linked to space. Until the advent of electroacoustic technologies, the idea of a sound decoupled from the imprint given to it by the space in which it was performed was technically and intellectually unimaginable. The voices and whispers of early stories were as much the product of the vocal cords of those who spoke or sang as they were the outcome of those vibrations bouncing on the width, height, and even humidity of caves. The whistles of birds and people were produced as much by the air flowing through their mouths as by the way they echoed through valleys and mountains. Mothers' voices entered their progeny's ears, mediated by the damped intimacy of curtains and pillows or the amplifying reverb of long corridors. The sounds that people heard in the premiere of Beethoven's String Quartet no. 11 were as much the outcome of the score and the strings of the instruments as they were the effects of the wood and drapery defining the walls of the room in which the musicians played. Every sound that shaped human and non-human existence, providing grounds for transferring knowledge, stories, joy, suffering, myth, and culture through different channels of sonic communication, was inevitably a byproduct of sounds *in* space, of sounds *as* space. Until the first decades of the twentieth century, nobody ever experienced a space without *its* sounds or a sound without *its* space. Space and sounds were always intrinsically linked by the phenomenon of reverberation. Sound was the time-sensible imprint of geometry and materiality on air as it moved.

During the first decades of the twentieth century, a well-researched break took place in the historical unity of sounds with space. That schism has been attributed to the scientific research pursued by engineer

Wallace C. Sabine in the United States, which turned reverberation into an objectively measurable by-product of the effects of sounds in space. Shaping a new sonic rationality, Sabine's work led to the dismantling of the millennia-long bond linking sonic vibrations with space: once mathematically isolated, reverberation could be controlled and eventually designed. The consequences of that break for architecture were enormous. Affected by this new sonic episteme, it did not take long until reverberation was associated with pure noise and understood as something to be eradicated from human sonic interactions.[44] As sound was understood as *signal* and reverberation as *noise,* specially engineered dampening materials were invented, and electroacoustics emerged, apparently giving the *coup de grâce* to the sound of space. Through a new mechanic and electroacoustic materiality, the "modern soundscape" was technically and epistemologically constructed as one of pure signal with zero noise. According to the historian of technology Emily Thompson, "the sound of space was effectively eliminated from the new modern sound as reverberation came to be considered an impediment, a noise that only interfered with the successful transmission and reception of the desired sound signal."[45]

If the first three decades of the twentieth century have been the historical locus to study the cultural and scientific understanding of noise and its later banishment from human sonic interactions in space, post-1933 Berlin is the precise context to expand on that understanding and eventually destabilize the linear narrative that explained the progressive separation and distinction of sound from space. This book sets out to retrace the relationship between sound, space, and electroacoustic technologies by situating the ear and the gaze after the periods of pioneering inventions, in a politically contested territory. In an agitated Berlin, the idea of an "ideal reverberation time" was abandoned, only to be replaced by a multiplicity of reverberation times fabricated by buildings and disseminated and overlapped throughout the city.

44 This reading is indebted to Emily Thompson's *The Soundscape of Modernity*, one of the most influential studies published on the interaction of sound, technology, and space. For more, see Emily Thompson, *The Soundscape of Modernity: Architectural Acoustics and the Culture of Listening in America, 1900–1933* (Cambridge, MA: MIT Press, 2002).

45 Thompson, *The Soundscape of Modernity*, 234.

Far from being abolished or avoided, noise was treated as a differential political factor and eventually understood as a signal itself, as in the case of jammed transmissions. Moreover, domestic as well as urban spaces were no longer considered as an exclusive by-product of built artifacts, but rather as a construction composed of interactions among buildings, sounds, and a series of electronic technologies that mediated those interactions. If we accept that the modern acoustic space during the first decades of the twentieth century was "dried out," emptied of reverberation and "noise," this book attempts to show how acoustic space in and from Berlin became "wet" again, so to speak, by becoming a globally contested battleground of political, technological, and cultural practices through the complex—and equally contested—construction of the architecture of radio.

The MIT Press Cambridge, Massachusetts London, England

RADIO-ACTIVITIES: ARCHITECTURE AND BROADCASTING IN COLD WAR BERLIN

ALFREDO
THIERMANN RIESCO

TABLE OF CONTENTS

PART 1
THE CITY
SOURCE MATERIAL

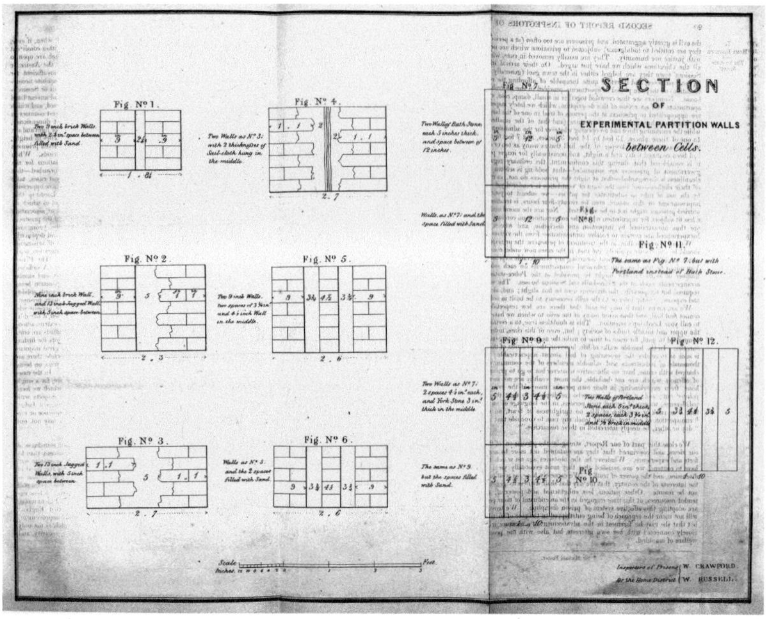
SECTION
OF
EXPERIMENTAL PARTITION WALLS
between Cells.
Fig. No 1.
Fig. No 2.
Fig. No 3.
Fig. No 4.
Fig. No 5.
Fig. No 6.
Fig. No 9.
Fig. No 11.
Fig. No 12.
W. CRAWFORD.
W. RUSSELL.

1

2

3

TELE
FUN
KEN

4

5

36

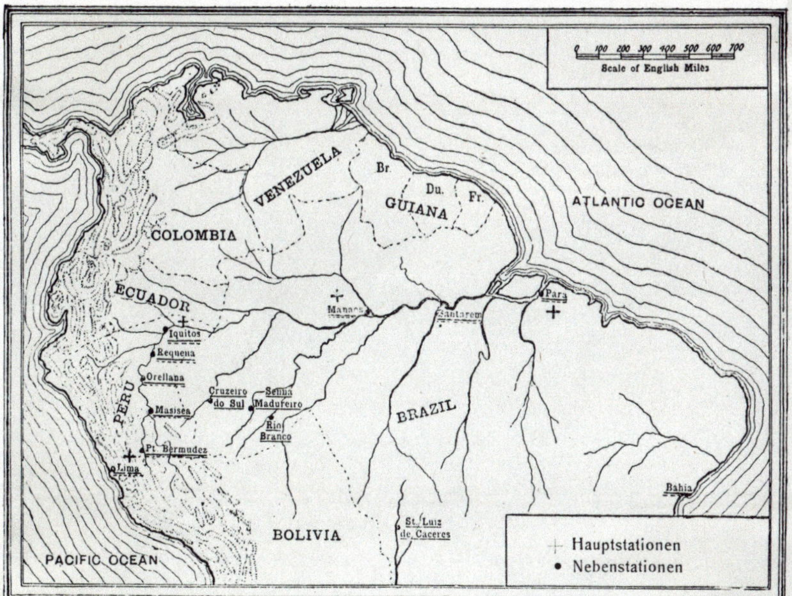

B40900

Bild 23

1905-1911 – Südamerika – Während der Knallfunkenperiode (bis 1909) wurde der peruanische Teil der Funkverbindung vom Pazifik zum Atlantik, die Verbindung von Puerto Bermudez bis Iquitos geschaffen. Die Fortsetzung durch Brasilien bis Parà wurde 1911 fertiggestellt. (Seite 28).

42

B 41094 L

Bild 29 -

1912 - Iquitos (Peru) - Die zweite von Telefunken in Iquitos errichtete ~~feste~~ Funkstation kurz vor ihrer Fertigstellung. Sie hatte einen Tönenden Löschfunkensender ~~für~~ ~~10~~ 5 kW Leistung ~~in der Schirmantenne~~ erhalten. Energiequelle war ein 40-PS-Rohölmotor. (S.28)

HB 565

Bild 24

1908 - Iquitos - Die Stationen Iquitos, Orellana und Requeña erhielten die ersten Schirmantennen mit einem 65 m hohen Eisengittermast, der durch einen Marmorblock von der Erde isoliert war und von drei ebenfalls isolierten Drahtseilen gehalten wurde. Der Sender war zuerst in der abgebildeten provisorischen Hütte aufgestellt. (S. 28)

9

10

11

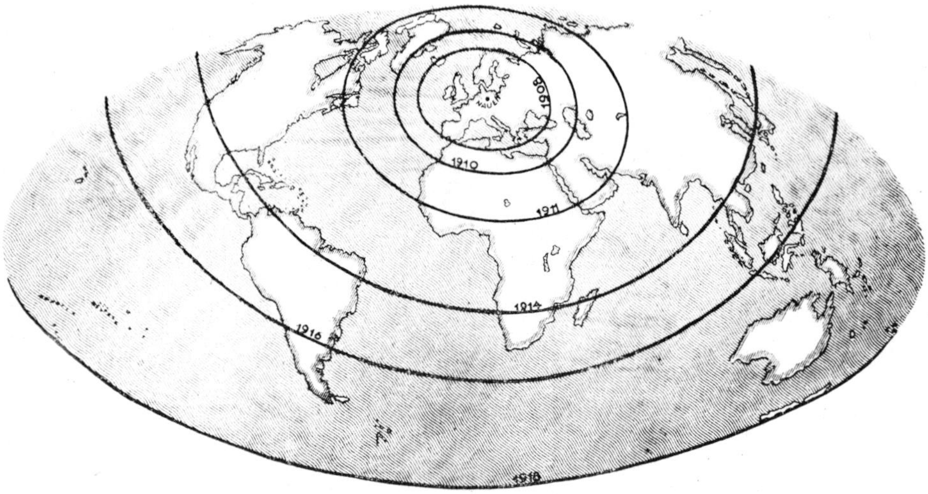
1908
1910
1911
1914
1916
1918

12

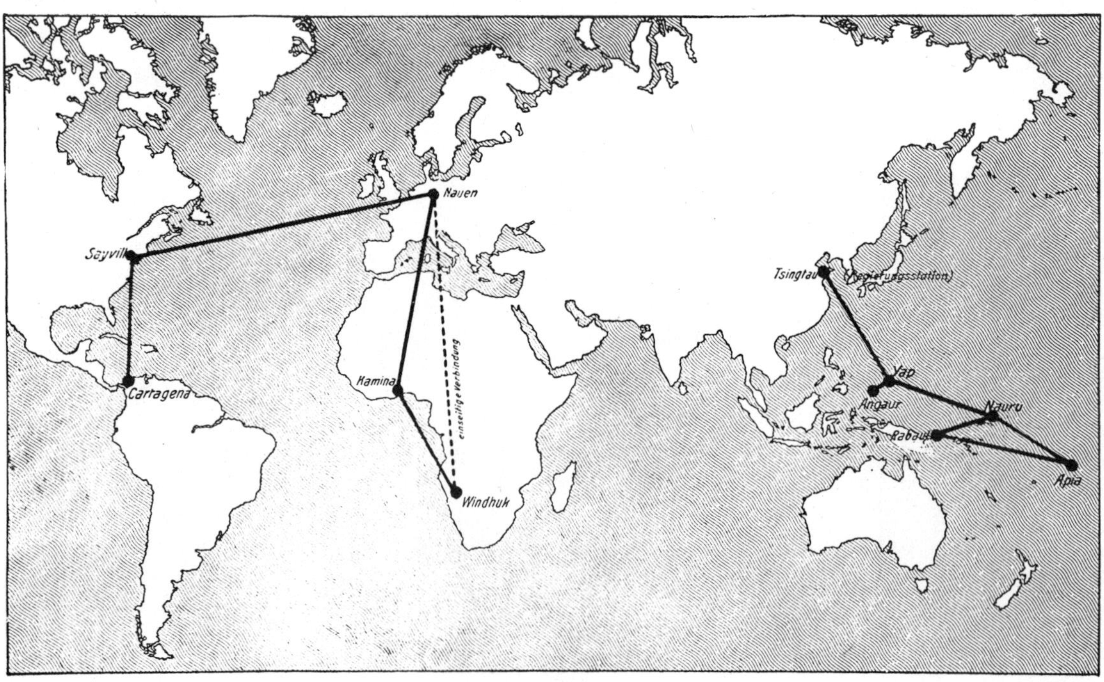
Nauen
Sayvill
Cartagena
Kamina
einseitige Verbindung
Windhuk
Tsingtau
(Regierungsstation)
Yap
Angaur
Nauru
Rabaul
Apia

13

Fig. 1 Study of wall sections conducted by Michael Faraday and Abel Blouet for Millbank Penitentiary. Source: W. Crawford and W. Russell, "Section of Experimental Partition Walls between Cells," in *Second Report of the Inspectors Appointed ... to Visit the Different Prisons of Great Britain,* vol. 1, *Home District*. London: W. Clowes & Sons for HMSO, 1837, Plate A.

Fig. 2 Nauen transmission station with its 100 m tower, ca. 1906. Source: Stiftung Deutsches Technikmuseum Berlin, Photo: Historisches Archiv. I.2.060 C 04270.

Fig. 3 In 1911, the Nauen tower was doubled in height, to 200 m, to extend the station's transmission capacities. Source: Stiftung Deutsches Technikmuseum Berlin, Photo: Historisches Archiv. I.2.060 C 04270.

Fig. 4 Interior view of the Nauen transmitter building. Source: Deutsches Dokumentationszentrum für Kunstgeschichte—Bildarchiv Foto Marburg. Aufnahme-Nr. 1.144.632.

Fig. 5 Aerial view of the Nauen transmitter building designed by Hermann Muthesius. Source: Deutsches Dokumentationszentrum für Kunstgeschichte—Bildarchiv Foto Marburg. Aufnahme-Nr. 1.144.628.

Fig. 6 Map showing Telefunken stations built between 1905 and 1911 in locations spanning from the Pacific to the Atlantic Ocean in Latin America. Source: Stiftung Deutsches Technikmuseum Berlin, Photo: Historisches Archiv. I.2.060 C 04270.

Fig. 7 Telefunken transmission station in Iquitos, Peru, under construction. Source: Stiftung Deutsches Technikmuseum Berlin, Photo: Historisches Archiv. I.2.060 C 04270.

Fig. 8 Construction view of Telefunken transmission station in Iquitos. The machinery is protected under an improvised tent, and the 65 m transmission tower sits on a marble block for electrical isolation. Source: Stiftung Deutsches Technikmuseum Berlin, Photo: Historisches Archiv. I.2.060 C 04270.

Fig. 9 Telefunken transmission station atop San Cristóbal hill in Lima, Peru. Source: *Telefunken-Zeitung* 2, no. 12 (1913): 156. Scan by PTT-Archiv, Köniz.

Fig. 10 Transmission station built by Telefunken in Beijing, China. Source: Hans Bredow, "Telefunken in China," *Telefunken-Zeitung* 1, no. 4 (1912): 41. Scan by PTT-Archiv, Köniz.

Fig. 11 An expedition of Telefunken technicians into the Bolivian *altiplano* was supported by the German army. Source: "Bolivien," *Telefunken-Zeitung* 3, no. 13 (1914): 30. Scan by PTT-Archiv, Köniz.

Fig. 12 Map showing the increasing coverage of the Nauen station's signals over the years. Source: Artur Fürts, *Im Bannkreis von Nauen: Die Eroberung der Erde Durch Drahtlose Telegraphie* (Stuttgart; Berlin: Deutsche Verlags-Anstalt, 1923), 213.

Fig. 13 Map showing Telefunken transmission and reception stations in the Americas, Africa, and Southeast Asia. Source: Artur Fürts, *Im Bannkreis von Nauen: Die Eroberung der Erde Durch Drahtlose Telegraphie* (Stuttgart; Berlin: Deutsche Verlags-Anstalt, 1923), 212.

7C

№ A S37

Bild 52

HB 550 (Wüst, Hann)

1914 - Kamina (Togo) - Fotografische Wiedergabe eines Gemäldes von Prof. Vollbehr, die die Gesamtanlage der Großfunkstelle erkennen läßt. Die Großfunkstelle Kamina gilt mit Recht als ein Markstein in der Entwicklung des kolonialen Nachrichtenverkehrs. (S. 45)

74

A 2008

Bild 56

1914 - Kamina - Links das Gebäude für die Empfangsanlage, in der auch bereits Kathodenröhren nach von Lieben als Niederfrequenz- und Hochfrequenzverstärker verwendet waren. Rechts das Gästehaus der Großfunkstelle. (S. 45)

71

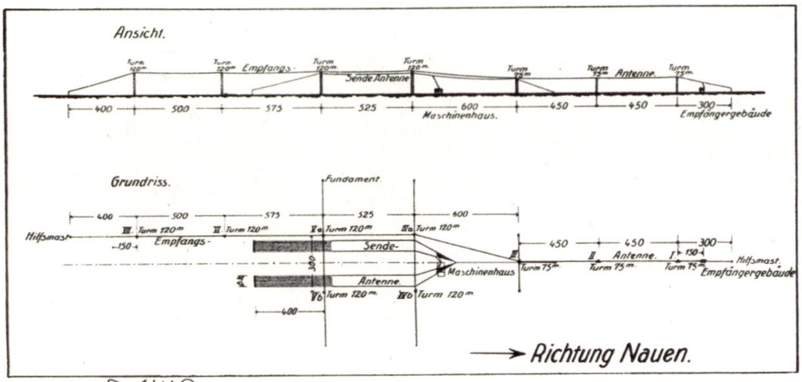

B 41119

Bild ~~28~~ 53

1914 – Kamina – Die Großstation hatte eine Dachantenne nach dem Muster von Nauen erhalten, die in der Richtung auf Nauen orientiert war. Die Senderantenne wurde von vier 120 m hohen Masten getragen. Weitere zwei Masten von 120 m Höhe und drei Masten von 75 m Höhe standen für die 3000 m lange Empfangsantenne zur Verfügung. (S. 45)

68

Bild ~~48~~ ~~49~~ 50

1913 - Küstenfunkstelle Daressalam (Deutsch-Ostafrika) - Der 100 m hohe Mast trug eine Schirmantenne. Der Sender führte ihr 7,5 kW Leistung zu. Der Wellenbereich betrug 300 bis 3500 m. Die Energie lieferte ein Rohölmotor von 40 PS. (S. 45)

18

B 36839

Bild 54

1914 - Kamina - Als Energiequelle der Großstation Kamina dienten Dampfturbinen, die mit 500-Periodengeneratoren und den dazugehörigen Erregermaschinen direkt gekuppelt waren. (S. 45)

20

79

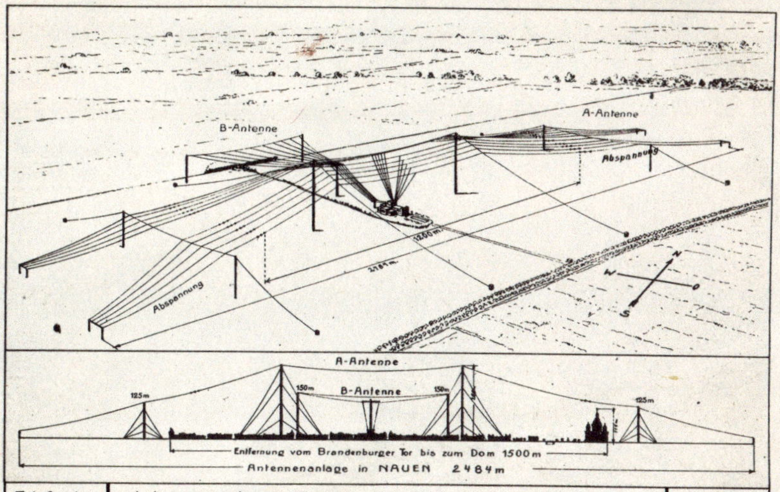

Telefunken | Antennenanlage Nauen 1920 für Langwellensender | D 3564

Bild ~~69~~ 59

1916 – <u>Nauen</u> – Schema der beiden Dachantennen in Nauen. Die langgestreckte A-Antenne mit einer Länge von 2484 m ohne die Abspannungen diente dem transozeanischen Verkehr. Ihr standen zwei Masten zu 260 m und vier von je 125 m zur Verfügung. Die B-Antenne in Gestalt eines gleichschenkligen Dreiecks war zwischen 150 m hohen Masten ausgespannt. Sie diente dem Europaverkehr. Da beide Antennen gleichzeitig zum Senden benutzt werden konnten, war zum ersten Male ein ~~Duplexverkehr~~ Doppelsendebetrieb möglich. (~~siehe~~ S.49)

80

Die Antennenanlagen der deutschen Großfunkstation
Nauen im Jahre 1916.
Sie ist noch gekennzeichnet durch ein Gewirr von
Antennendrähten.

B42146 o.

Fig. 14 Telefunken transmission station in Kamina, Togo. Source: Stiftung Deutsches Technikmuseum Berlin, Foto: Historisches Archiv. I.2.060 C 04270.

Fig. 15 Telefunken transmission station in Kamina, Togo. On the left is the building dedicated to receiving signals, and on the right is the guesthouse. The construction and operation of the station involved forced labor, and these workers were housed elsewhere. Source: Stiftung Deutsches Technikmuseum Berlin, Foto: Historisches Archiv. I.2.060 C 04270.

Fig. 16 Diagram in plan and section of the reception and transmission roof antenna in the Telefunken station in Kamina. The diagram shows how the antenna was oriented toward Nauen with a 3,000 m extension. Source: Stiftung Deutsches Technikmuseum Berlin, Foto: Historisches Archiv. I.2.060 C 04270.

Fig. 17 Telefunken transmission station, Dar es Salaam, Tanzania, 100 m tower. Source: "Der Bau der Telefunken-Station Kamina (Togo)," *Telefunken-Zeitung* 2, no. 12 (1913): 180.

Fig. 18 Telefunken transmission station in Kamina, Togo, interior view of boiler plant. Source: Stiftung Deutsches Technikmuseum Berlin, Foto: Historisches Archiv. I.2.060 C 04270. This image was previously published in "Die Grossstation Kamina und der Beginn des Weltkrieges," *Telefunken-Zeitung* 3, no. 16 (1919): 31. PTT-Archiv, Köniz.

Fig. 19 Telefunken transmission station in Kamina, Togo. Interior view showing the steam turbines fabricated by AEG. Source: Stiftung Deutsches Technikmuseum Berlin, Photo: Historisches Archiv. I.2.060 C 04270.

Fig. 20 Interior view of the turbine hall designed by Peter Behrens for AEG in 1909 depicting the fabrication process of steam turbines. Source: Allgemeine Elektricitäts-Gesellschaft, ed., *25 Jahre AEG-Dampfturbinen* (Berlin: VDI-Verlag, 1928), 99.

Fig. 21 Nauen transmission station, perspective and section of the antenna comparing its extent to the length of the boulevard Unter den Linden. Source: Stiftung Deutsches Technikmuseum Berlin, Photo: Historisches Archiv. I.2.060 C 04270.

Fig. 22 Nauen transmission station depicted under the web of roof antennas constructed after the final expansion of its transmitting capacities in 1916. Source: Stiftung Deutsches Technikmuseum Berlin, Photo: Historisches Archiv. I.2.060 C 04270.

Köln 1928.

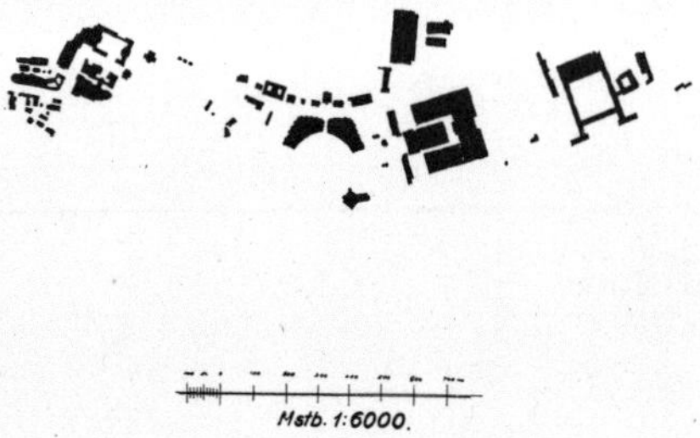

Düsseldorf 1926.

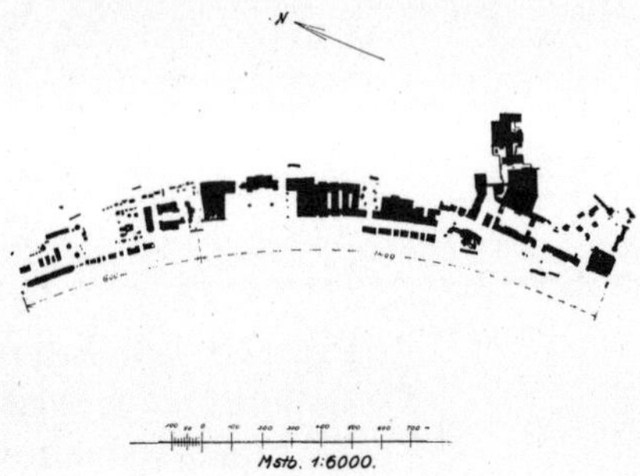

Brüssel 1910.

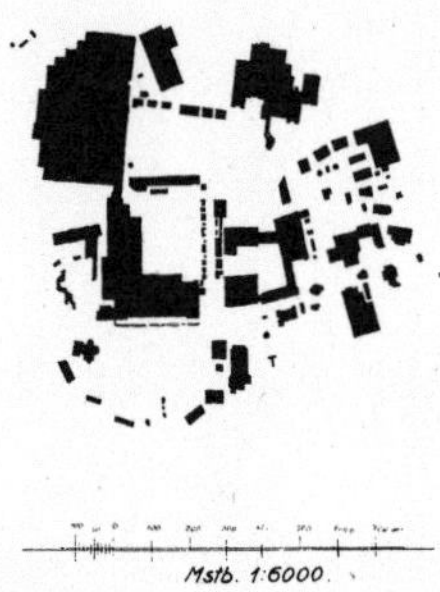

Paris 1900.

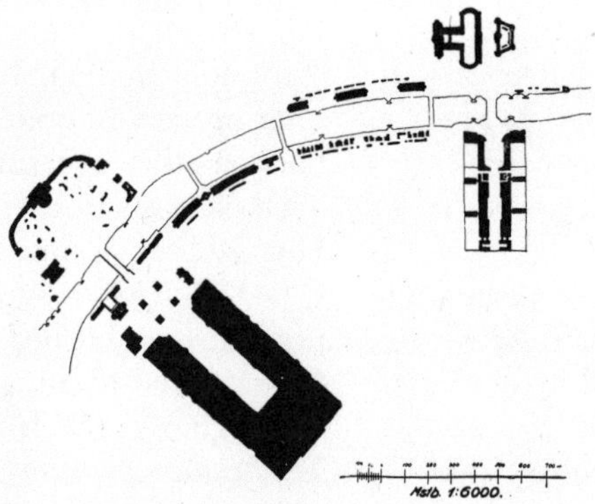

Mailand 1906.

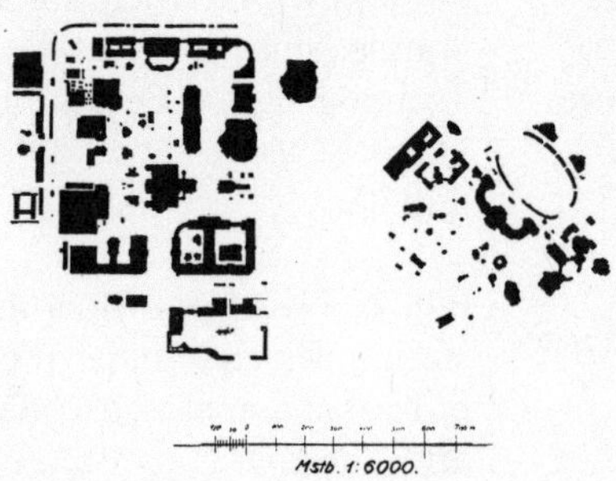

Berlin 1930.

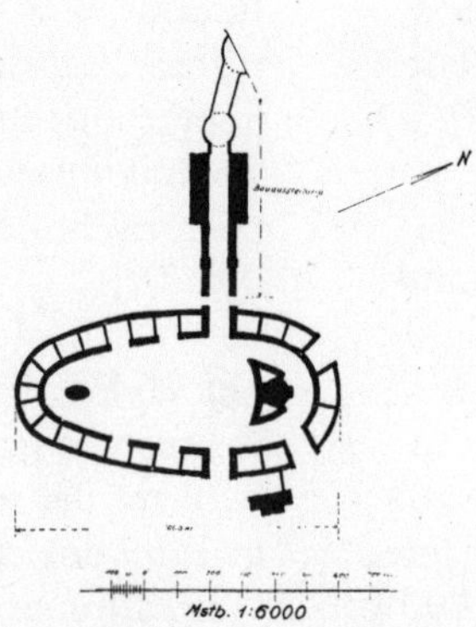

24

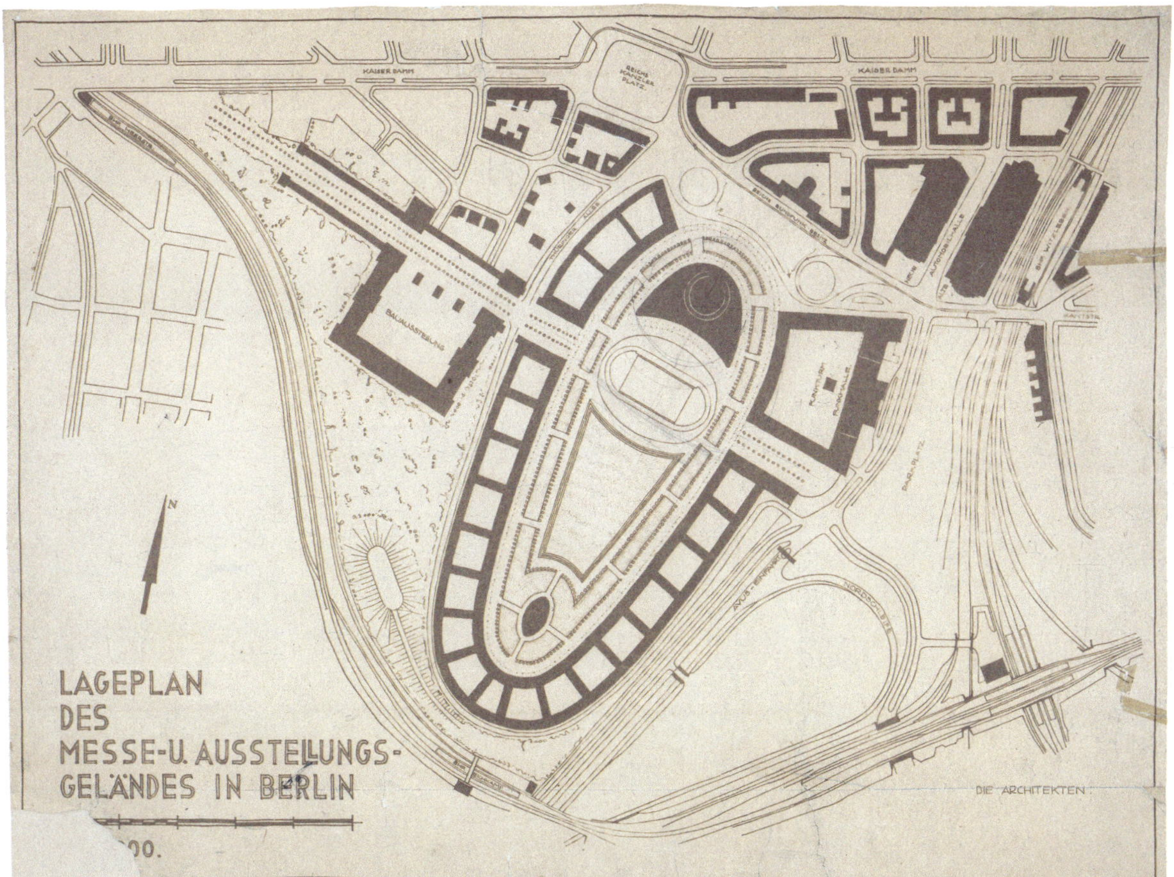
LAGEPLAN
DES
MESSE-U. AUSSTELLUNGS-
GELÄNDES IN BERLIN
DIE ARCHITEKTEN

25

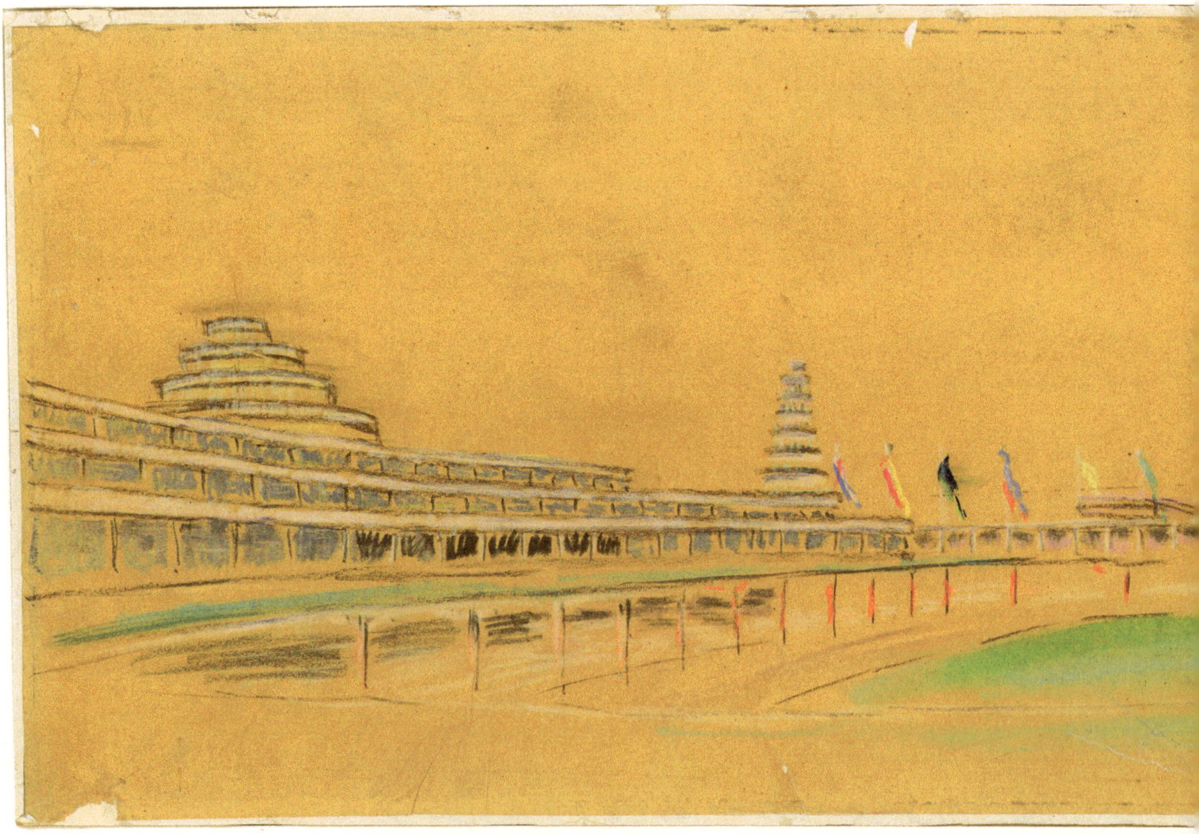

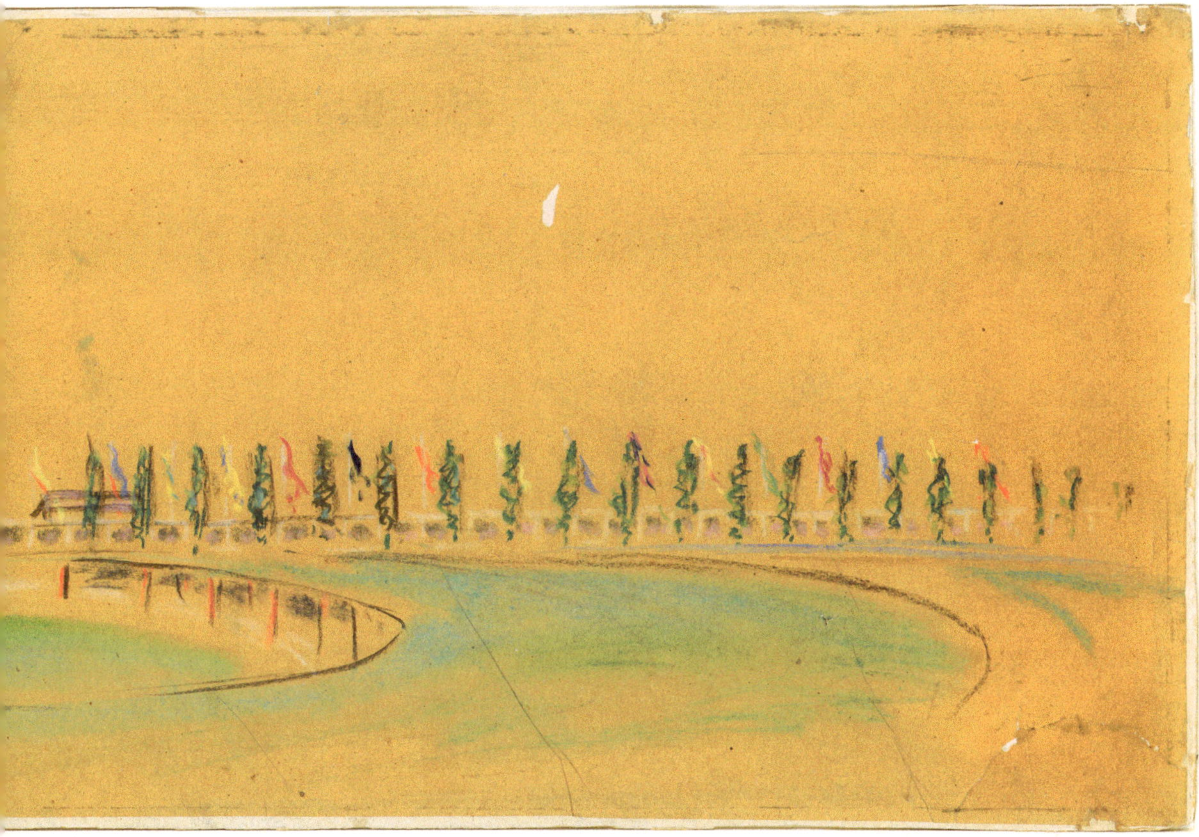

26

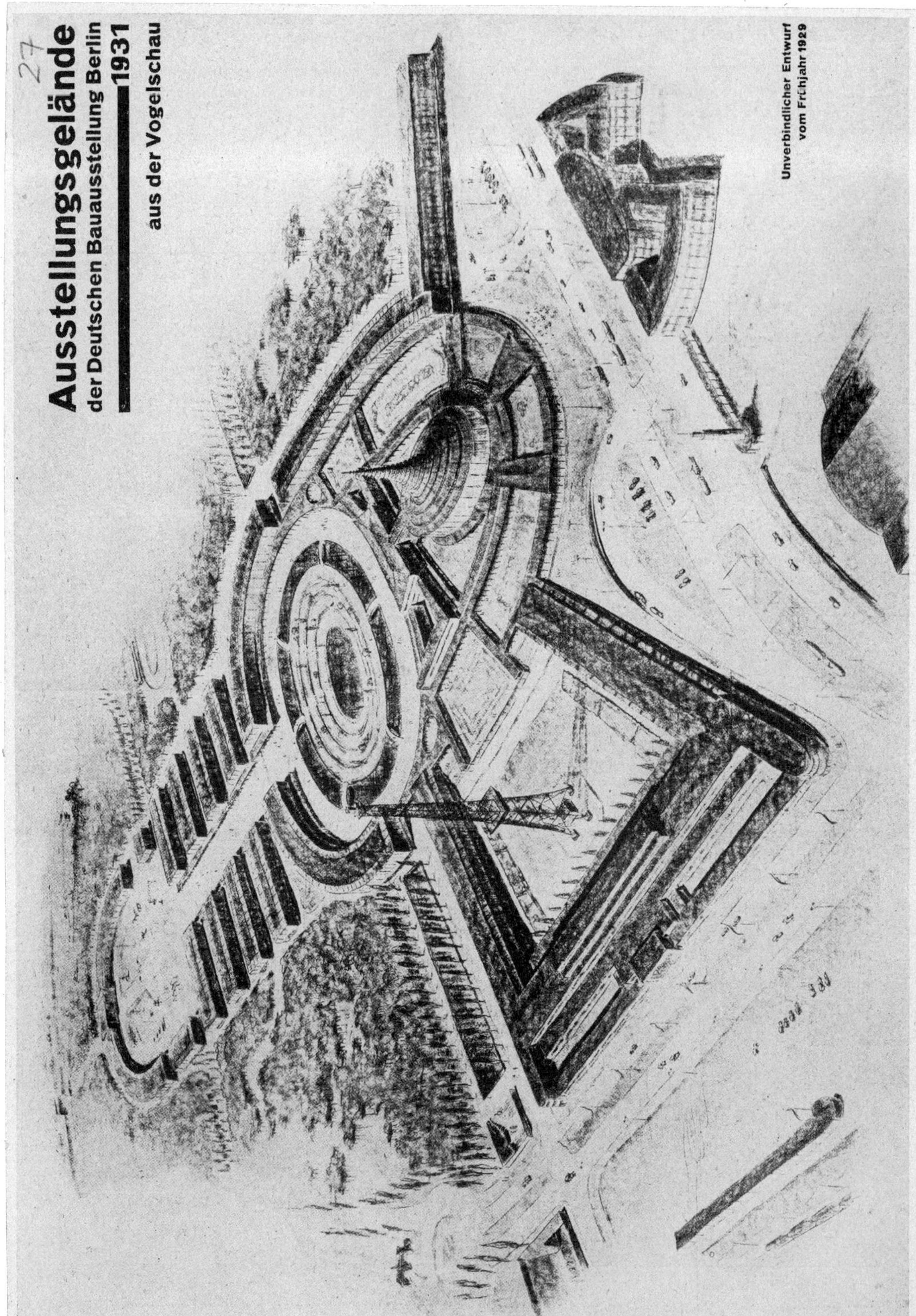
Ausstellungsgelände
der Deutschen Bauausstellung Berlin
1931
aus der Vogelschau
Unverbindlicher Entwurf
vom Frühjahr 1929

27

Inneres der Halle.

Das Haus der Deutfchen Funkinduftrie.
Berlin-Witzleben.

Architekt: Prof. Heinrich Straumer, B. D. A.
Berlin.

28

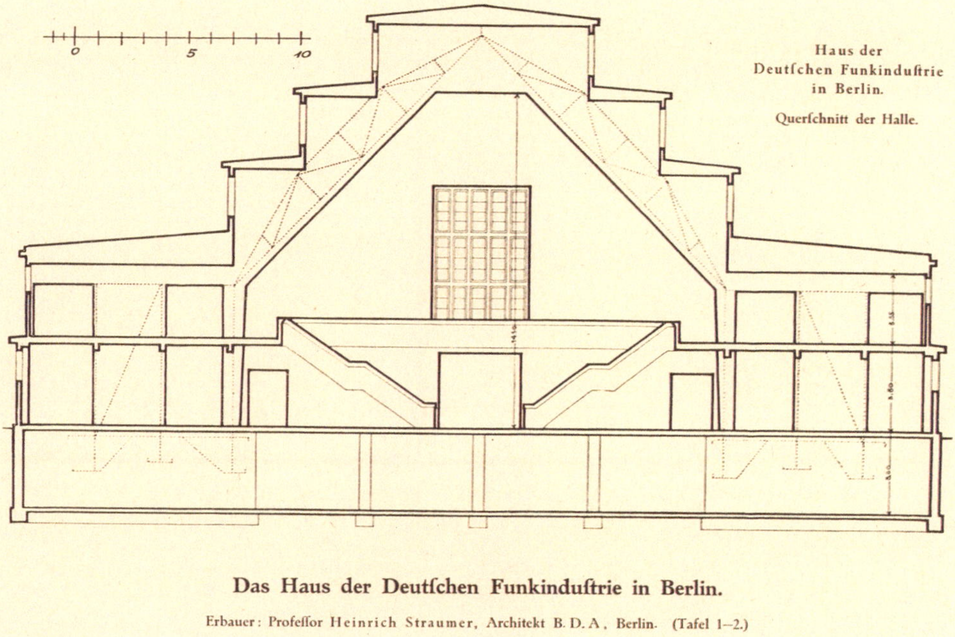

Haus der
Deutfchen Funkinduftrie
in Berlin.

Querfchnitt der Halle.

Das Haus der Deutfchen Funkinduftrie in Berlin.

Erbauer: Profeffor Heinrich Straumer, Architekt B. D. A., Berlin. (Tafel 1–2.)

29

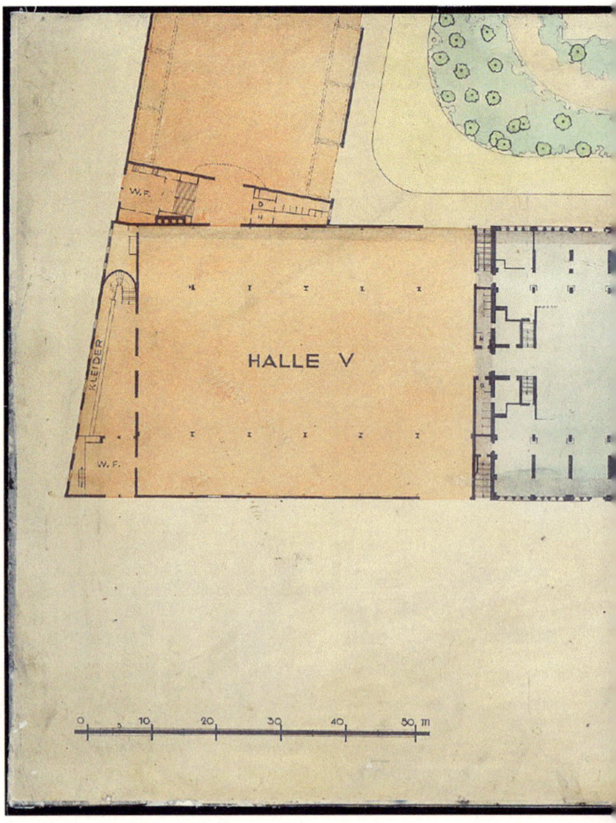
HALLE V
KLEIDER
W.F.
W.F.
0 5 10 20 30 40 50 m

30

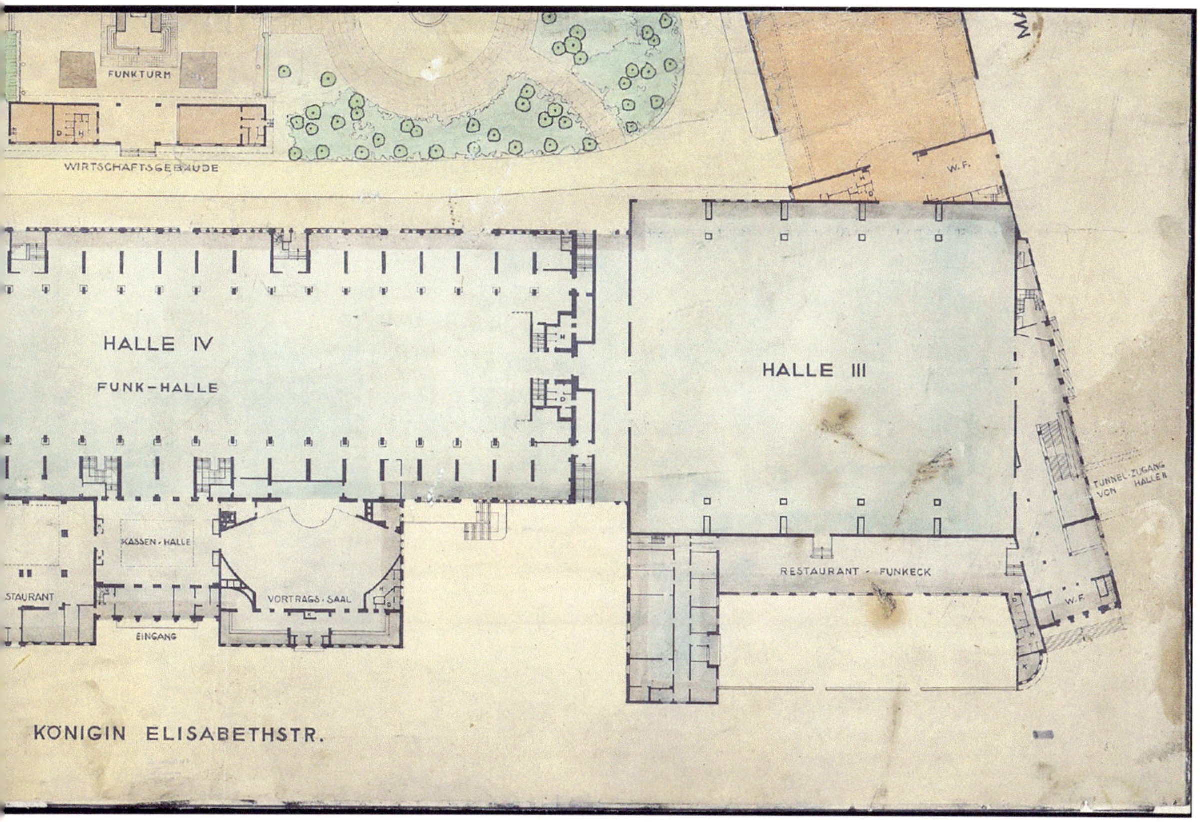
FUNKTURM
WIRTSCHAFTSGEBÄUDE
W.F.
HALLE IV
FUNK-HALLE
HALLE III
TUNNEL-ZUGANG VON HALLEN
KASSEN-HALLE
VORTRAGS-SAAL
RESTAURANT - FUNKECK
W.F.
EINGANG
KÖNIGIN ELISABETHSTR.

31

32

33

34

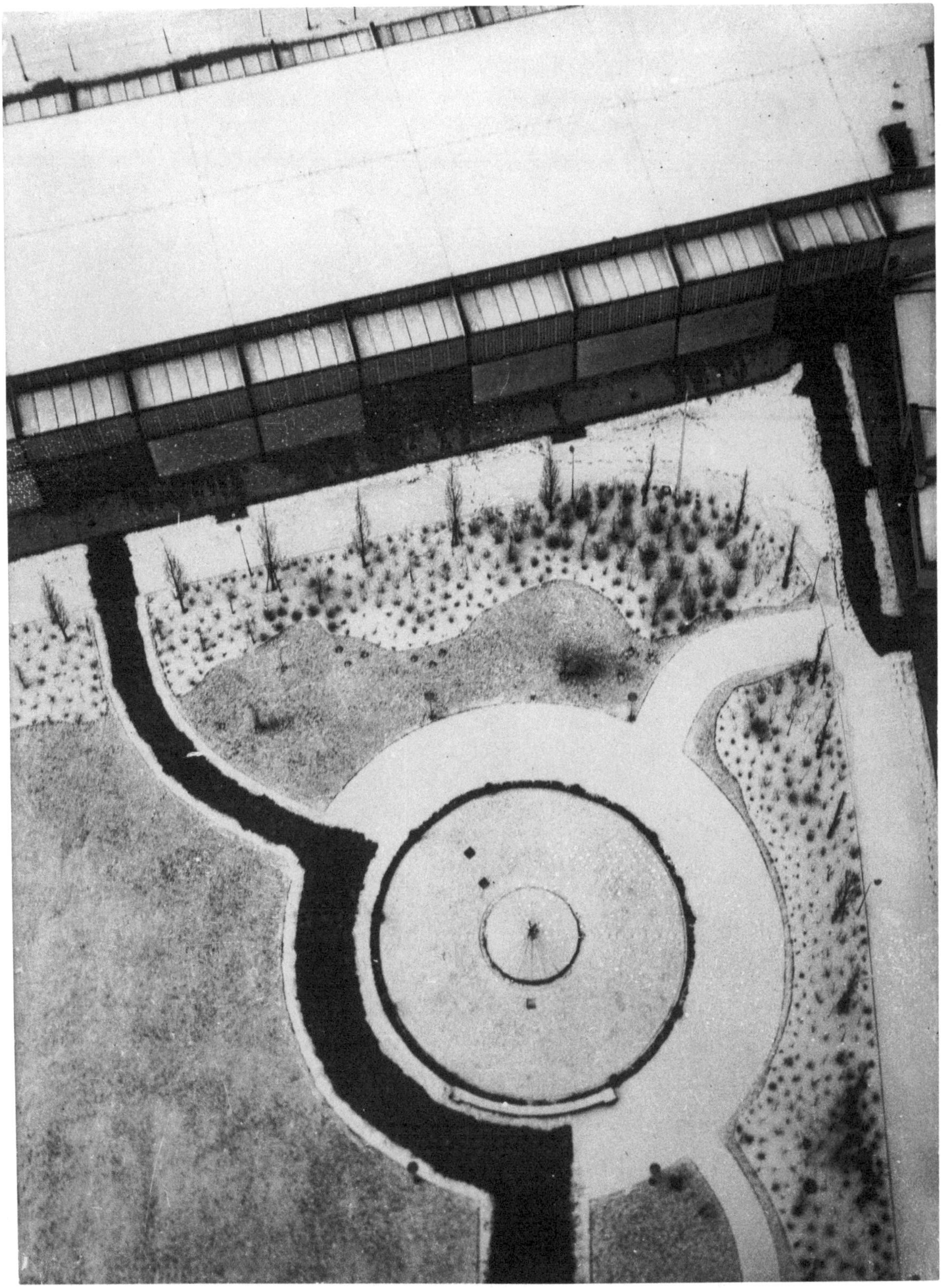

35

36

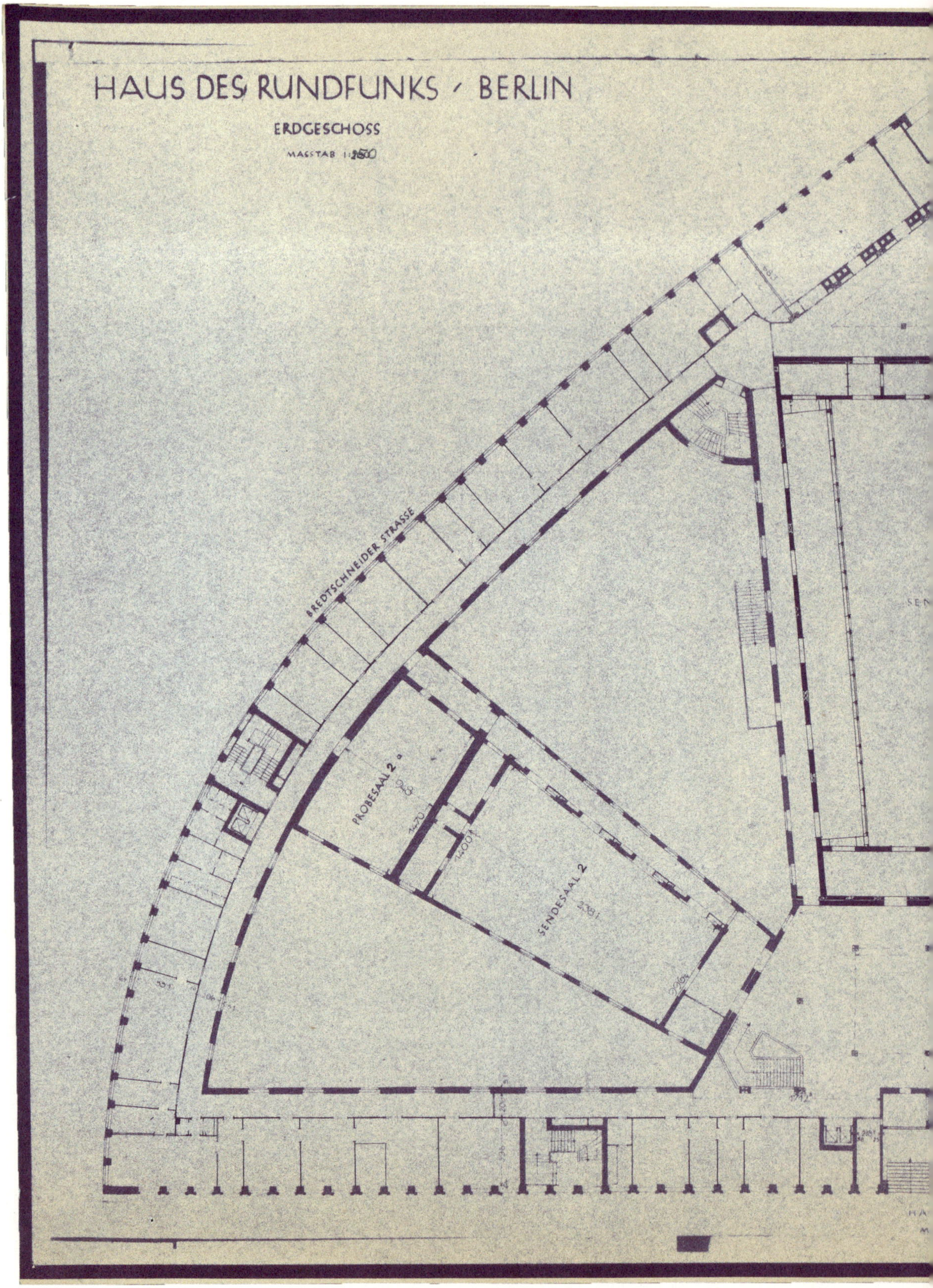
HAUS DES RUNDFUNKS / BERLIN
ERDGESCHOSS
MASSTAB 1:250
BREDTSCHNEIDER STRASSE
PROBESAAL 2 a
SENDESAAL 2

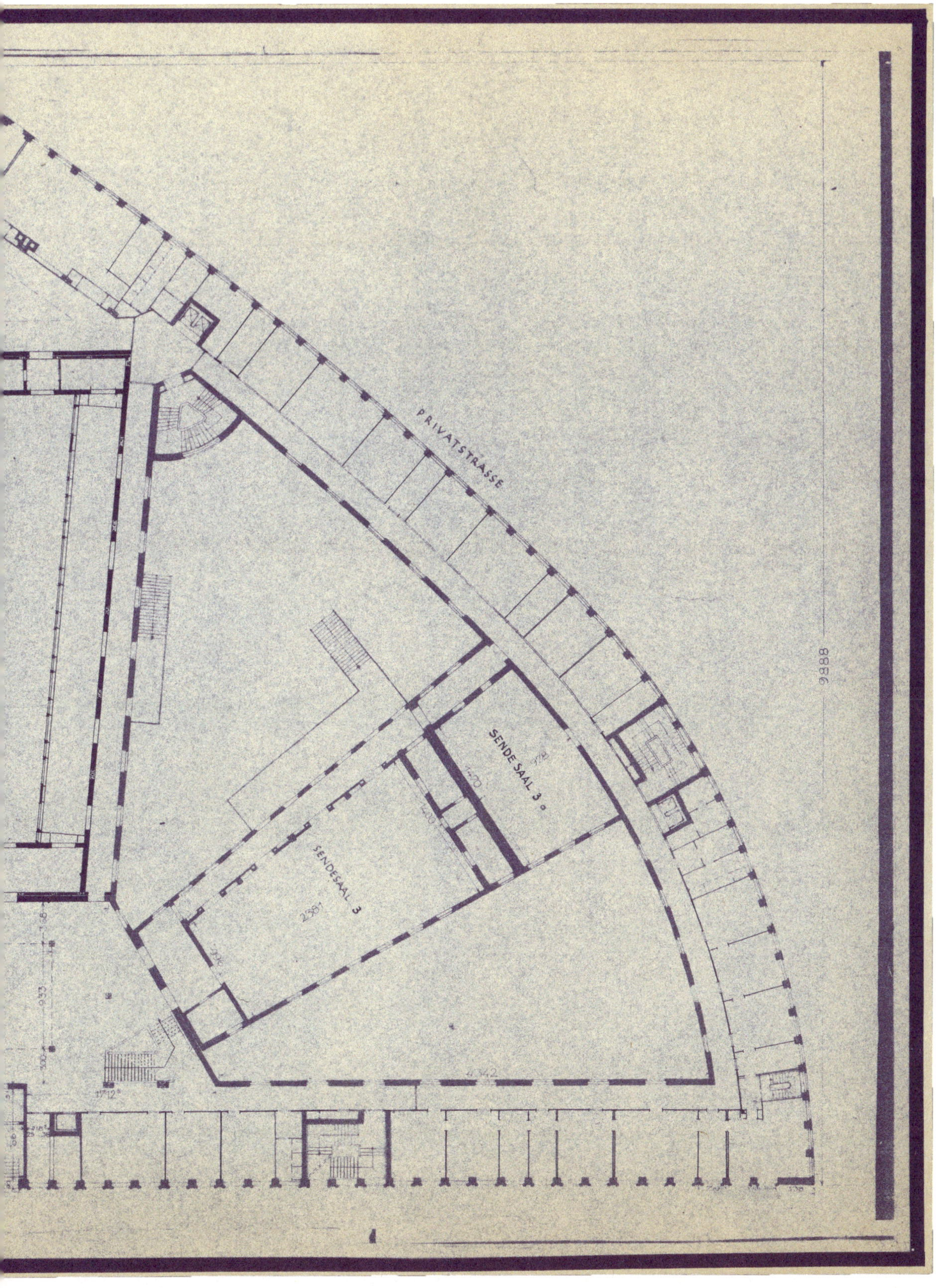

37

38

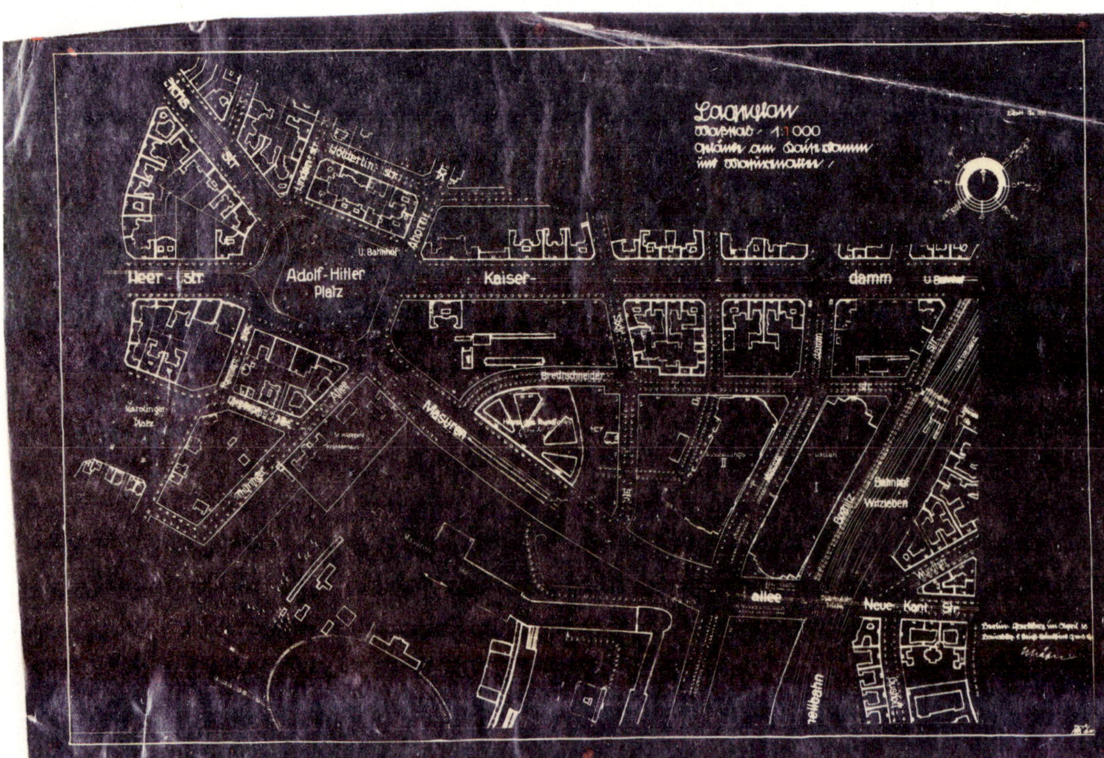

Heer - Str.
Adolf-Hitler Platz
U Bahnhof
Kaiser-
damm
allee
Neue Kant Str.

39

Fig. 23 Series of plans comparing the exhibition grounds of different cities. Source: Martin Wagner, "Die Reichsausstellungsstadt Berlin," *Das Neue Berlin* 1 (1929): 12.

Fig. 24 Berlin radio tower designed by Heinrich Straumer. Photograph by the author.

Fig. 25 Plan of the exhibition grounds designed by Hans Poelzig and Martin Wagner. Source: TU Berlin Architekturmuseum, Inv. Nr. 3451.

Fig. 26 Berlin exhibition grounds, perspective showing the congress hall and main restaurant on the left and, in the center, marked in red, the radio tower designed by Heinrich Straumer. Source: TU Berlin Architekturmuseum, Inv. Nr. 3512.

Fig. 27 Berlin exhibition grounds, bird's-eye view showing the Haus des Rundfunks in the bottom-right corner. Source: Frances Loeb Library, Harvard University Graduate School of Design. Papers of Martin Wagner, 1924–1965 (inclusive), 1937–1957 (bulk). *Neue Berlin: Monatshefte für Probleme der Grossstadt.* [New Berlin: Monthly Journal on the Issues of the Metropolis]. DES-1971-0001-000603299 27.

Fig. 28 Funkhalle/Haus der Deutschen Funkindustrie, interior perspective. Architect: Heinrich Straumer. Source: Erich Blunck, "Das Haus der deutschen Funkindustrie in Berlin-Charlottenburg," *Deutsche Bauzeitung* 59, no. 14 (1925): 107.

Fig. 29 Funkhalle, cross section. Source: Erich Blunck, "Das Haus der deutschen Funkindustrie in Berlin-Charlottenburg," *Deutsche Bauzeitung* 59, no. 14 (1925): 108.

Fig. 30 Aerial view of the exhibition grounds showing the radio tower and the Funkhalle/Haus der Deutschen Funkindustrie. Source: TU Berlin Architekturmuseum, Inv. Nr. 3535.

Fig. 31 Funkhalle, plan of level 0 inserted within the Hans Poelzig and Martin Wagner master plan. Source: TU Berlin Architekturmuseum, Inv. Nr. 3461.

Fig. 32 Exterior perspective of Hans Poelzig's competition design for the Funkhalle. Source: TU Berlin Architekturmuseum, Inv. Nr. 3543.

Fig. 33 Interior perspective of Hans Poelzig's competition design for the Funkhalle. Source: TU Berlin Architekturmuseum, Inv. Nr. 3541.

Fig. 34–35 László Moholy-Nagy's photographs from the top of the radio tower designed by Heinrich Straumer. Source: Gelatin silver print, Thomas Walther Collection, Museum of Modern Art, New York. Object numbers 1793.2001 and 485.1939. The sequence of images was published in *Der Querschnitt 11, no. 5 (1931): 303.*

Fig. 36 Haus des Rundfunks, main facade onto Masurenallee. Source: TU Berlin Architekturmuseum, Inv. Nr. 4918.

Fig. 37 Haus des Rundfunks, level 0. Source: TU Berlin Architekturmuseum, Inv. Nr. 4635.

Fig. 38 Heinrich-Hertz-Institut für Schwingungsforschung, exterior view. Source: Karl Willy Wagner, "Das Heinrich-Hertz-Institut für Schwingungsforschung," *Elektrische Nachrichten-Technik* 7, no. 5 (1930): 174.

Fig. 39 Site plan showing the Haus des Rundfunks and the exhibition grounds as central elements of the east–west axis of Hitler's Berlin. Source: Deutsche Rundfunkarchiv Frankfurt am Main. A48 RRG Bauzeichnungen, 411-531.

40

41

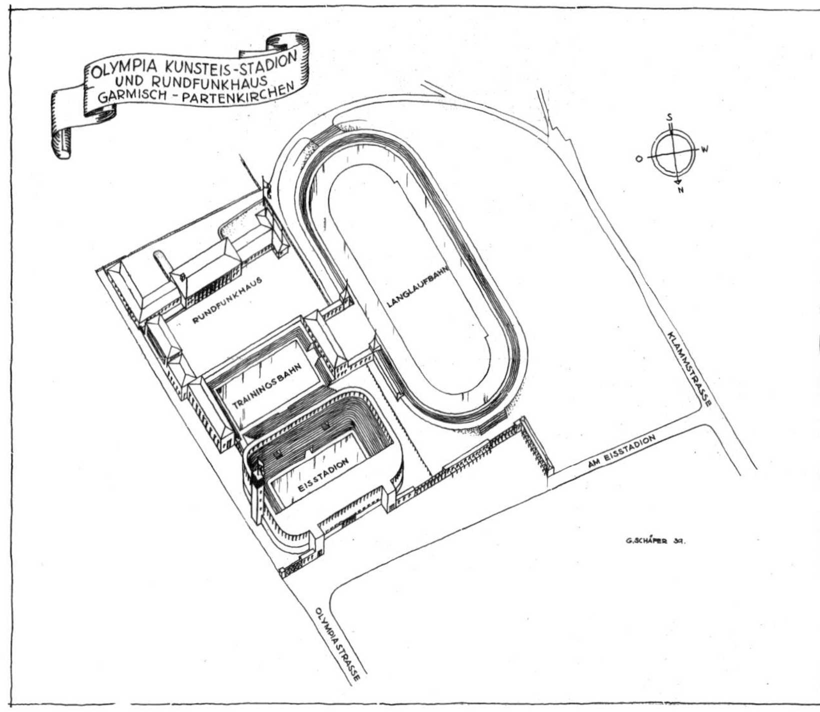
OLYMPIA KUNSTEIS-STADION
UND RUNDFUNKHAUS
GARMISCH - PARTENKIRCHEN
RUNDFUNKHAUS
LANGLAUFBAHN
TRAININGSBAHN
EISSTADION
KLAMMSTRASSE
AM EISSTADION
OLYMPIASTRASSE
S
W
O
N

42

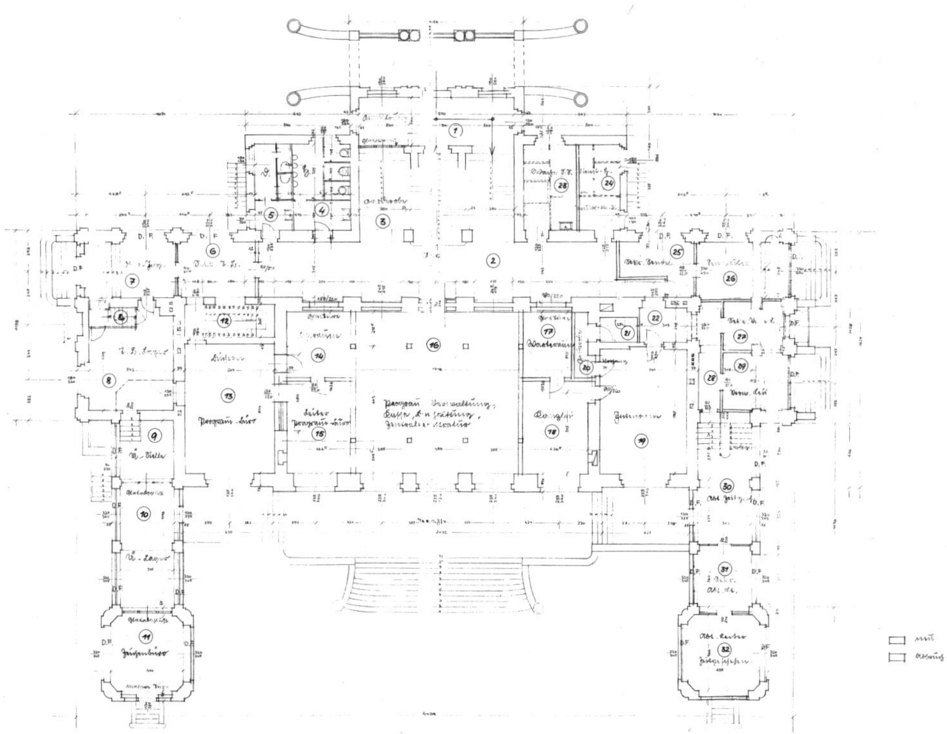

43

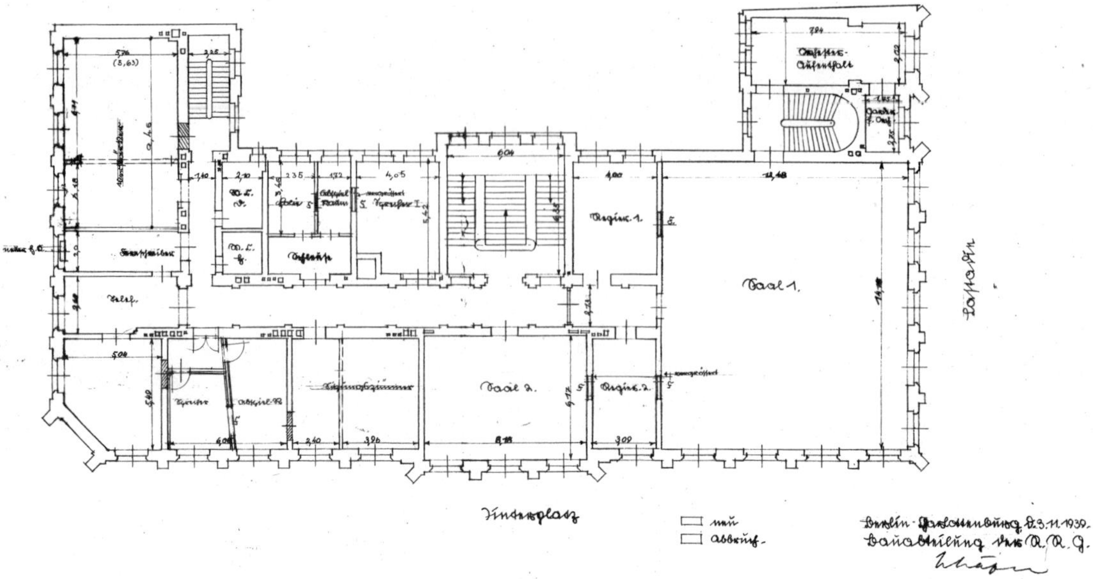

44

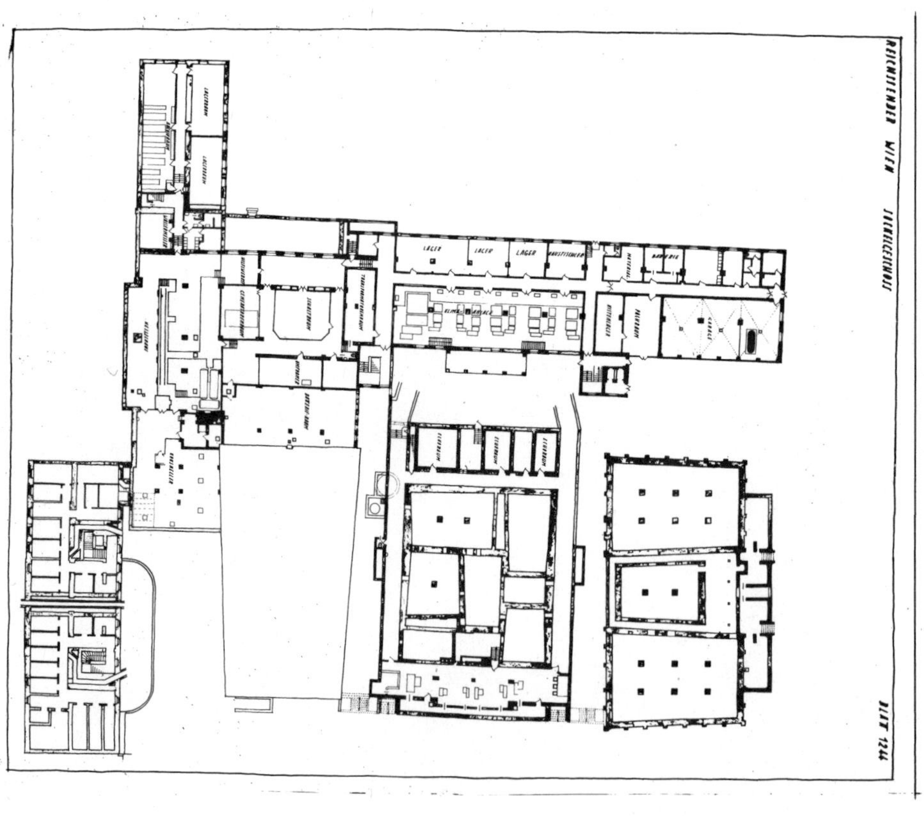

45

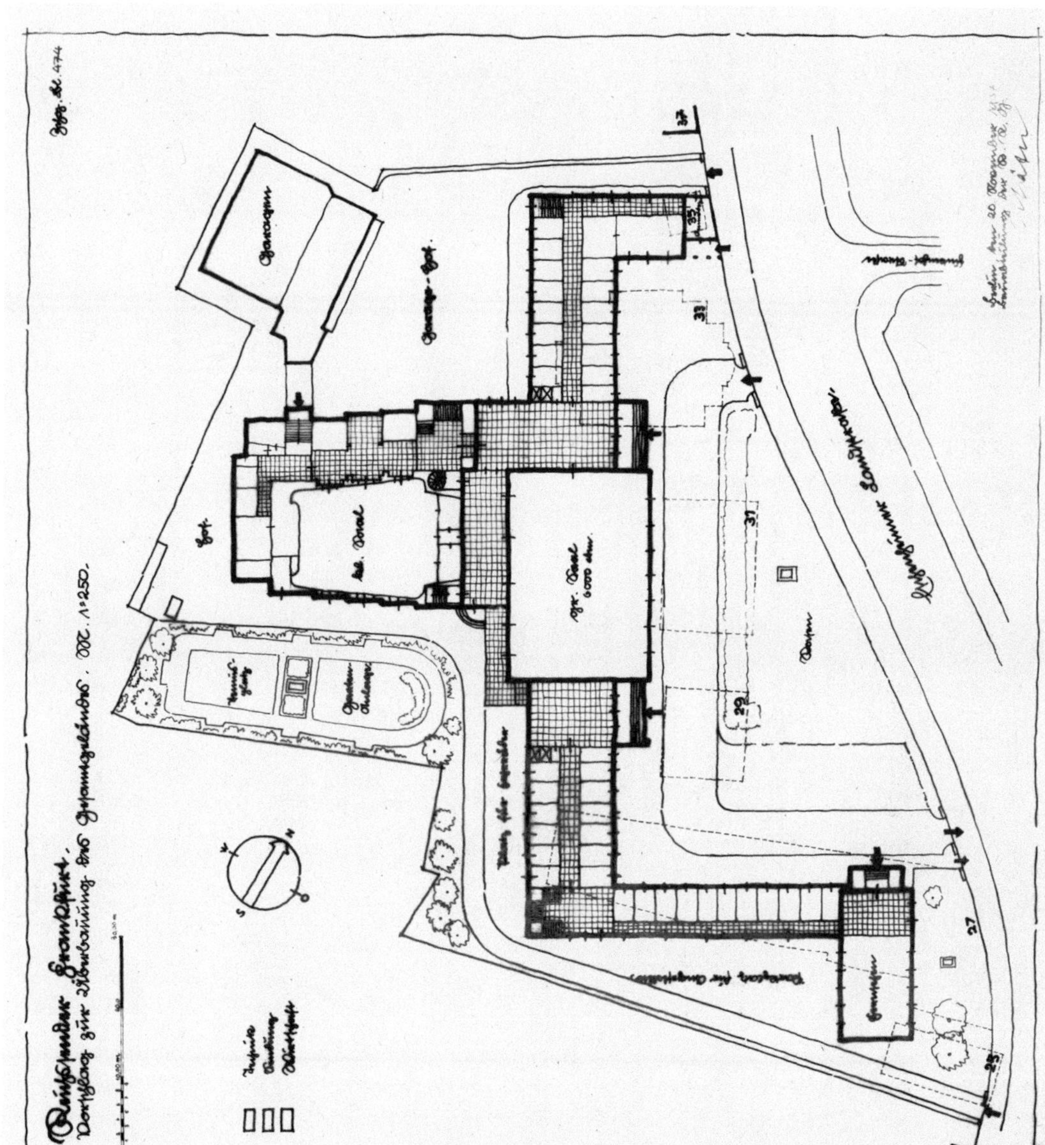

46

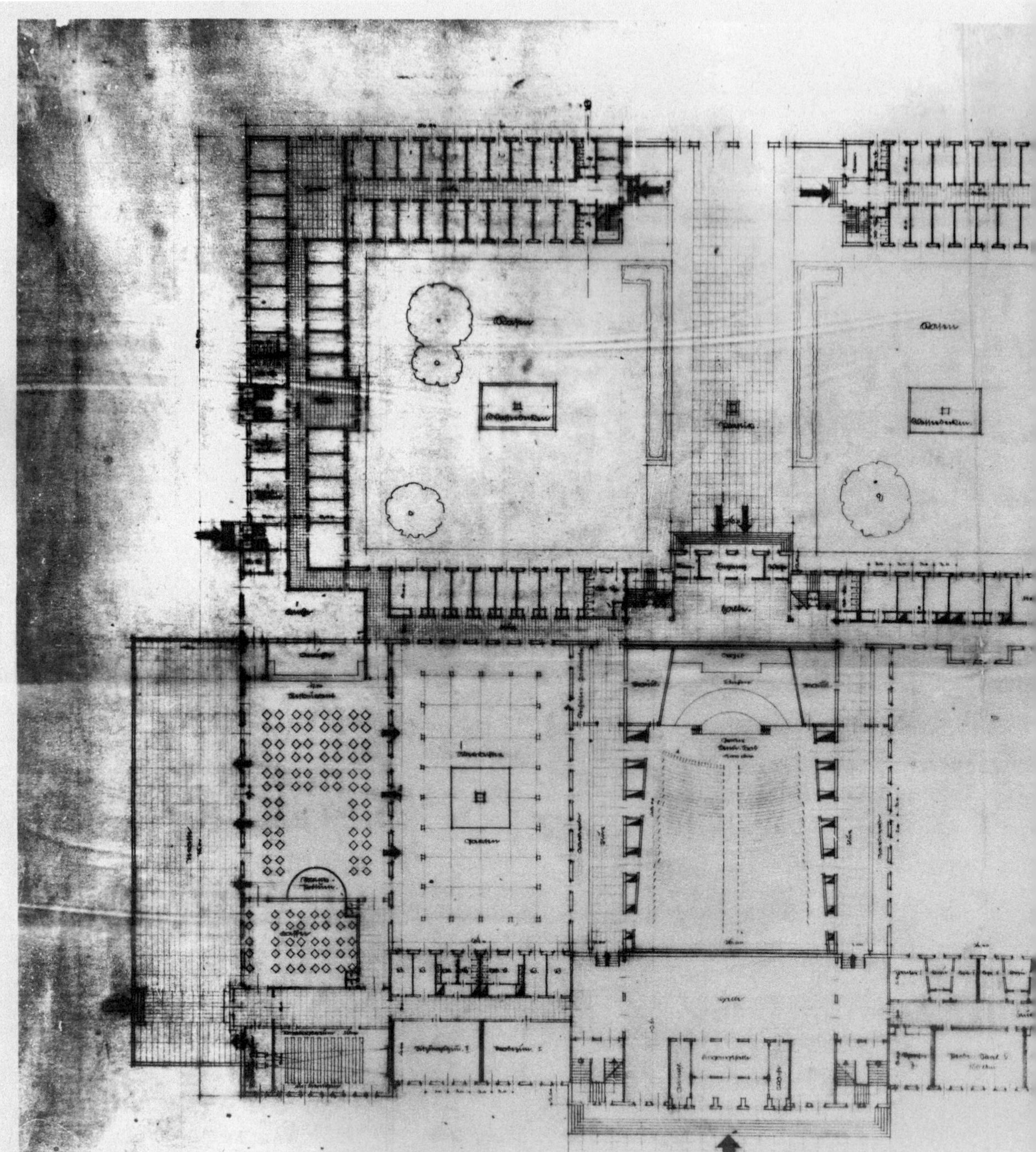

47

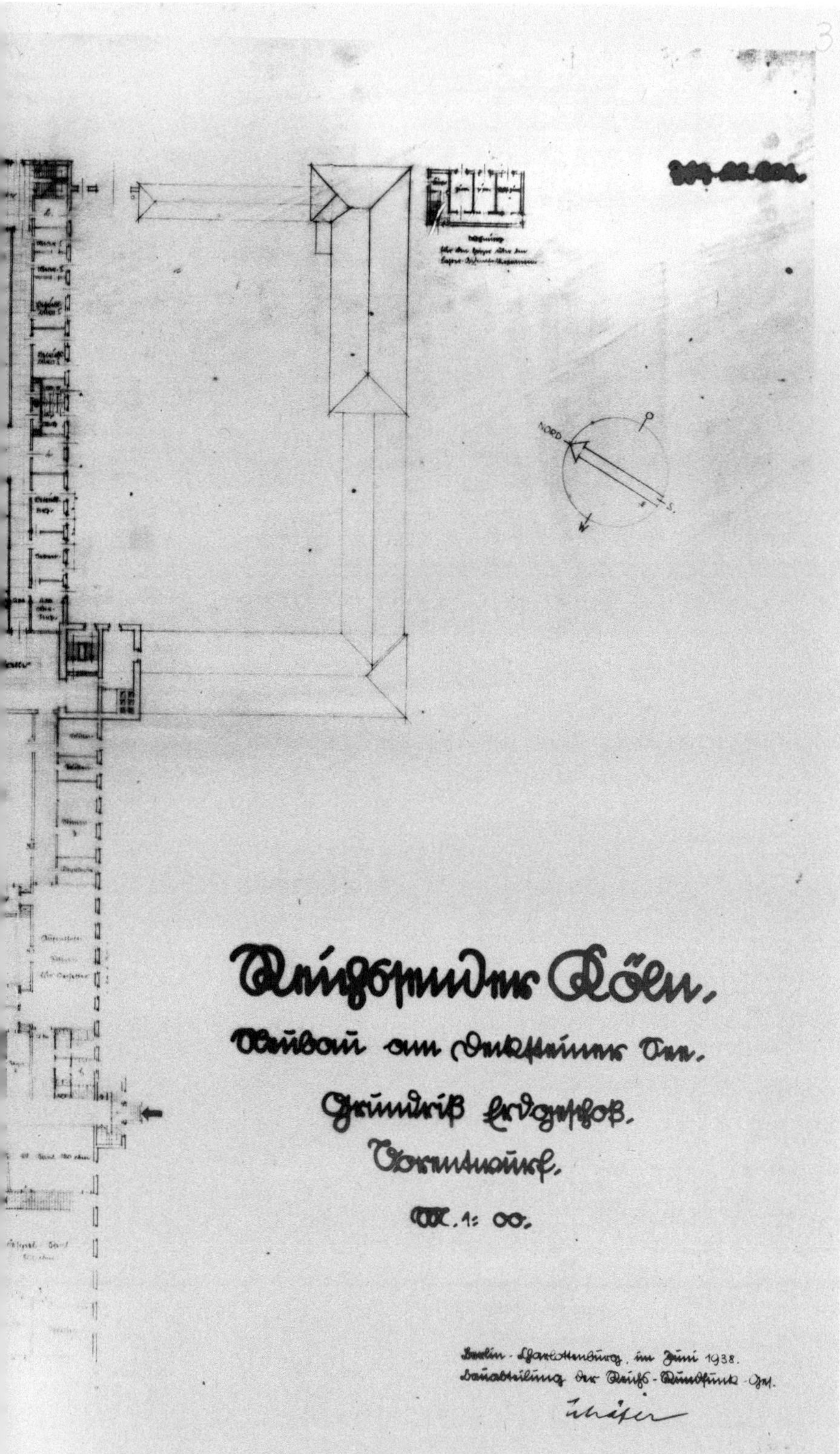
NORD
Reichssender Köln.
Grundriß Erdgeschoß.
Vorentwurf.
Berlin-Charlottenburg, im Juni 1938.
Bauabteilung der Reichs-Rundfunk-Ges.

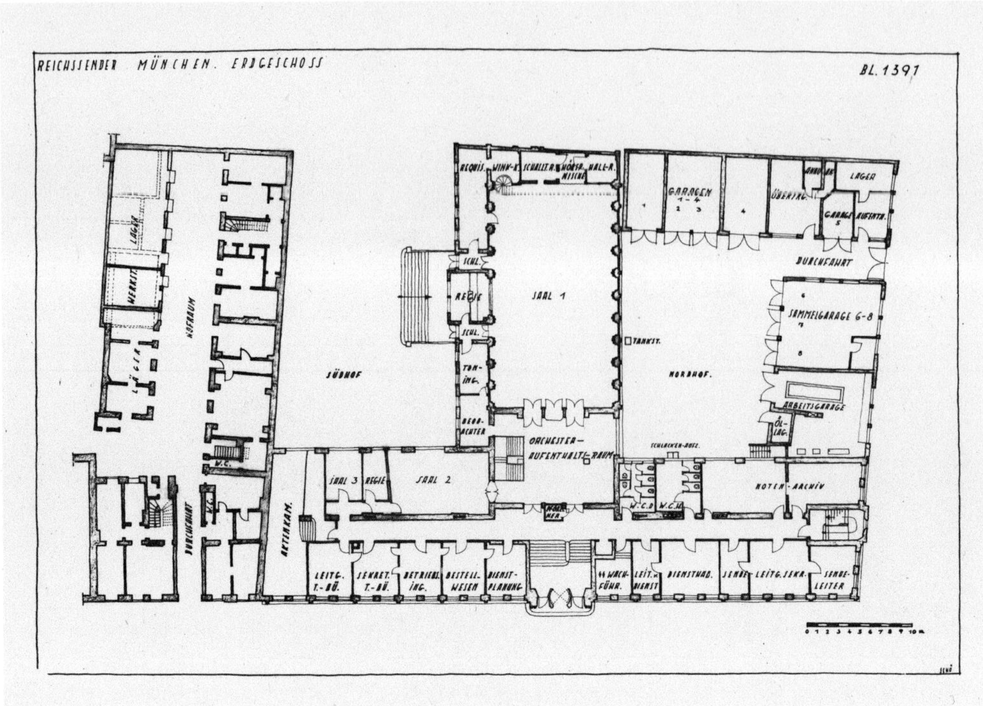
REICHSSENDER MÜNCHEN. ERDGESCHOSS
BL. 1391
GARAGEN
SAAL 1
SÜDHOF
NORDHOF
DURCHFAHRT
SAMMELGARAGE 6-8
ARBEITSGARAGE
SAAL 2
SAAL 3
REGIE
ORCHESTER-AUFENTHALTS-RAUM
NOTEN-ARCHIV
AKTENKAM.
LAGER

48

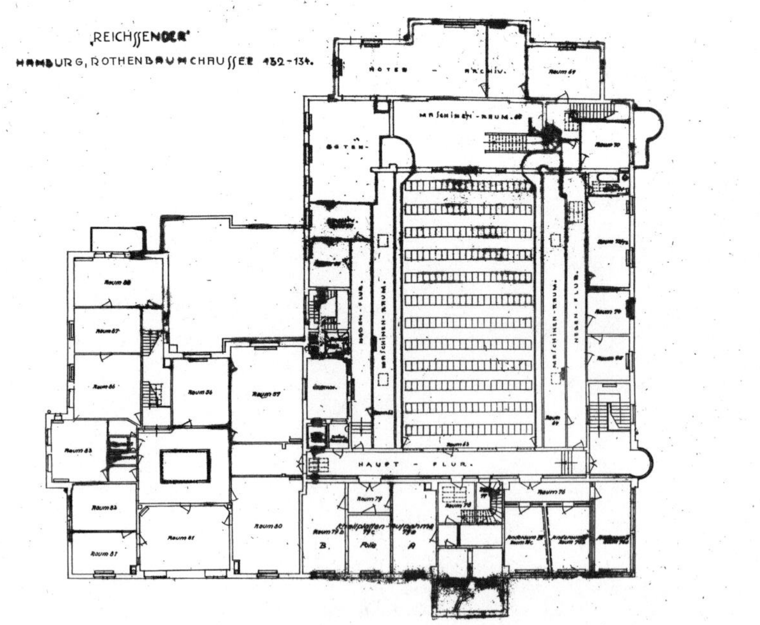
REICHSSENDER
HAMBURG, ROTHENBAUMCHAUSSEE 132-134.
HAUPT-FLUR.
I. OBERGESCHOSS.

49

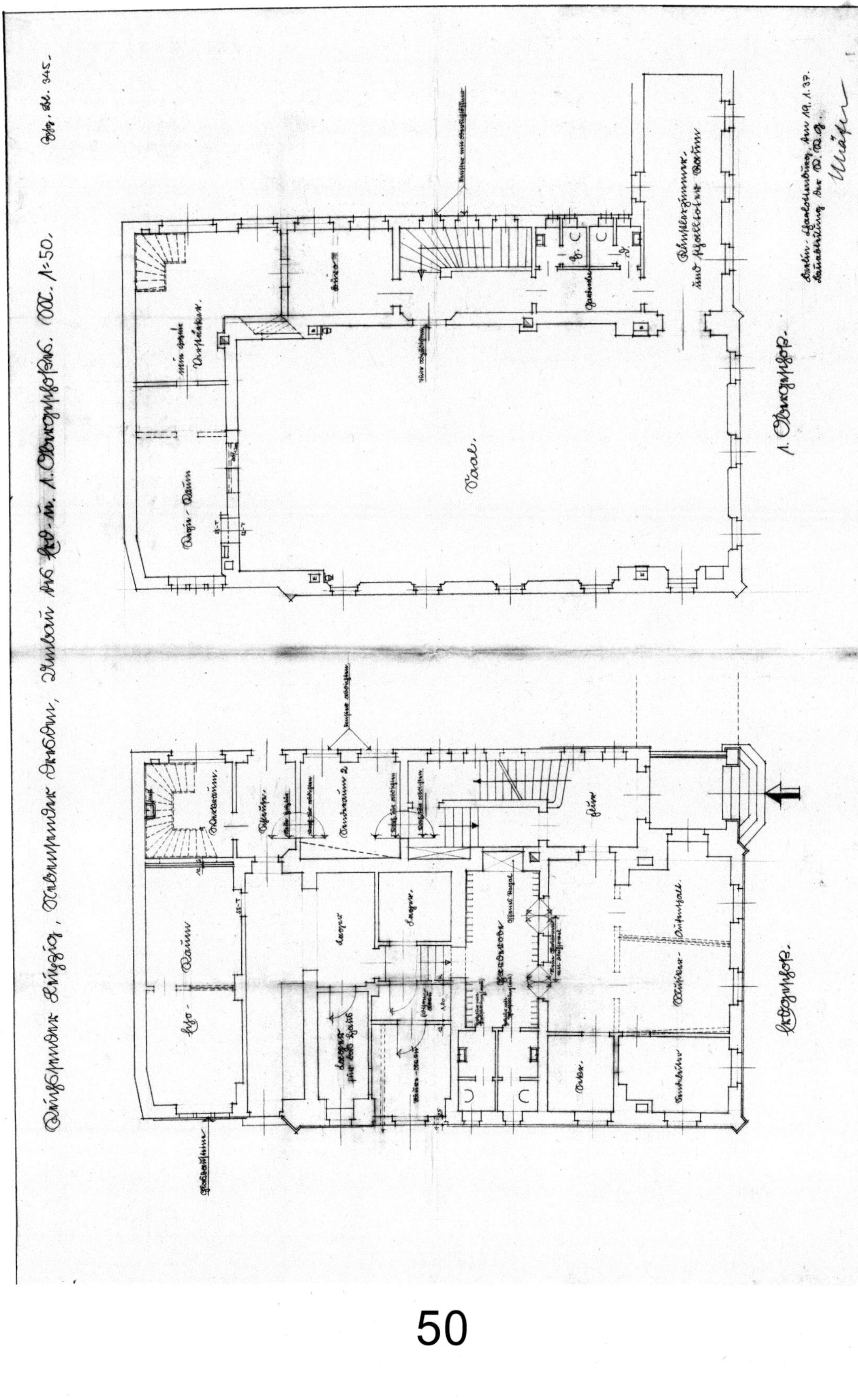

50

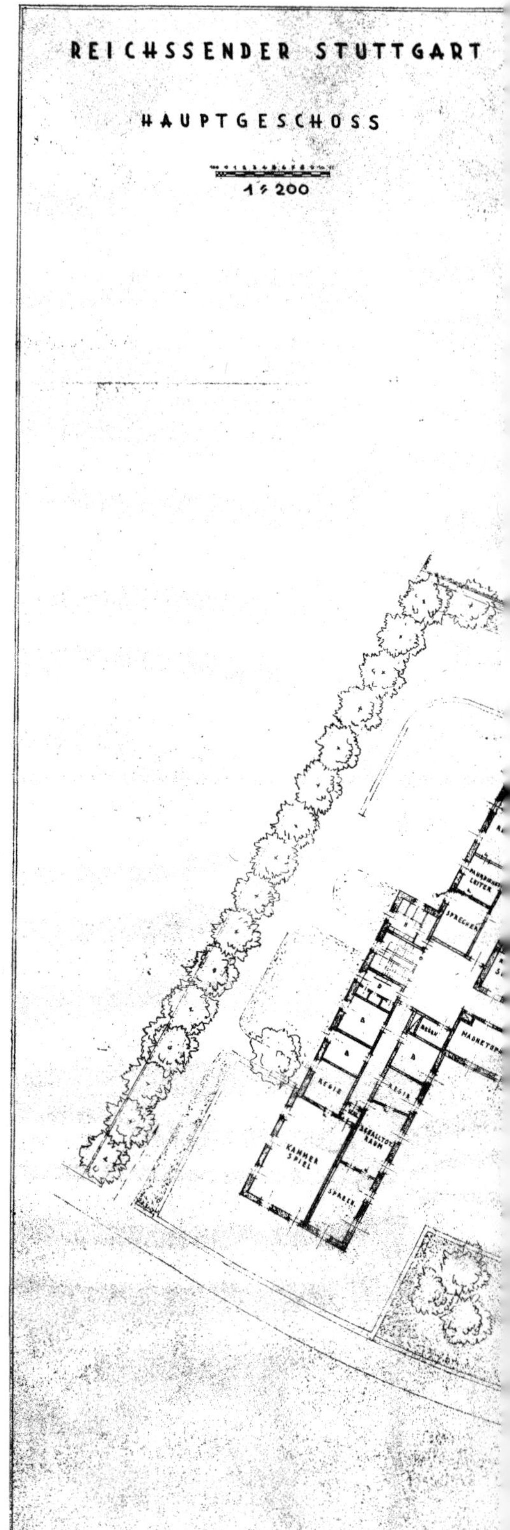
REICHSSENDER STUTTGART
HAUPTGESCHOSS
1:200

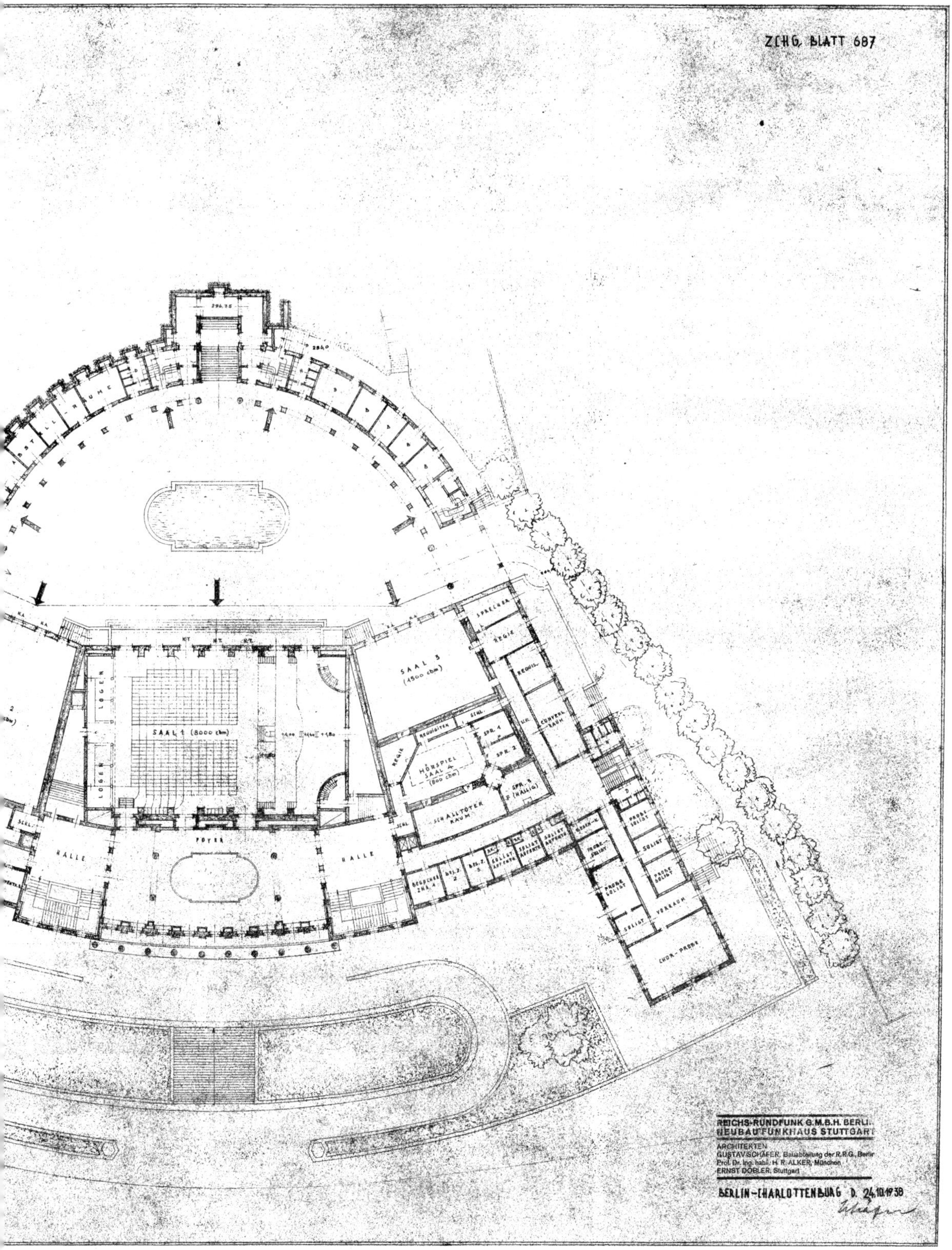
ZCHG. BLATT 687
SAAL 1 (8000 cbm)
SAAL 3 (1500 cbm)
HÖRSPIEL SAAL (800 cbm)
FOYER
HALLE
HALLE
LOGEN
LOGEN
SPRECHER
REGIE
CHOR-PROBE
VORRAUM
REICHS-RUNDFUNK G.M.B.H. BERLI
NEUBAU FUNKHAUS STUTTGART
ARCHITEKTEN
GUSTAV SCHÄFER, Bauabteilung der R.R.G., Berlin
Prof. Dr. Ing. habil. H. R. ALKER, München
ERNST DOBLER, Stuttgart
BERLIN-CHARLOTTENBURG D. 24.10.1938

51

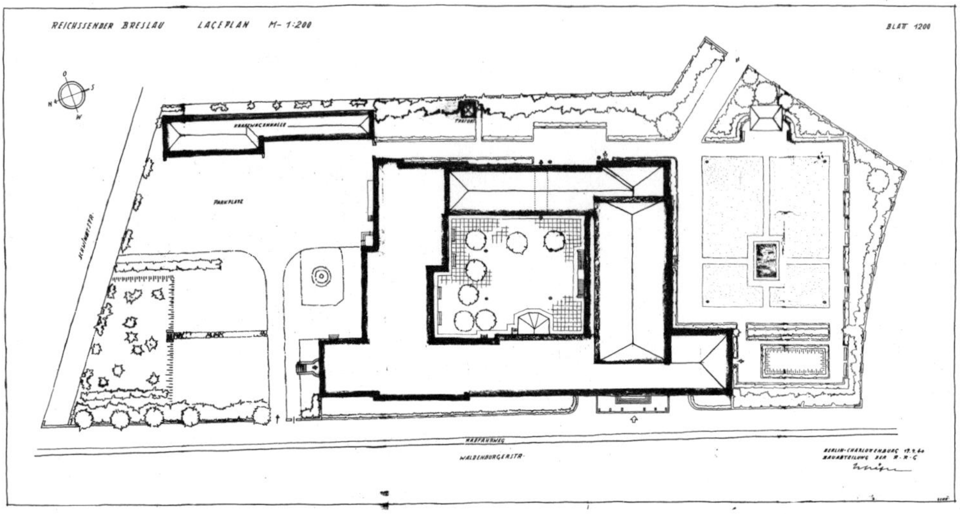
REICHSSENDER BRESLAU LAGEPLAN M- 1:200
BLATT 1200
PARKPLATZ
WALDENBURGERSTR.

52

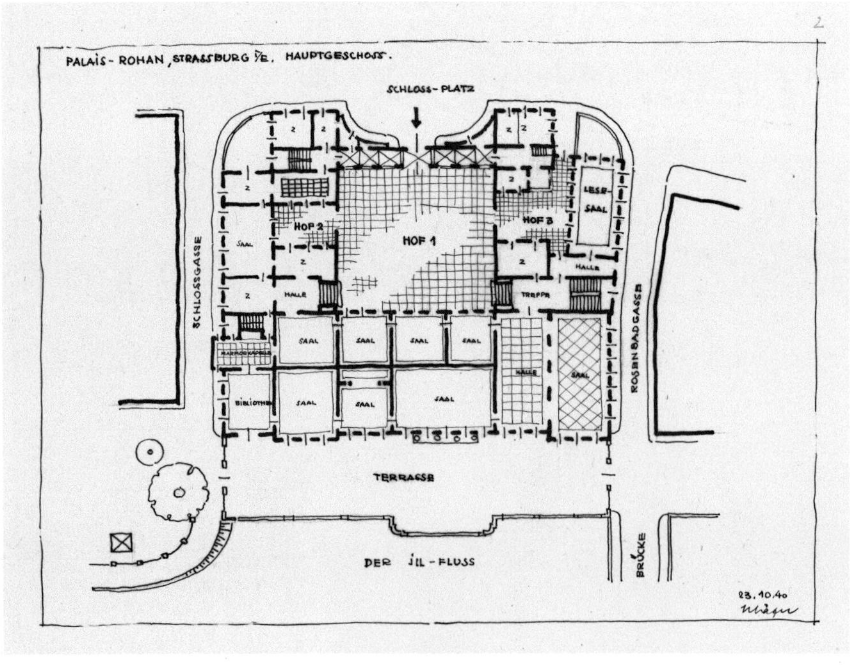
PALAIS-ROHAN, STRASSBURG I/E. HAUPTGESCHOSS.
SCHLOSS-PLATZ
HOF 2
HOF 1
HOF 3
LESE-SAAL
SAAL
HALLE
TREPPE
SCHLOSSGASSE
ROSENBADGASSE
BIBLIOTH.
TERRASSE
DER ILL-FLUSS
BRÜCKE
23.10.40

53

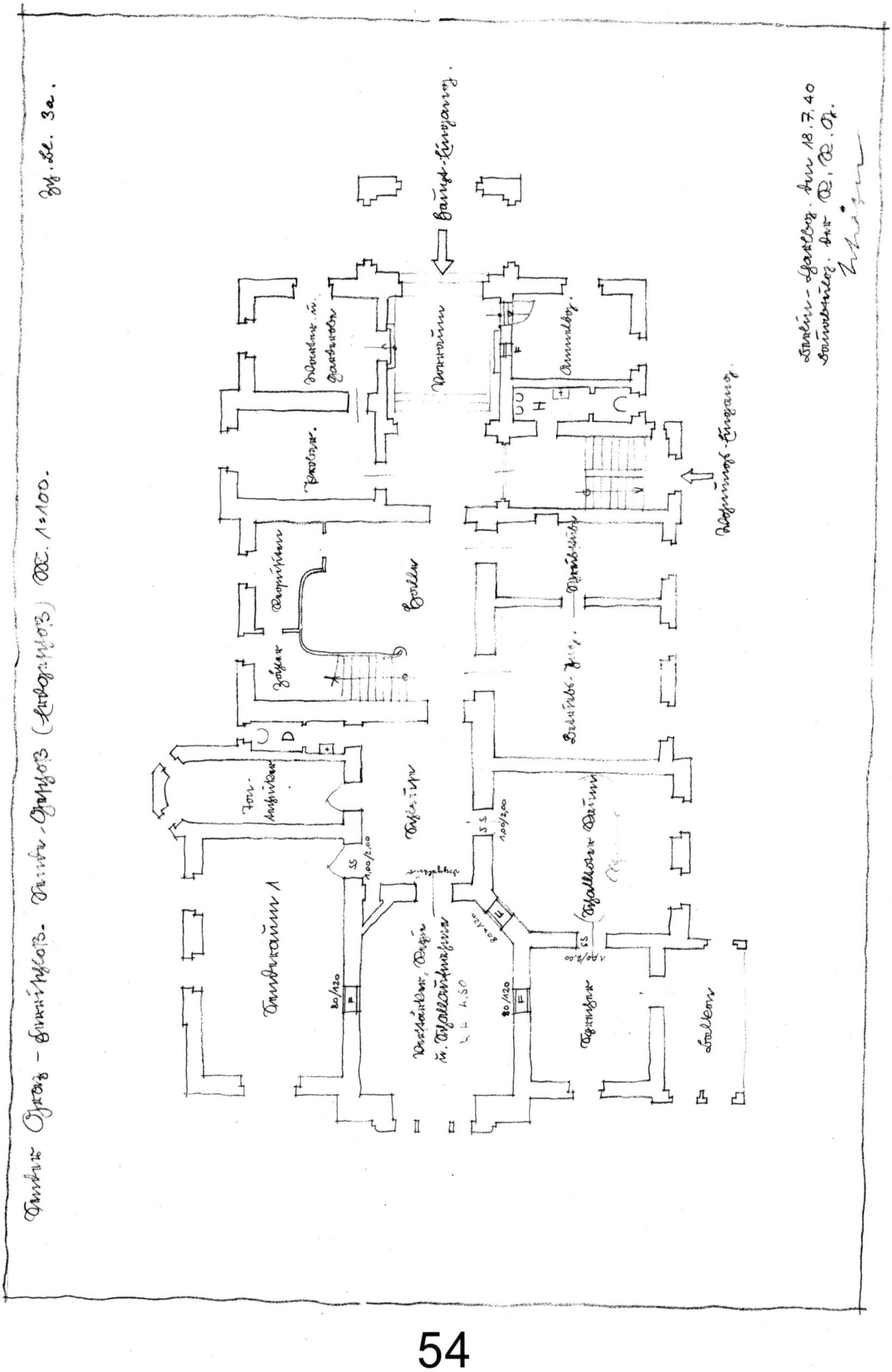
M. 1:100
80/120
80/120
1,00/2,00
1,00/2,00
SS
F
D
H
18.7.40

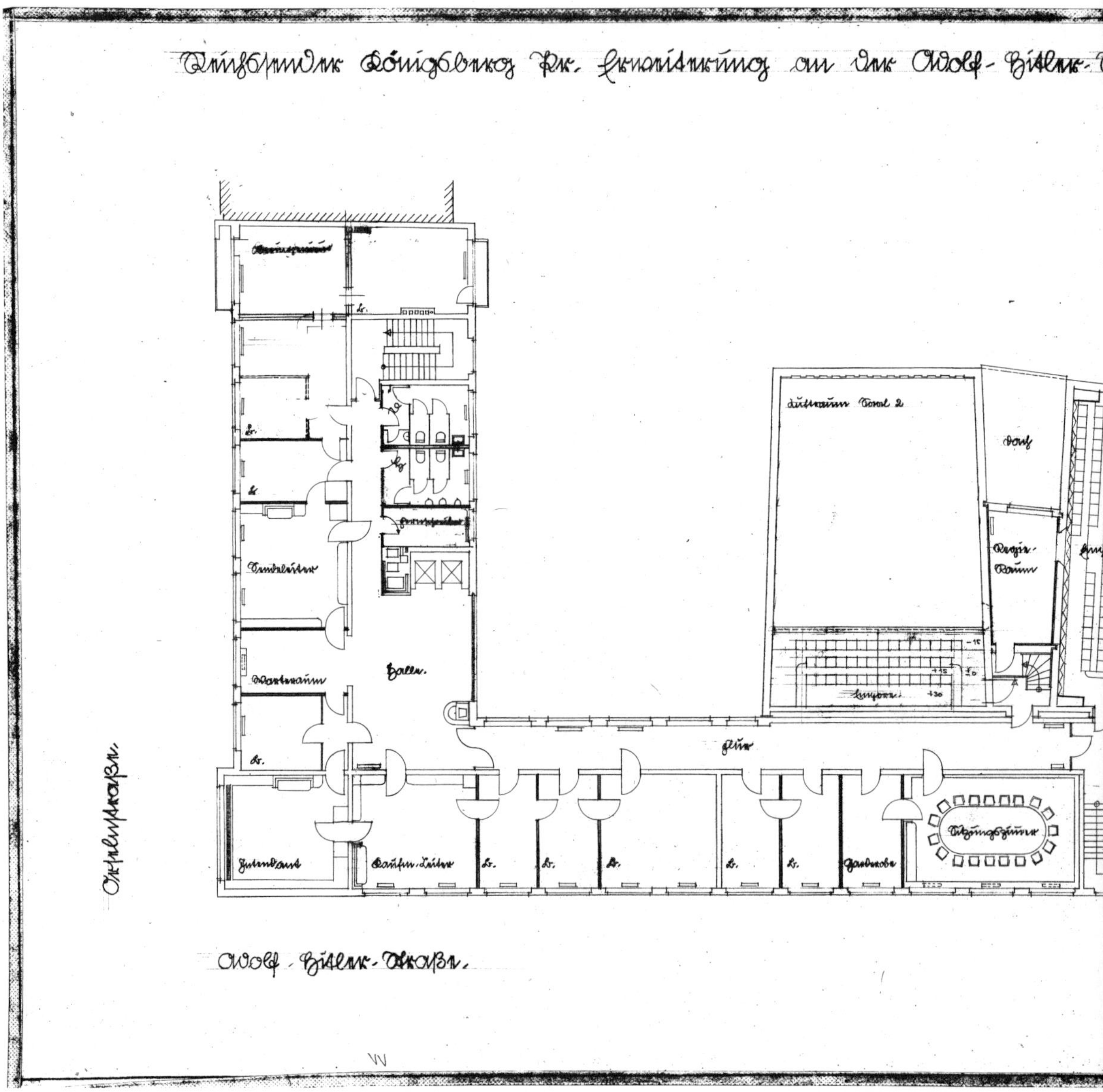

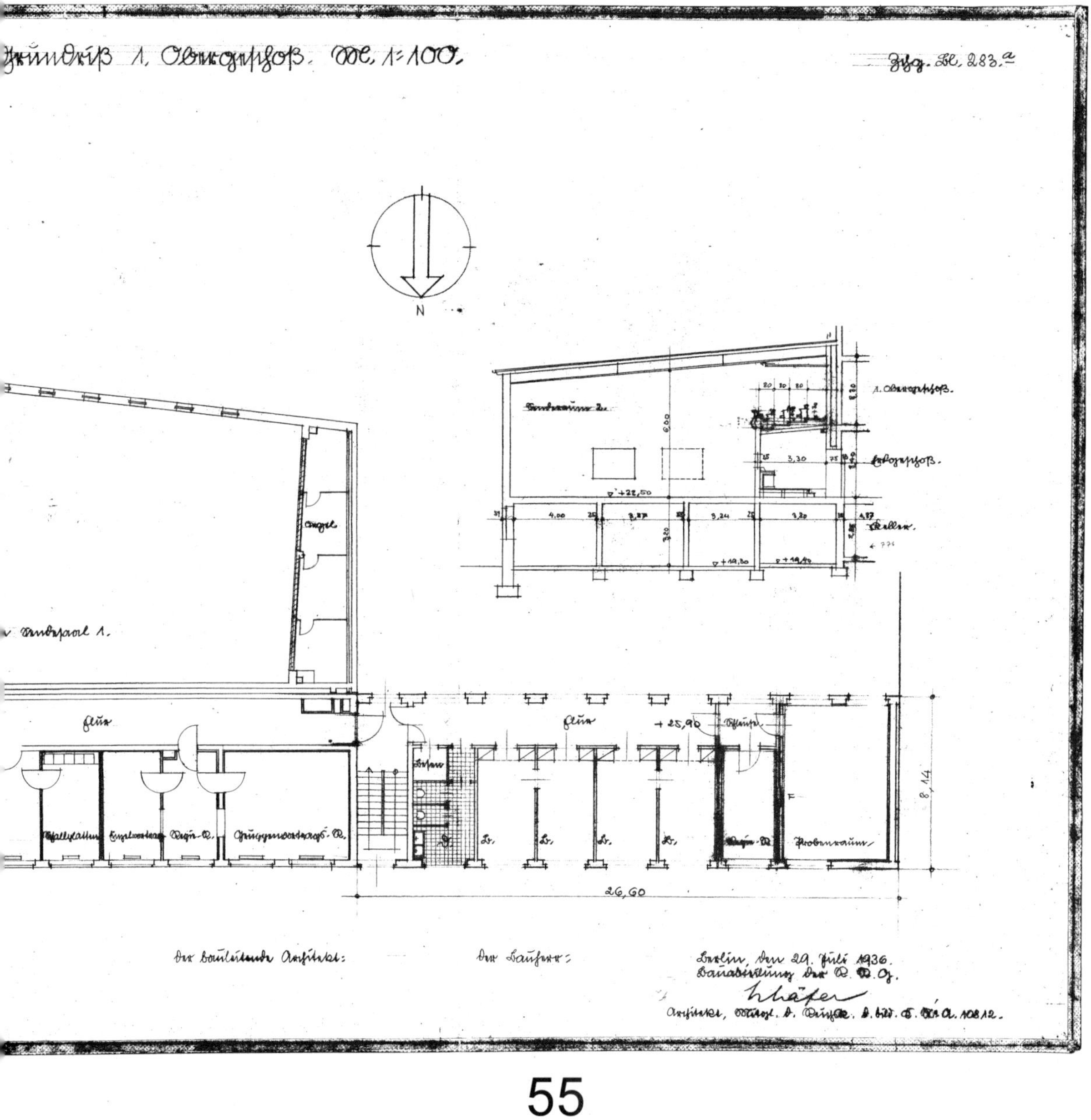

55

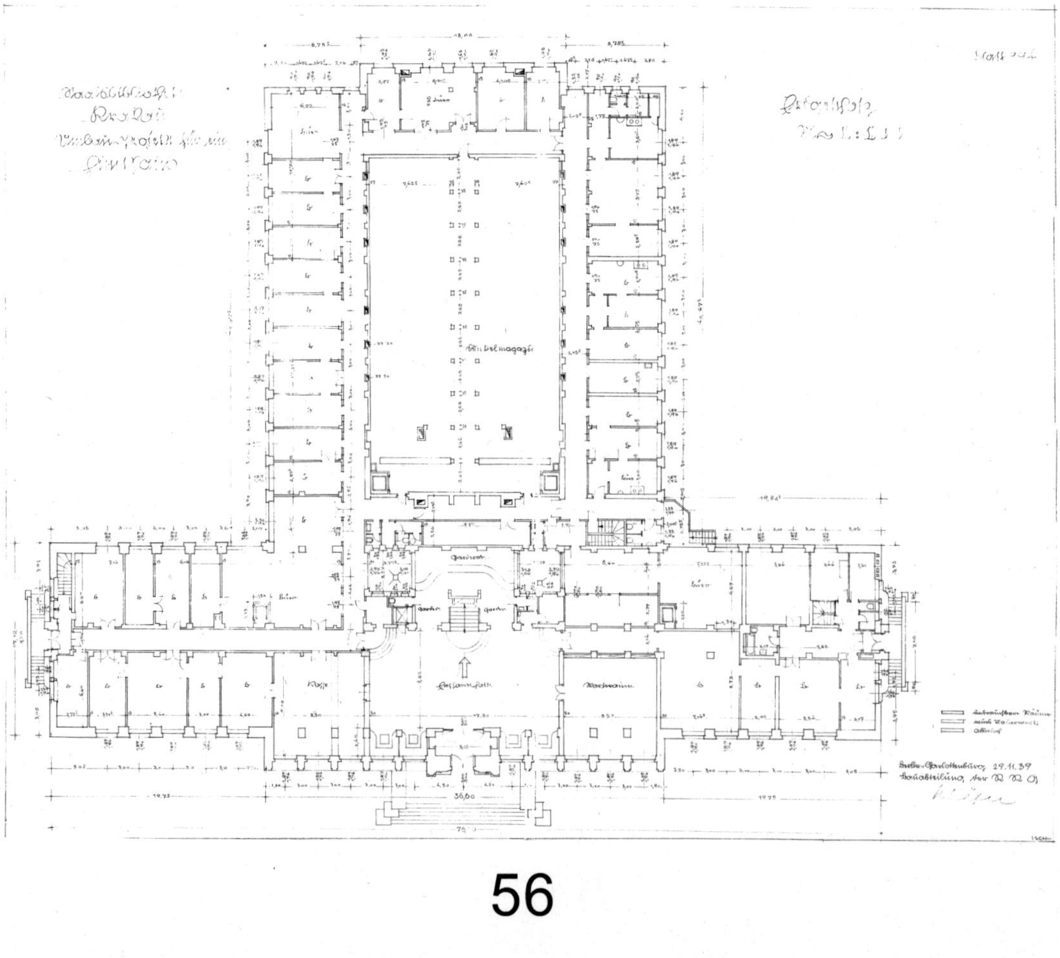

56

Fig. 40 Perspective, section, and photograph depicting the extinguishing loudspeaker (*Löschstrahler*) developed by Erwin Meyer in collaboration with Telefunken. Source: Erwin Meyer, *Electro-Acoustics* (London: G. Bell and Sons, 1939), 59.

Fig. 41 Extinguishing loudspeakers installed in the Olympic stadium in Berlin. Source: Thomas Schmidt, *Werner March: Der Architekt des Olympia-Stadions 1894–1976* (Basel: Birkhäuser, 1992), 77.

Fig. 42 Olympic ice rink and adjacent broadcasting station in Garmisch-Partenkirchen, axonometric view. Source: Deutsches Rundfunkarchiv Frankfurt am Main. A48 RRG Bauzeichnungen, 402.

Fig. 43 Broadcasting house in Prague, level 0. Source: Deutsches Rundfunkarchiv Frankfurt am Main. A48 RRG Bauzeichnungen, 393-893.

Fig. 44 Reich station in Gdansk, level 1. Source: Deutsches Rundfunkarchiv Frankfurt am Main. A48 RRG Bauzeichnungen, 390-972.

Fig. 45 Reich station in Vienna, level -1. Source: Deutsches Rundfunkarchiv Frankfurt am Main. A48 RRG Bauzeichnungen, 392-1244.

Fig. 46 Reich station in Frankfurt am Main, site plan. Source: Deutsches Rundfunkarchiv Frankfurt am Main. A48 RRG Bauzeichnungen, 388-474.

Fig. 47 Reich station in Cologne, level 0. Source: Deutsches Rundfunkarchiv Frankfurt am Main. A48 RRG Bauzeichnungen, 388-604.

Fig. 48 Reich station in Munich, level 0. Source: Deutsches Rundfunkarchiv Frankfurt am Main. A48 RRG Bauzeichnungen, 389-1391.

Fig. 49 Reich station in Hamburg, level 2. Source: Deutsches Rundfunkarchiv Frankfurt am Main. A48 RRG Bauzeichnungen, 389.

Fig. 50 Reich station in Leipzig, levels 0 and 1. Source: Deutsches Rundfunkarchiv Frankfurt am Main. A48 RRG Bauzeichnungen, 388-345.

Fig. 51 Reich station in Stuttgart, level 0. Source: Deutsches Rundfunkarchiv Frankfurt am Main. A48 RRG Bauzeichnungen, 390-687.

Fig. 52 Reich station in Breslau, site plan. Source: Deutsches Rundfunkarchiv Frankfurt am Main. A48 RRG Bauzeichnungen, 391-827.

Fig. 53 Reich station in Strasbourg, level 0. Source: Deutsches Rundfunkarchiv Frankfurt am Main. A48 RRG Bauzeichnungen, 398.

Fig. 54 Reich station in Graz, level 0. Source: Deutsches Rundfunkarchiv Frankfurt am Main. A48 RRG Bauzeichnungen, 392-3a.

Fig. 55 Reich station in Königsberg (now Kaliningrad), level 1. Source: Deutsches Rundfunkarchiv Frankfurt am Main. A48 RRG Bauzeichnungen, 390-283.

Fig. 56 Reich station in Krakow, level 0. Source: Deutsches Rundfunkarchiv Frankfurt am Main. A48 RRG Bauzeichnungen, 393-994.

57

58

59

60

574 kHz Mühlacker, Kassel
592 kHz Wien, München
658 kHz Leipzig, Breslau
740 kHz Berlin, Hamburg
785 kHz Langenberg, Frankfurt/M.
841 kHz Heilsberg
886 kHz Nürnberg, Linz, Kempten, Graz Zell a. See
904 kHz Magdeburg, Pirna, Sonneberg Karlsbad, Troppau
950 kHz Reichenberg, Gleiwitz, Mähr-Schönberg, Grünberg/Schl.
1031 kHz Bremen, Hannover, Flensburg, Stendal, Schwerin, Stralsund Frankfurt/O. Stettin, Stolp, Schneidemühl
1195 kHz Saarbrücken, Trier, Freiburg, Konstanz, Osnabrück Koblenz, Würzburg
1204 kHz Innsbruck, Lienz, Salzburg, Regensburg, Passau, Eisenerz, Klagenfurt, Nordling
Reine
Osnabrück
Münster
Paderborn
Essen
Dortmund
Duisburg
Wuppertal
Meschede
Düsseldorf
Köln
Siegen
Bonn
Gießen
Prüm
Koblenz
Wiesbaden
Frankfurt
Mainz
Bad-Kreuznach
Darmstadt
Trier
Mannheim
Saarbrücken
Karlsruhe
Heilbronn
Stuttgart
Ulm
Freiburg
Überlingen
Bregenz

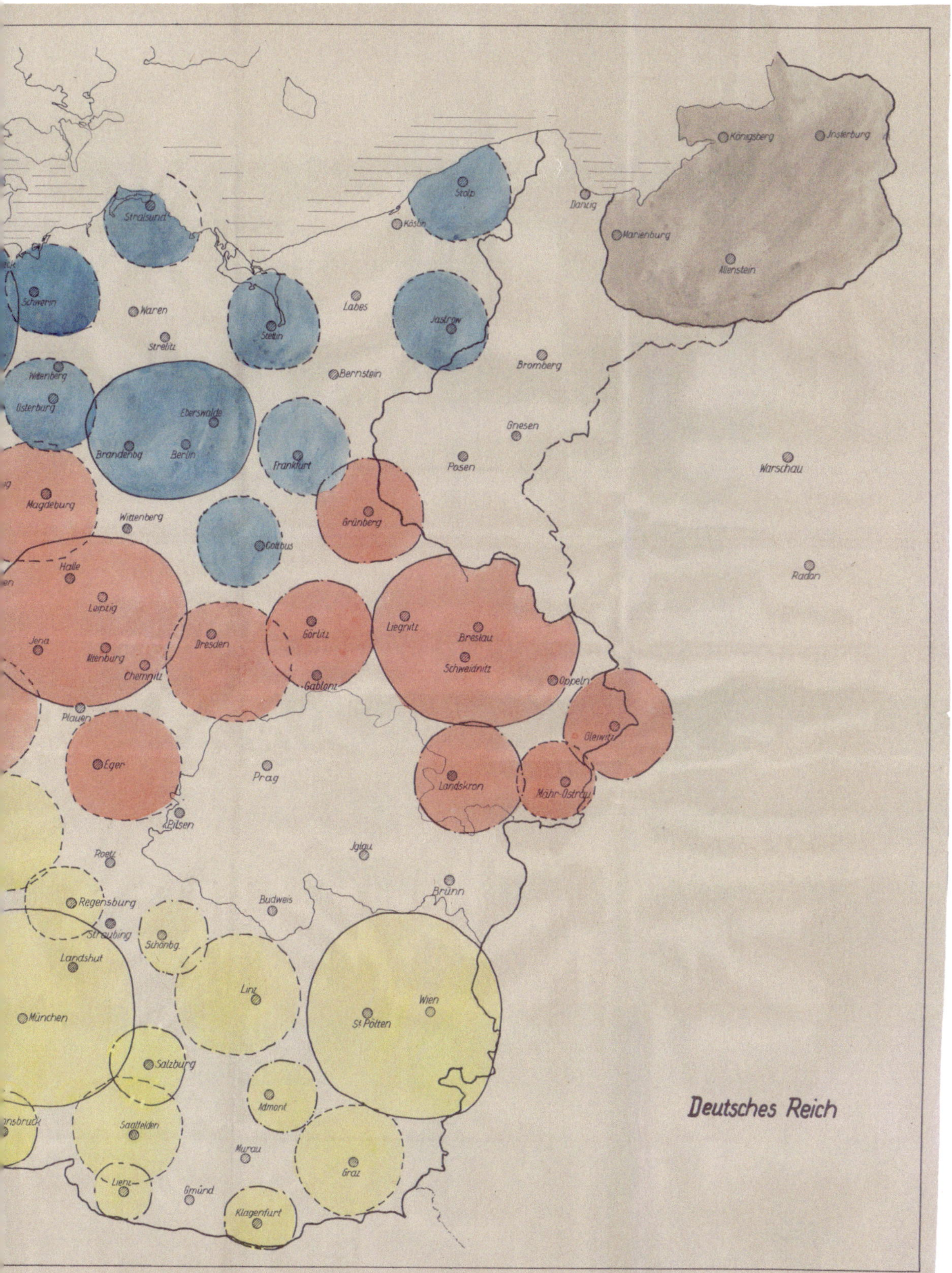
Königsberg
Insterburg
Stolp
Danzig
Köslin
Stralsund
Marienburg
Allenstein
Schwerin
Waren
Labes
Jastrow
Stettin
Strelitz
Bromberg
Bernstein
Wittenberg
Osterburg
Eberswalde
Gnesen
Brandenbg
Berlin
Frankfurt
Posen
Warschau
Magdeburg
Wittenberg
Grünberg
Cottbus
Radom
Halle
Leipzig
Liegnitz
Breslau
Jena
Görlitz
Dresden
Altenburg
Schweidnitz
Chemnitz
Oppeln
Gablonz
Plauen
Gleiwitz
Eger
Prag
Landskron
Mähr-Ostrau
Pilsen
Iglau
Brünn
Regensburg
Budweis
Straubing
Schönbg.
Landshut
Linz
Wien
München
St. Pölten
Salzburg
Admont
Saalfelden
Murau
Graz
Lienz
Gmünd
Klagenfurt
Deutsches Reich

RIAS
BERLIN
EINE
FREIE
STIMME
DER
FREIEN
WELT

62

ACHTUNG !
DIES IST
KEIN
WESTBERLINER SENDER
SFB HEIDELBERGER PL.3
RIAS KUFSTEINER STR.69

63

Richtung
SPANDAU

64

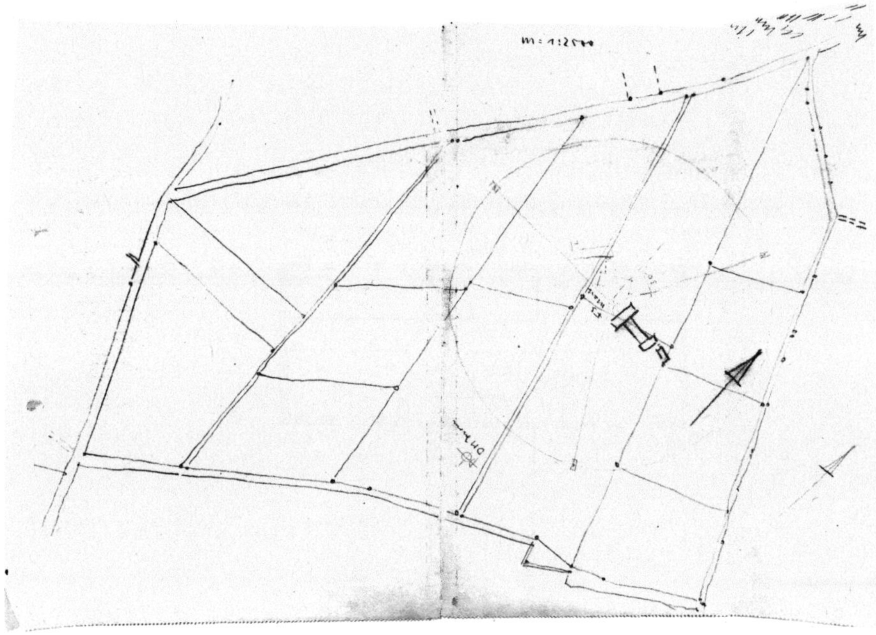

65

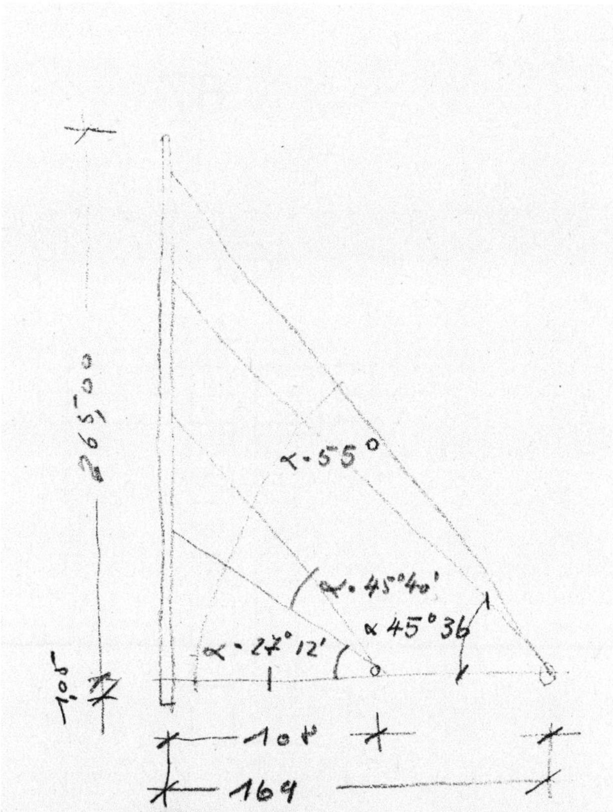

66

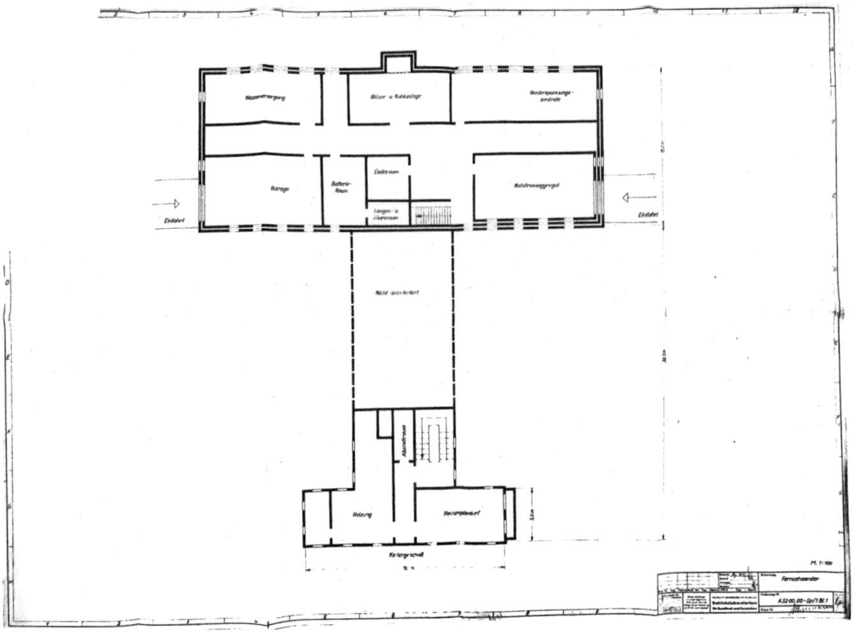
Garage
Batterie-Raum
Laderaum
Notstromaggregat
Einfahrt
Einfahrt
Heizung
Kellergeschoß

67

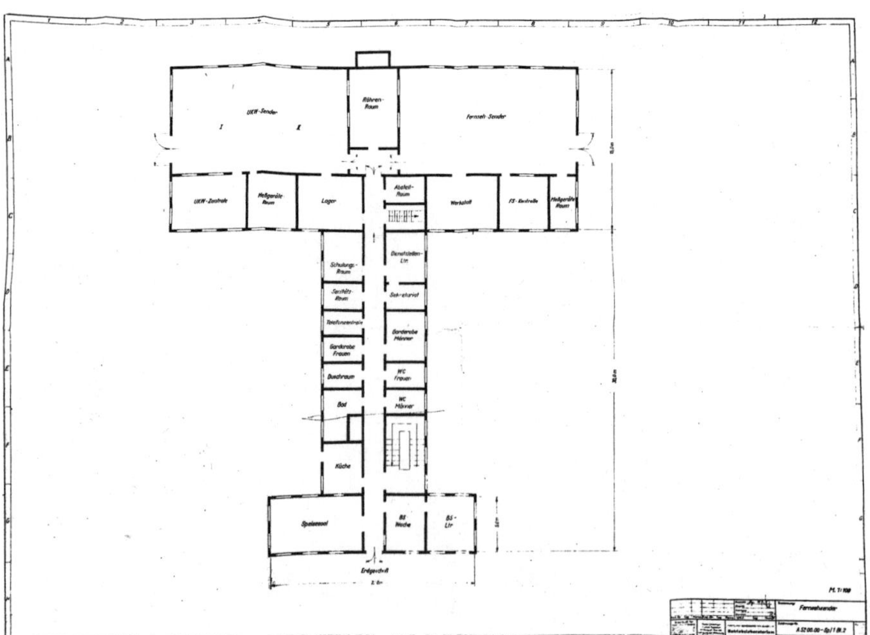
Röhren-Raum
UKW-Sender
Fernseh-Sender
UKW-Zentrale
Meßgeräte Raum
Lager
Werkstatt
Meßgeräte Raum
Schulungs-Raum
Sekretariat
Garderobe Männer
Garderobe Frauen
Duschraum
WC Frauen
Bad
WC Männer
Küche
Speisesaal
BS Wache
BS-Ltr
Erdgeschoß

68

Fig. 57 View of the Deutschlandsender III in Herzberg, the tallest building in Europe at the time of its construction. Source: Helmut Knuppe, *Ein Riese unter Riesen* (Leipzig: Leipziger Verlagsgesellschaft, 2003).

Fig. 58 View of the transmission tower in Tegel. Source: Gunter Erler, "Von Tegel nach Königs Wusterhausen: Die Geschichte eines Berliner Rundfunksenders," *Post und Telekommunikationsgeschichte* 2000, no. 5 (2000): 25.

Fig. 59 Street facade of the Fernmeldeamt (telephone exchange building) designed by Berlin architects Otto Spalding and Kurt Kuhlow in 1929. Source: Architekten- und Ingenieur-Verein zu Berlin, ed., *Berlin und seine Bauten*, pt. 10, vol. B: *Anlagen und Bauten für den Verkehr (4): Post und Fernmeldewesen* (Berlin: Ernst & Sohn, 1987), 105.

Fig. 60 Berliners listen to a RIAS broadcast through an improvised loudspeaker mounted on a military truck. Source: Bundesarchiv. Bild 183-Z1105-504.

Fig. 61 Map showing the coverage and frequency range of transmission stations during the Third Reich. Source: Deutsche Rundfunkarchiv Frankfurt am Main. A48 RRG Bauzeichnungen, 400.

Fig. 62 Postcard of the RIAS Broadcasting House. The postcard was sent to local and international audiences requesting feedback about the signal reception of short-, medium, and longwave transmissions. Source: Personal archive.

Fig. 63 Haus des Rundfunks, view from Masurenallee. The sign bluntly states: "This is NOT a West Berlin station. The West Berlin stations are SFR and RIAS." Source: Bundesarchiv. Bild F003013-0010.

Fig. 64 British military police enclosing the perimeter of the Haus des Rundfunks with barbed wire. Source: Landesarchiv Berlin.

Fig. 65 Schwerin transmission station, plan sketch showing position of the mast relative to the station building. Source: Bundesarchiv. DM 302/2832.

Fig. 66 Schwerin transmission station, sketch in section of the mast. Source: Bundesarchiv. DM 302/2832.

Fig. 67 Schwerin transmission station, level -1. Source: Bundesarchiv. DM 302/2832.

Fig. 68 Schwerin transmission station, level 0. Source: Bundesarchiv. DM 302/2832.

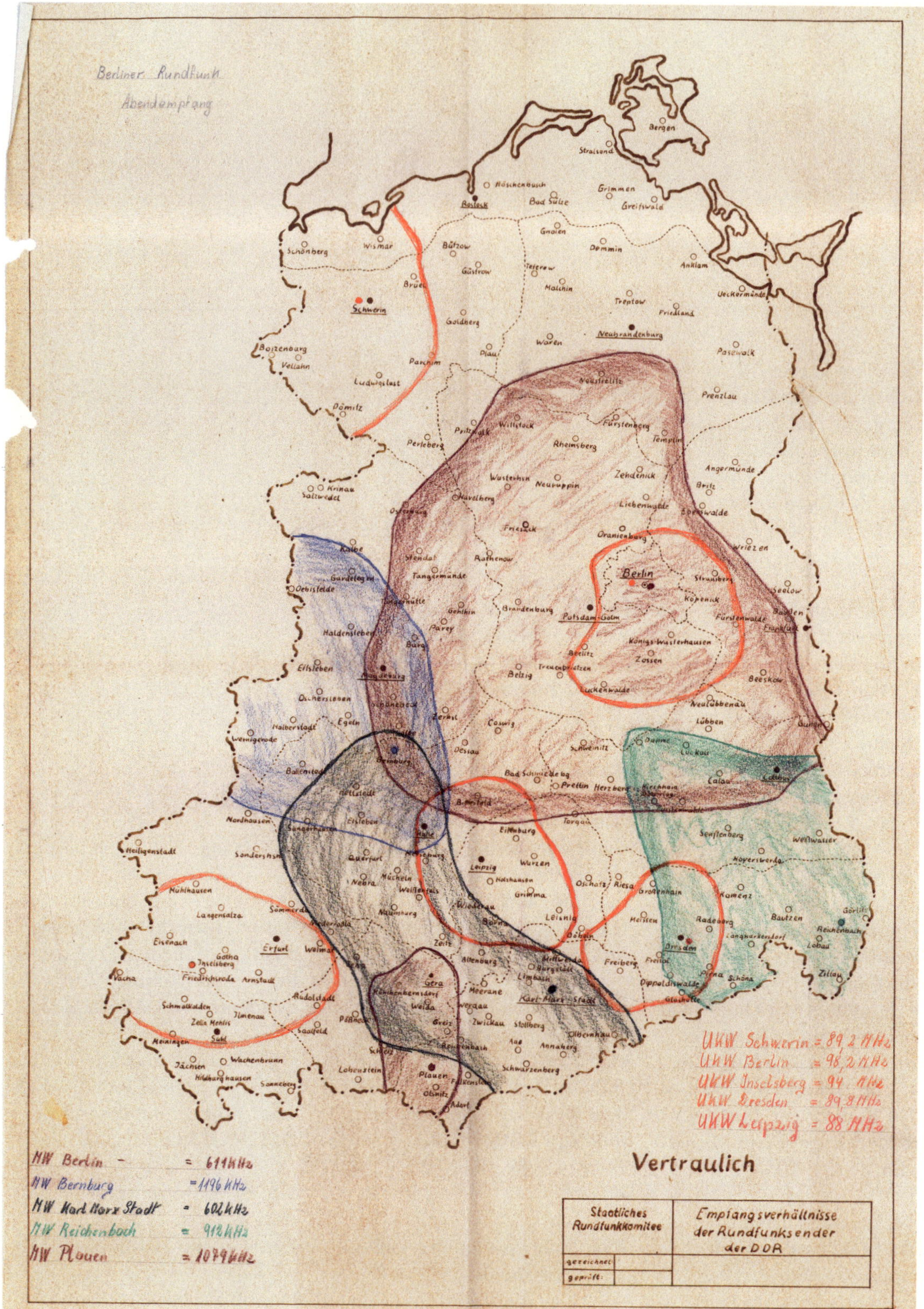
Berliner Rundfunk
Abendempfang
UKW Schwerin = 89,2 MHz
UKW Berlin = 98,2 MHz
UKW Inselsberg = 94 MHz
UKW Dresden = 89,8 MHz
UKW Leipzig = 88 MHz
MW Berlin = 611 kHz
MW Bernburg = 1196 kHz
MW Karl Marx Stadt = 602 kHz
MW Reichenbach = 912 kHz
MW Plauen = 1079 kHz
Vertraulich
Staatliches Rundfunkkomitee
Empfangsverhältnisse der Rundfunksender der DDR
gezeichnet:
geprüft:

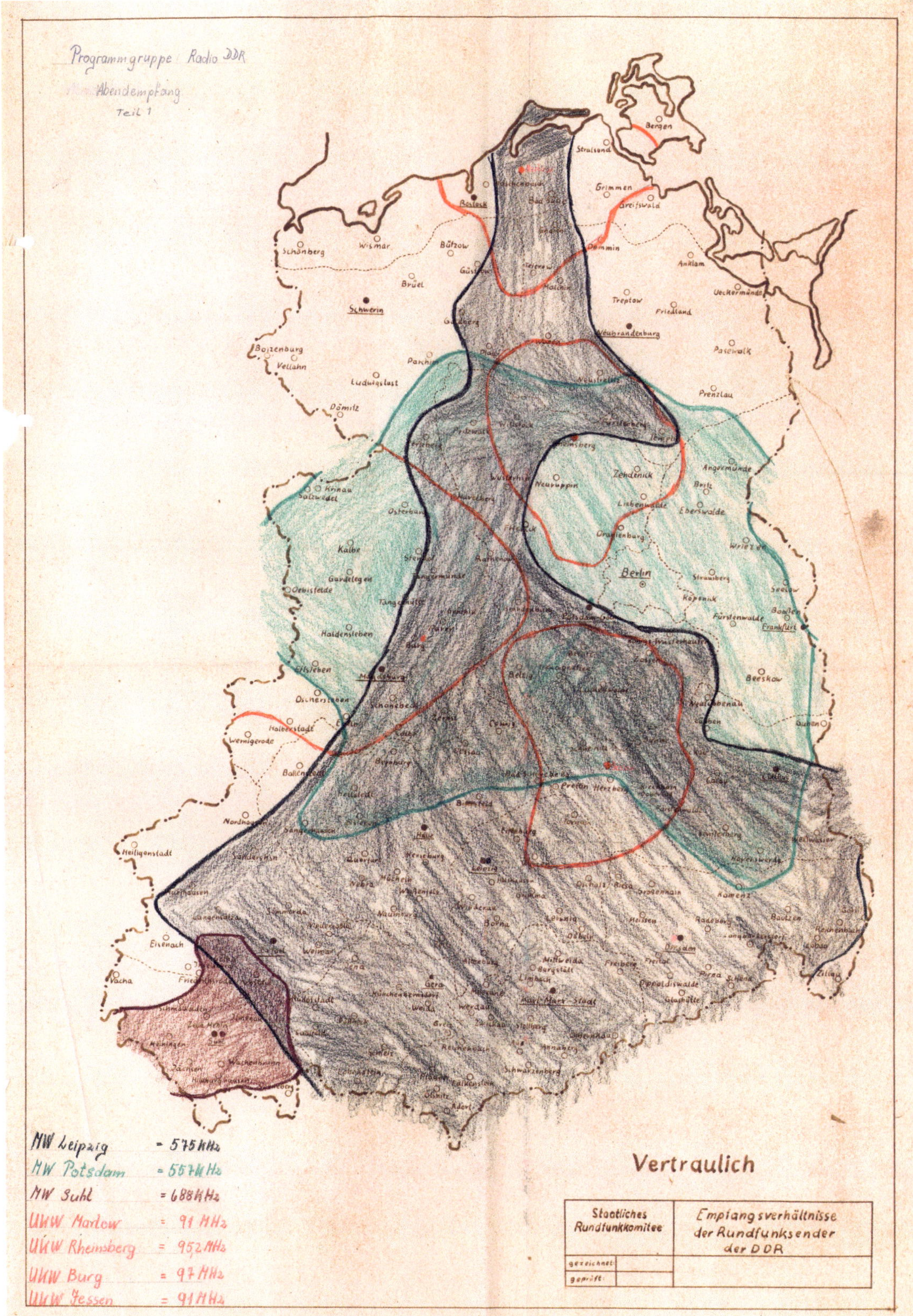
Programmgruppe Radio DDR
Abendempfang
Teil 1
MW Leipzig = 575 kHz
MW Potsdam = 557 kHz
MW Suhl = 688 kHz
UKW Marlow = 91 MHz
UKW Rheinsberg = 95,2 MHz
UKW Burg = 97 MHz
UKW Jessen = 91 MHz
Vertraulich
Staatliches Rundfunkkomitee
Empfangsverhältnisse der Rundfunksender der DDR
gezeichnet:
geprüft:

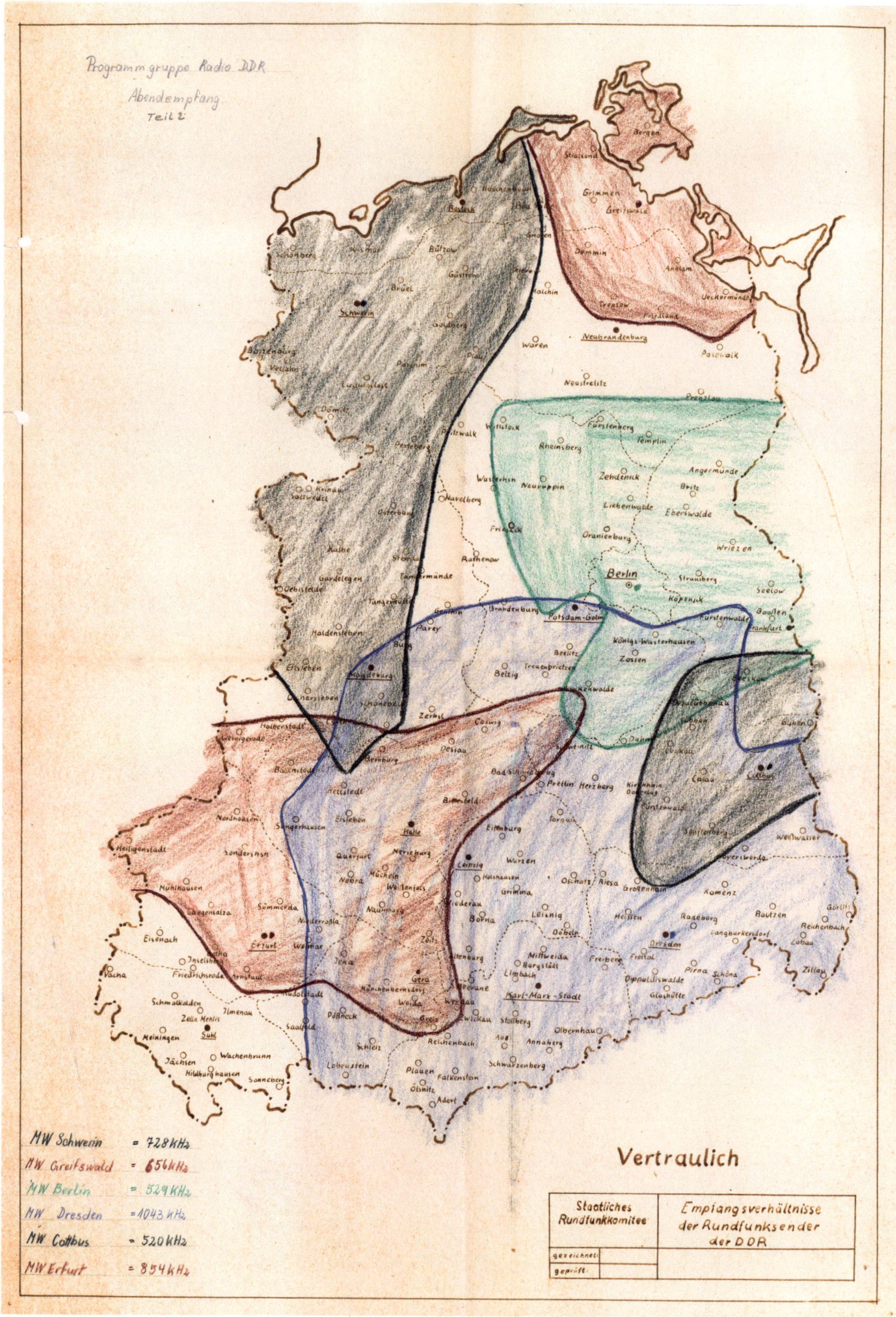
Programmgruppe Radio DDR
Abendempfang
Teil 2
MW Schwerin = 728 kHz
MW Greifswald = 656 kHz
MW Berlin = 529 kHz
MW Dresden = 1043 kHz
MW Cottbus = 520 kHz
MW Erfurt = 854 kHz
Vertraulich
Staatliches Rundfunkkomitee
Empfangsverhältnisse der Rundfunksender der DDR
gezeichnet:
geprüft:

71

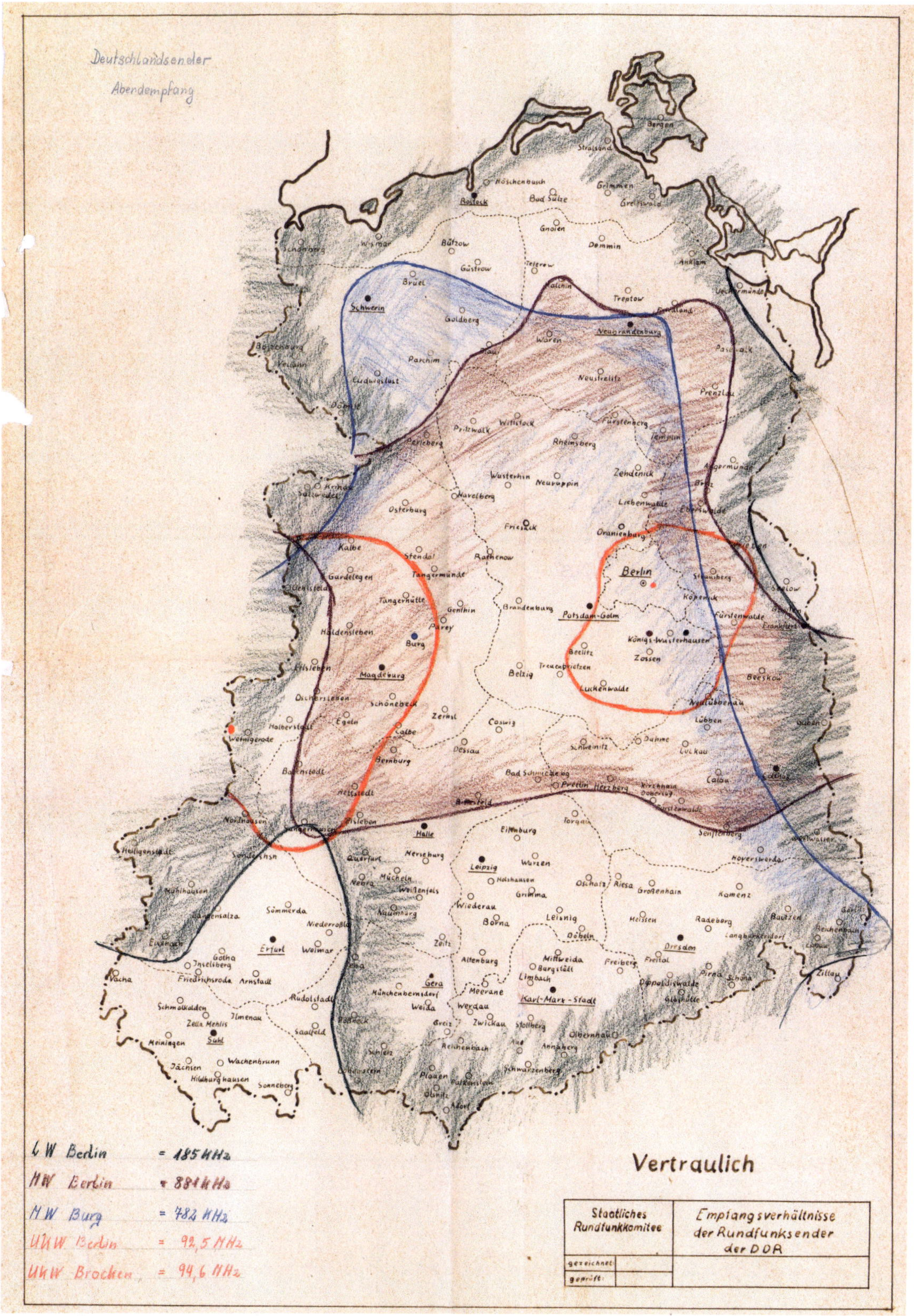
Deutschlandsender
Abendempfang
Rostock
Schwerin
Neubrandenburg
Berlin
Potsdam-Golm
Magdeburg
Halle
Leipzig
Erfurt
Gera
Karl-Marx-Stadt
Dresden
Suhl
LW Berlin = 185 kHz
MW Berlin = 881 kHz
MW Burg = 782 kHz
UKW Berlin = 92,5 MHz
UKW Brocken = 94,6 MHz
Vertraulich
Staatliches Rundfunkkomitee
Empfangsverhältnisse der Rundfunksender der DDR
gezeichnet:
geprüft:

72

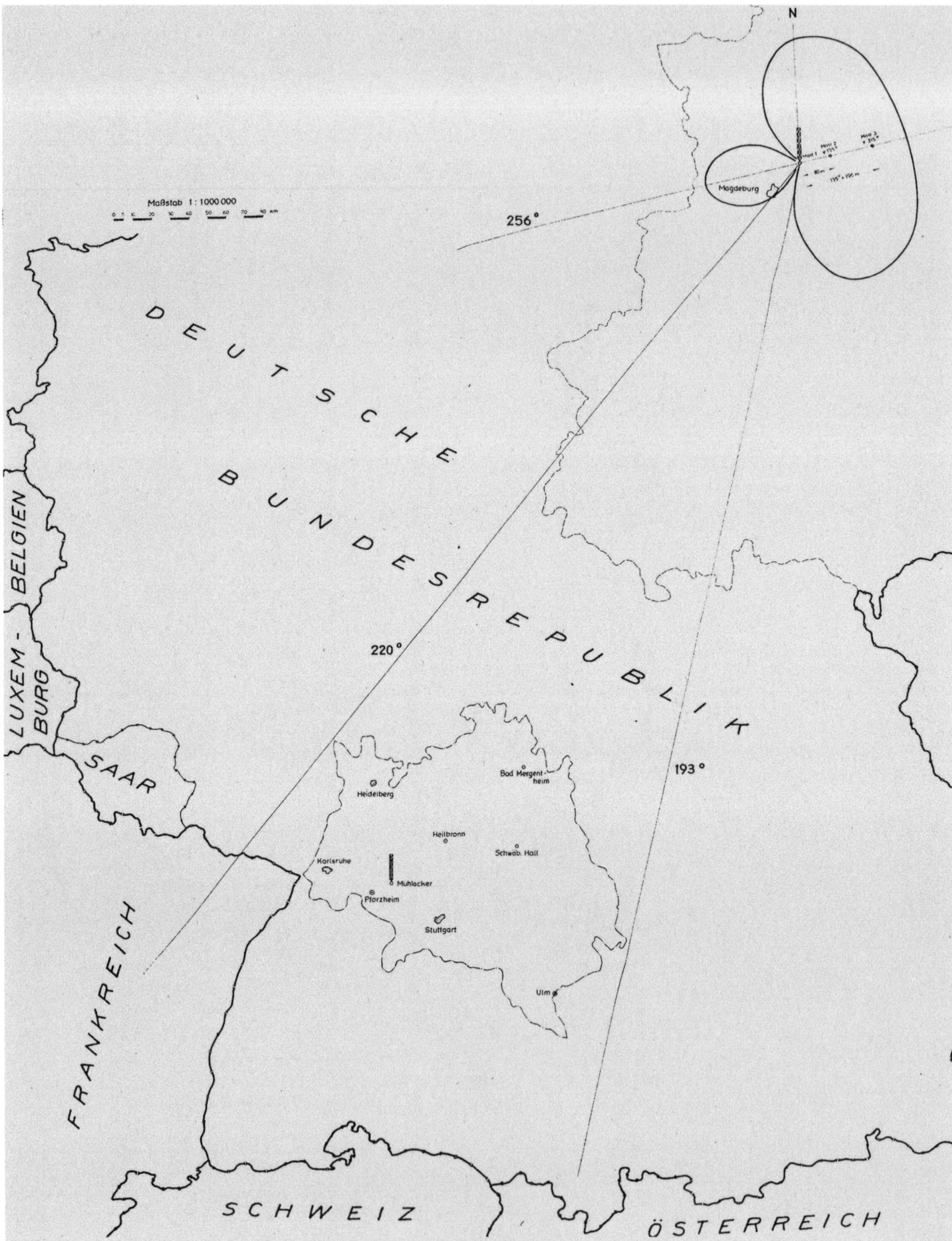
N
Maßstab 1 : 1000 000
Magdeburg
256°
DEUTSCHE BUNDESREPUBLIK
BELGIEN
LUXEMBURG
SAAR
FRANKREICH
220°
193°
Bad Mergentheim
Heidelberg
Heilbronn
Schwäb. Hall
Karlsruhe
Mühlacker
Pforzheim
Stuttgart
Ulm
SCHWEIZ
ÖSTERREICH

73

HAMBURG
BERLIN
Tagesempfang
Nachtempfang
Gebietsw. Tag- und Nacht empfang
Starke Interferenz
Hamburg: 19³⁰ - 21³⁰ Uhr
Essen: 20⁰⁰ - 24⁰⁰ Uhr
Stuttgart: 20⁰⁰ - 24⁰⁰ Uhr
Vertraulich
Staatliches Rundfunkkomitee
Empfangsverhältnisse der Rundfunksender
gezeichnet: 7.8.55
geprüft:
DS - Burg - 782 kHz

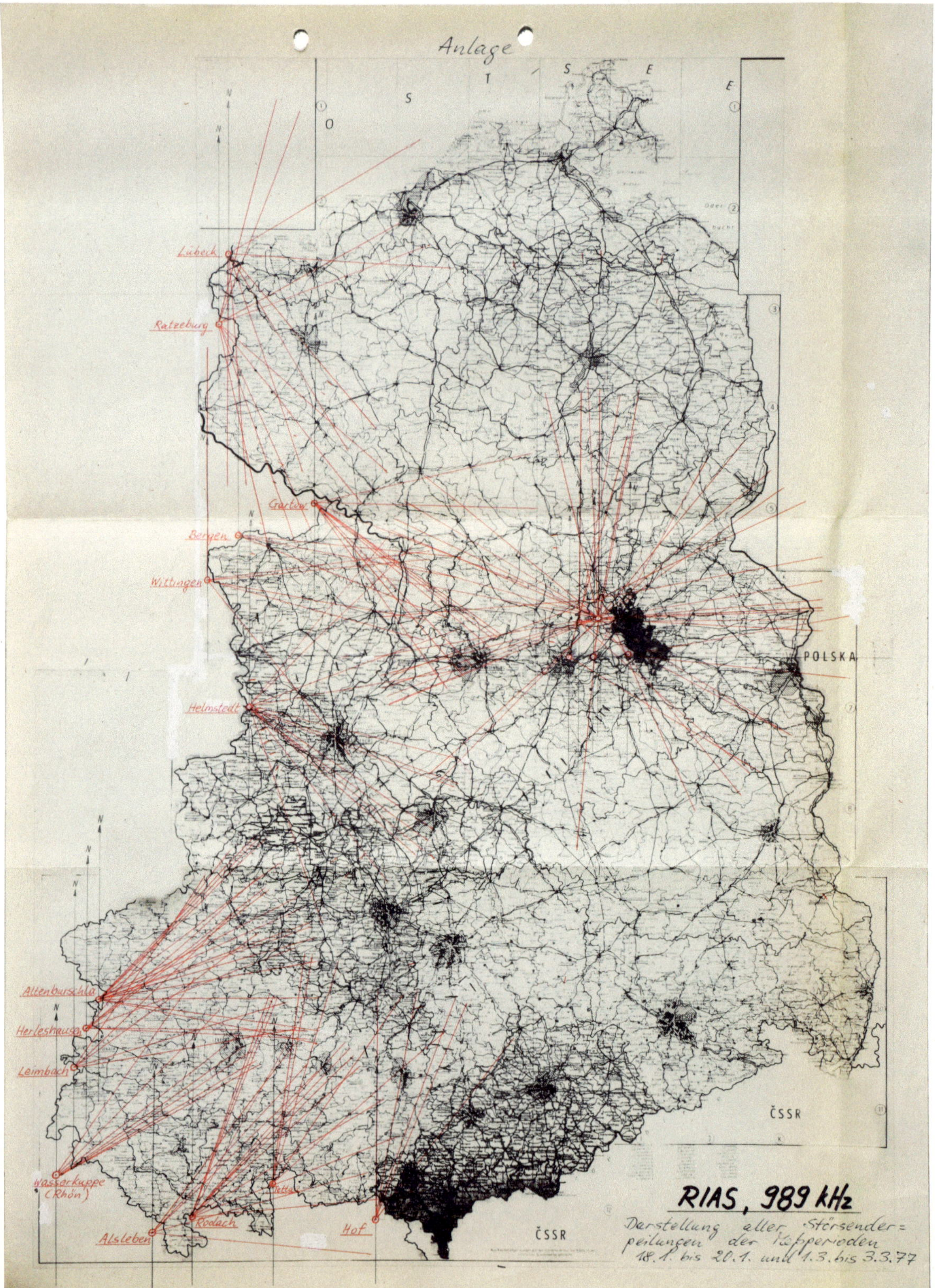
Anlage
O S T S E E
Lübeck
Ratzeburg
Gartow
Bergen
Wittingen
Helmstedt
POLSKA
Altenburschla
Herleshausen
Leimbach
Wasserkuppe
(Rhön)
Alsleben
Rodach
Hof
ČSSR
ČSSR
RIAS, 989 kHz
Darstellung aller Störsender=
peilungen der
18.1. bis 20.1. und 1.3. bis 3.3.77

75

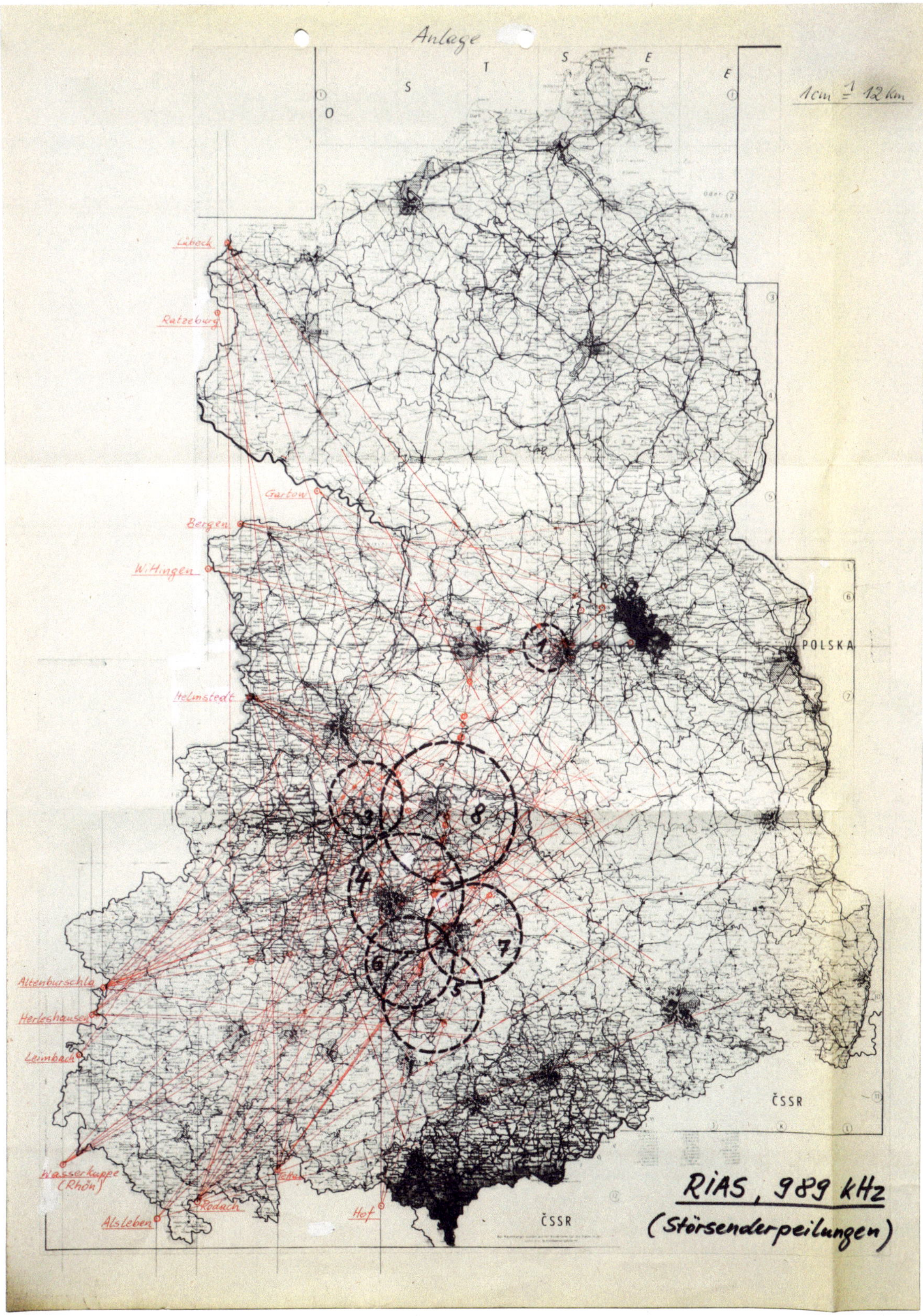
Anlage
O S T S E E
1cm ≙ 12 km
Lübeck
Ratzeburg
Gartow
Bergen
Wittingen
Helmstedt
POLSKA
Altenburschla
Herleshausen
Leimbach
Wasserkuppe (Rhön)
Rodach
Alsleben
Hof
ČSSR
ČSSR
RIAS, 989 kHz
(Störsenderpeilungen)

Anlage 1

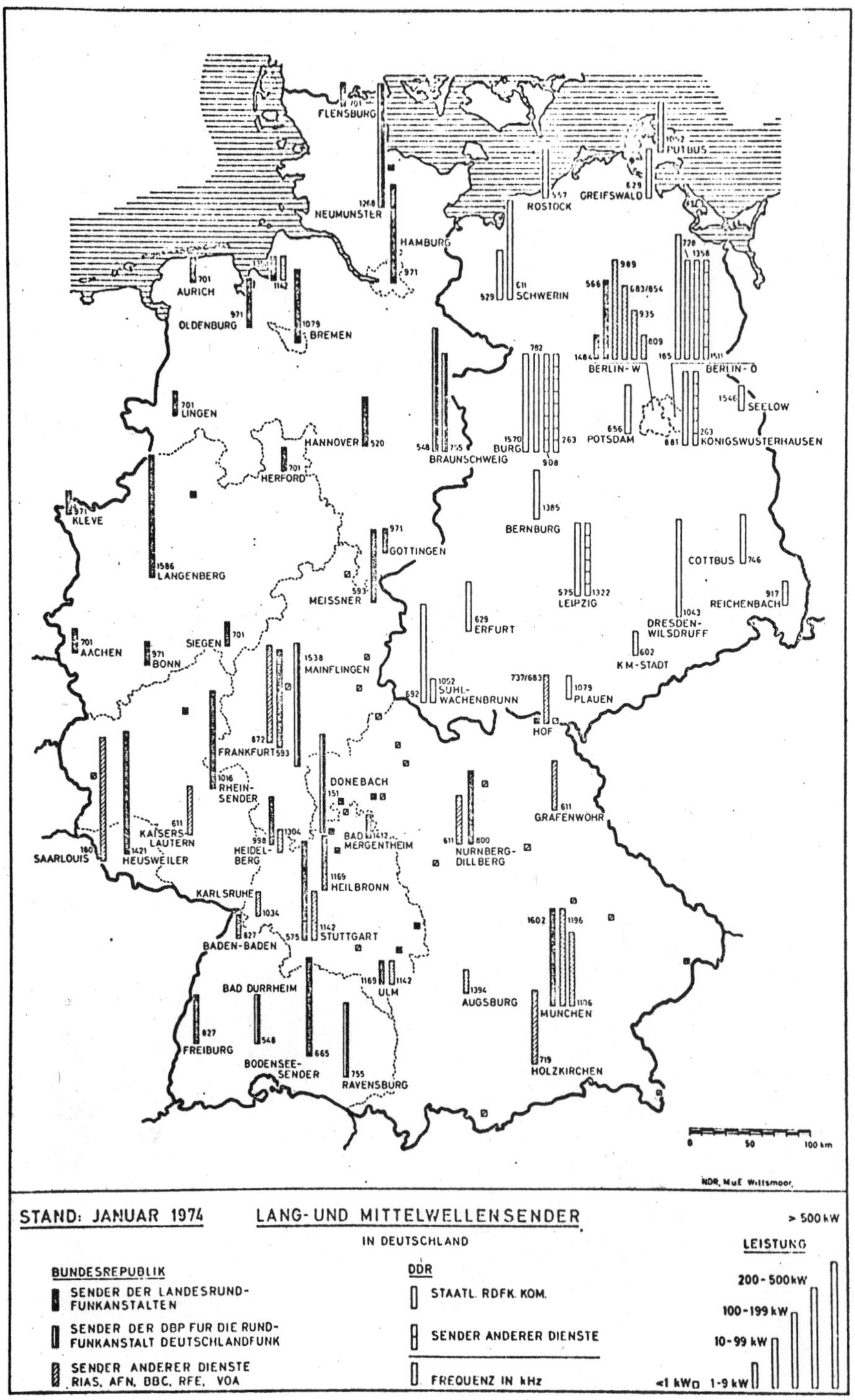
STAND: JANUAR 1974
LANG- UND MITTELWELLENSENDER
IN DEUTSCHLAND
BUNDESREPUBLIK
SENDER DER LANDESRUNDFUNKANSTALTEN
SENDER DER DBP FÜR DIE RUNDFUNKANSTALT DEUTSCHLANDFUNK
SENDER ANDERER DIENSTE RIAS, AFN, BBC, RFE, VOA
DDR
STAATL. RDFK. KOM.
SENDER ANDERER DIENSTE
FREQUENZ IN kHz
LEISTUNG
> 500 kW
200 - 500 kW
100 - 199 kW
10 - 99 kW
<1 kW
1 - 9 kW
0
50
100 km
NDR, MuE Wittsmoor
FLENSBURG
NEUMÜNSTER
HAMBURG
AURICH
OLDENBURG
BREMEN
LINGEN
HANNOVER
HERFORD
BRAUNSCHWEIG
KLEVE
LANGENBERG
GÖTTINGEN
MEISSNER
AACHEN
BONN
SIEGEN
MAINFLINGEN
FRANKFURT
RHEINSENDER
DONEBACH
KAISERSLAUTERN
SAARLOUIS
HEUSWEILER
HEIDELBERG
BAD MERGENTHEIM
HEILBRONN
KARLSRUHE
BADEN-BADEN
STUTTGART
BAD DÜRRHEIM
ULM
FREIBURG
BODENSEE-SENDER
RAVENSBURG
AUGSBURG
MÜNCHEN
HOLZKIRCHEN
NÜRNBERG-DILLBERG
GRAFENWÖHR
HOF
PLAUEN
SUHL-WACHENBRUNN
ERFURT
LEIPZIG
BERNBURG
BURG
POTSDAM
BERLIN-W
BERLIN-O
KÖNIGSWUSTERHAUSEN
SEELOW
COTTBUS
REICHENBACH
DRESDEN-WILSDRUFF
K M-STADT
SCHWERIN
ROSTOCK
GREIFSWALD
PUTBUS

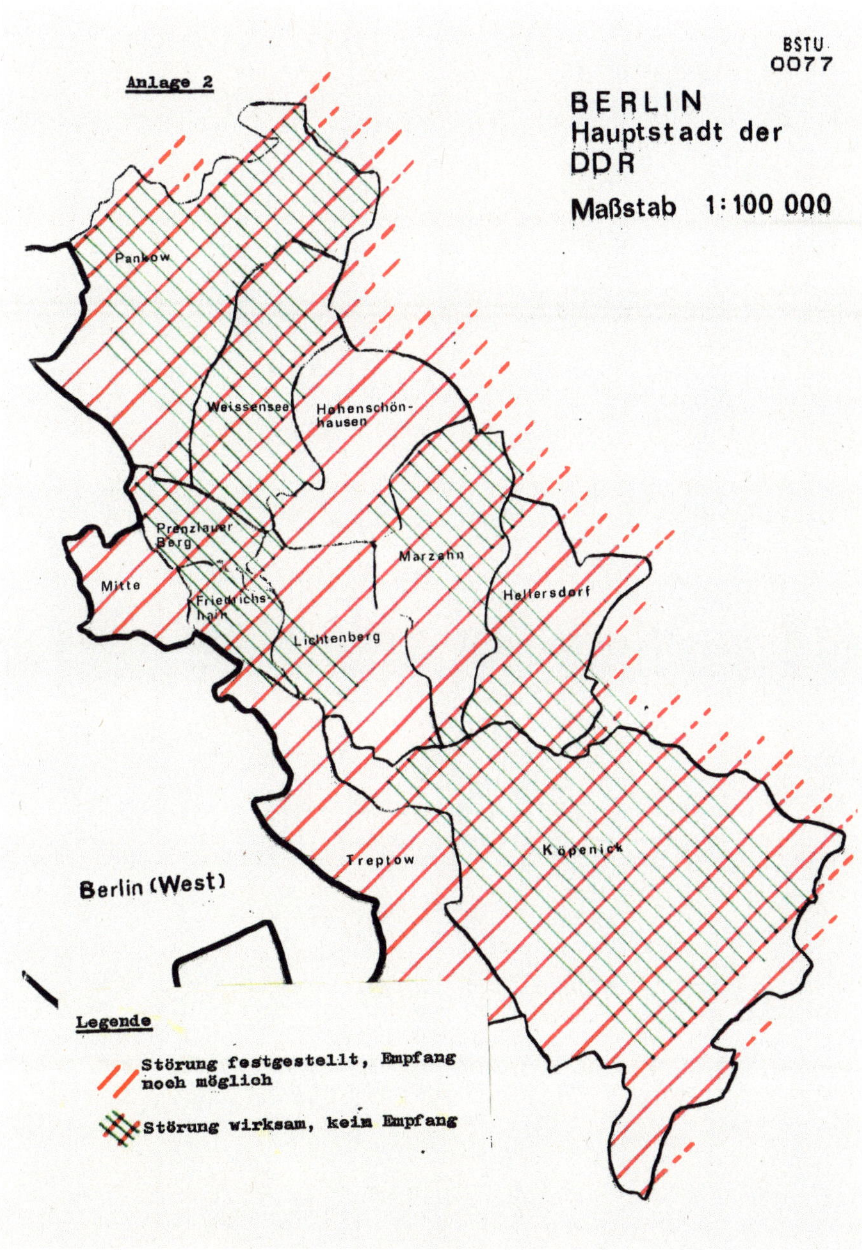
Anlage 2
BSTU
0077
BERLIN
Hauptstadt der
DDR
Maßstab 1:100 000
Pankow
Weissensee
Hohenschön-
hausen
Prenzlauer
Berg
Mitte
Friedrichs-
hain
Lichtenberg
Marzahn
Hellersdorf
Treptow
Köpenick
Berlin (West)
Legende
Störung festgestellt, Empfang
noch möglich
Störung wirksam, kein Empfang

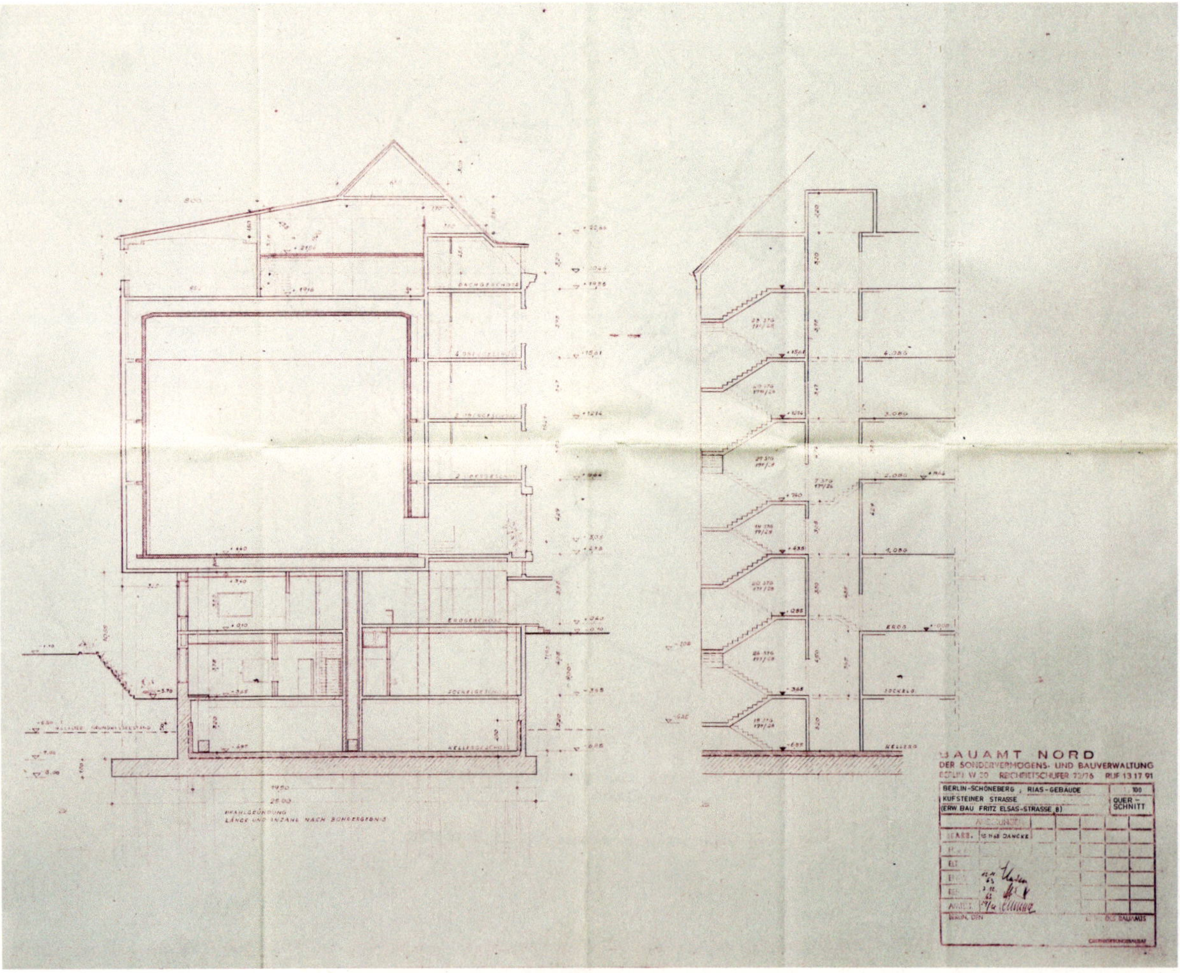
BAUAMT NORD
DER SONDERVERMÖGENS- UND BAUVERWALTUNG
RUF 13 17 91
BERLIN-SCHÖNEBERG ; RIAS-GEBÄUDE
KUFSTEINER STRASSE
(ERW. BAU FRITZ ELSAS-STRASSE 8)
100
QUER-SCHNITT
DANCKE

79

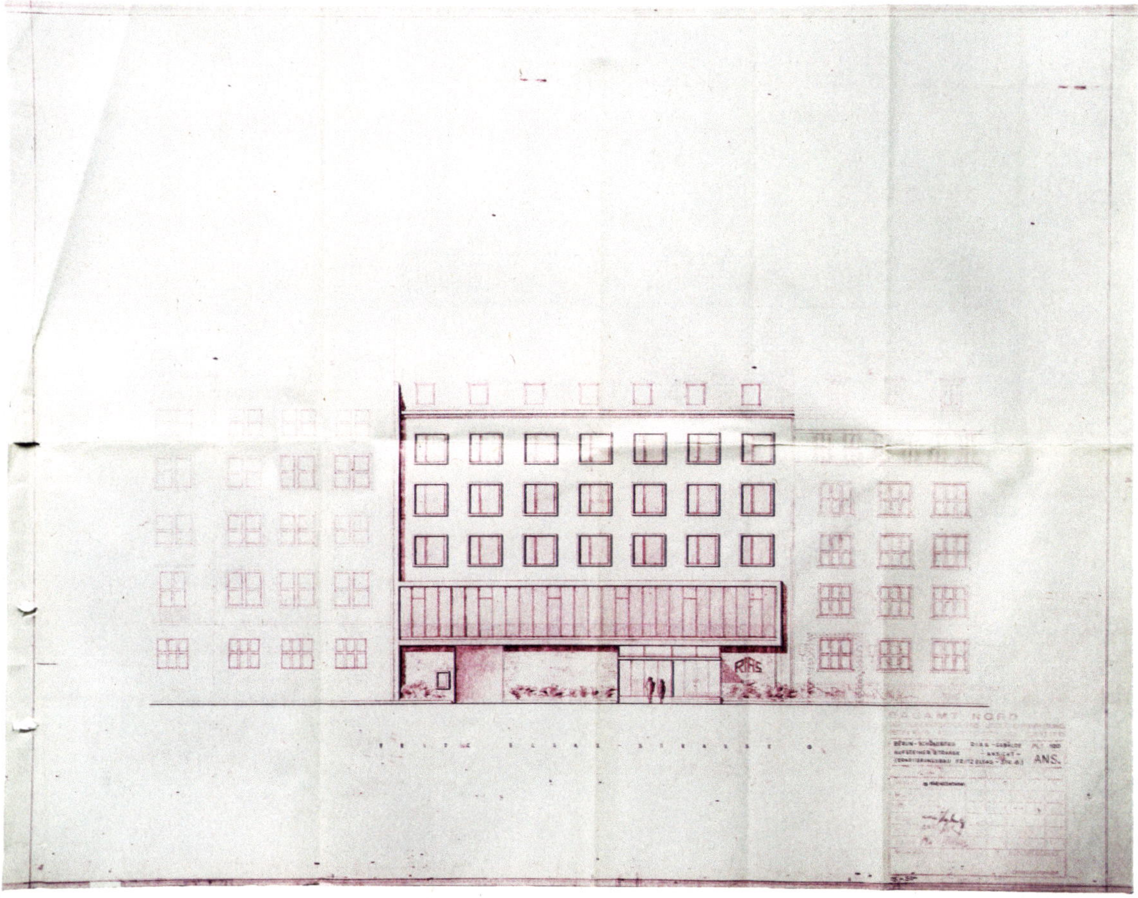

80

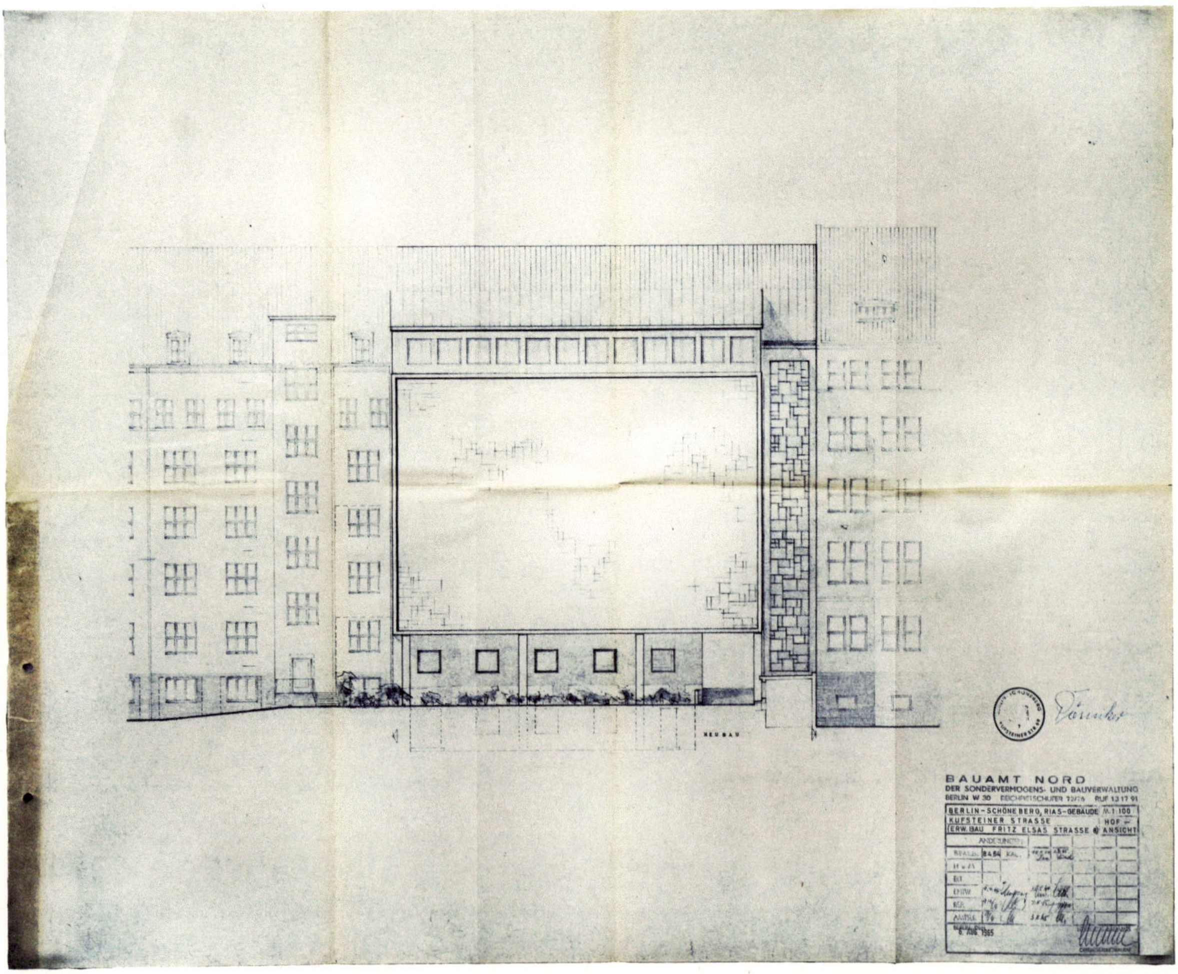
NEUBAU
BAUAMT NORD
DER SONDERVERMÖGENS- UND BAUVERWALTUNG
BERLIN W 30
RUF 13 17 91
BERLIN-SCHÖNEBERG, RIAS-GEBÄUDE
M.1:100
KUFSTEINER STRASSE
(ERW.BAU FRITZ ELSAS STRASSE 8)
HOF-ANSICHT
8484
6. AUG. 1965

81

82

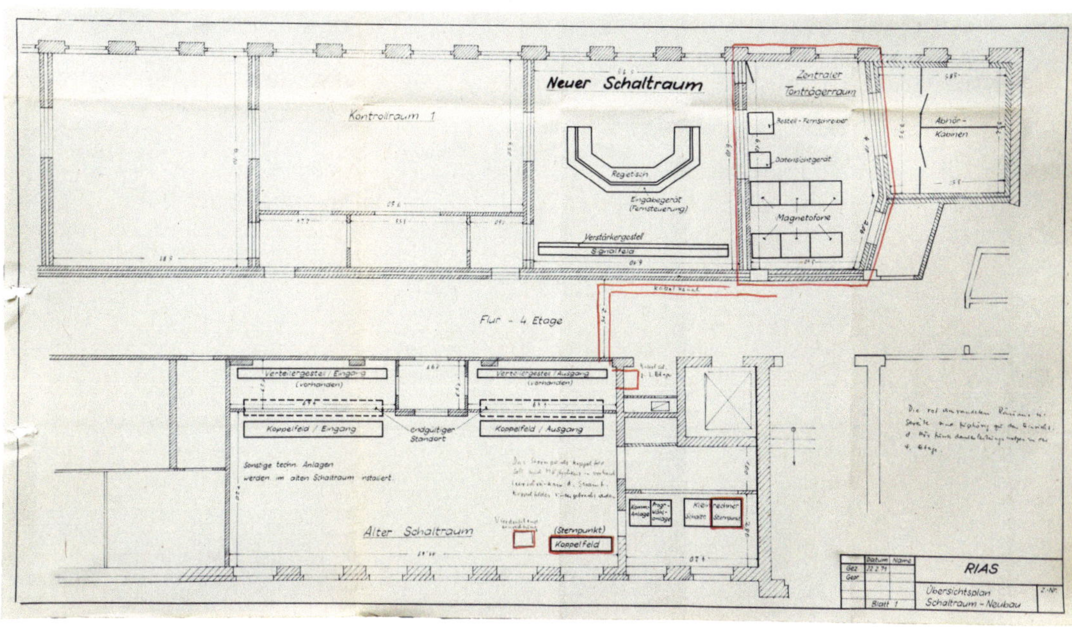
Neuer Schaltraum
Zentraler Tonträgerraum
Kontrollraum 1
Regietisch
Eingabegerät (Fernsteuerung)
Verstärkergestell
Signalfeld
Magnetofone
Abhör-Kabinen
Flur - 4 Etage
Verteilergestell / Eingang (vorhanden)
Verteilergestell / Ausgang (vorhanden)
Koppelfeld / Eingang
endgültiger Standort
Koppelfeld / Ausgang
Sonstige techn. Anlagen werden im alten Schaltraum installiert.
Alter Schaltraum
(Sternpunkt)
Koppelfeld
RIAS
Übersichtsplan Schaltraum - Neubau

83

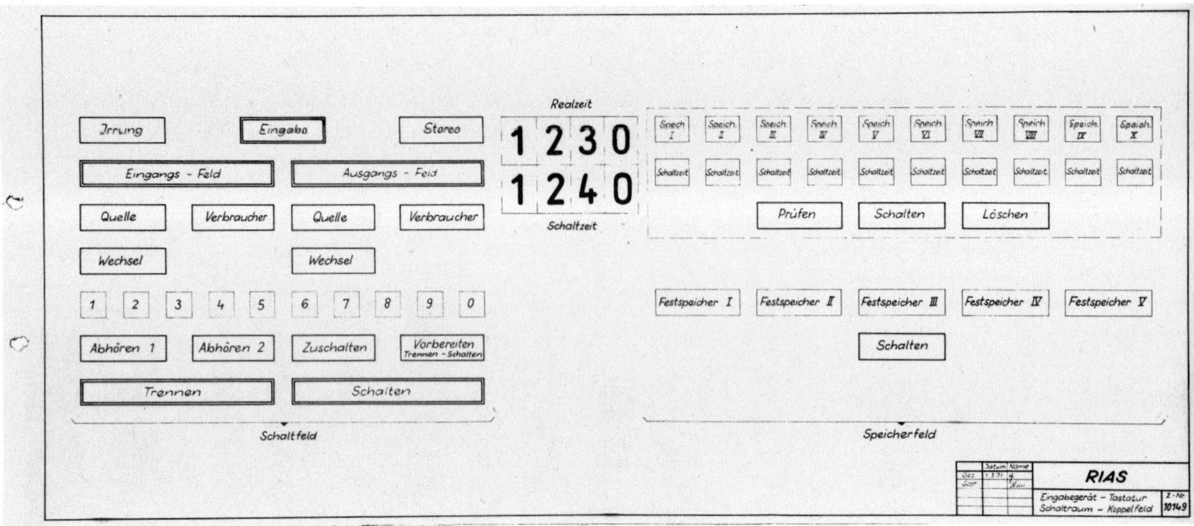
Realzeit
1230
1240
Schaltzeit
Jrrung
Eingabe
Stereo
Eingangs - Feld
Ausgangs - Feld
Quelle
Verbraucher
Quelle
Verbraucher
Wechsel
Wechsel
1 2 3 4 5 6 7 8 9 0
Abhören 1
Abhören 2
Zuschalten
Vorbereiten Trennen - Schalten
Trennen
Schalten
Schaltfeld
Speich I
Speich II
Speich III
Speich IV
Speich V
Speich VI
Speich VII
Speich VIII
Speich IX
Speich X
Schaltzeit
Prüfen
Schalten
Löschen
Festspeicher I
Festspeicher II
Festspeicher III
Festspeicher IV
Festspeicher V
Schalten
Speicherfeld
RIAS
Eingabegerät – Tastatur
Schaltraum – Koppelfeld
10149

84

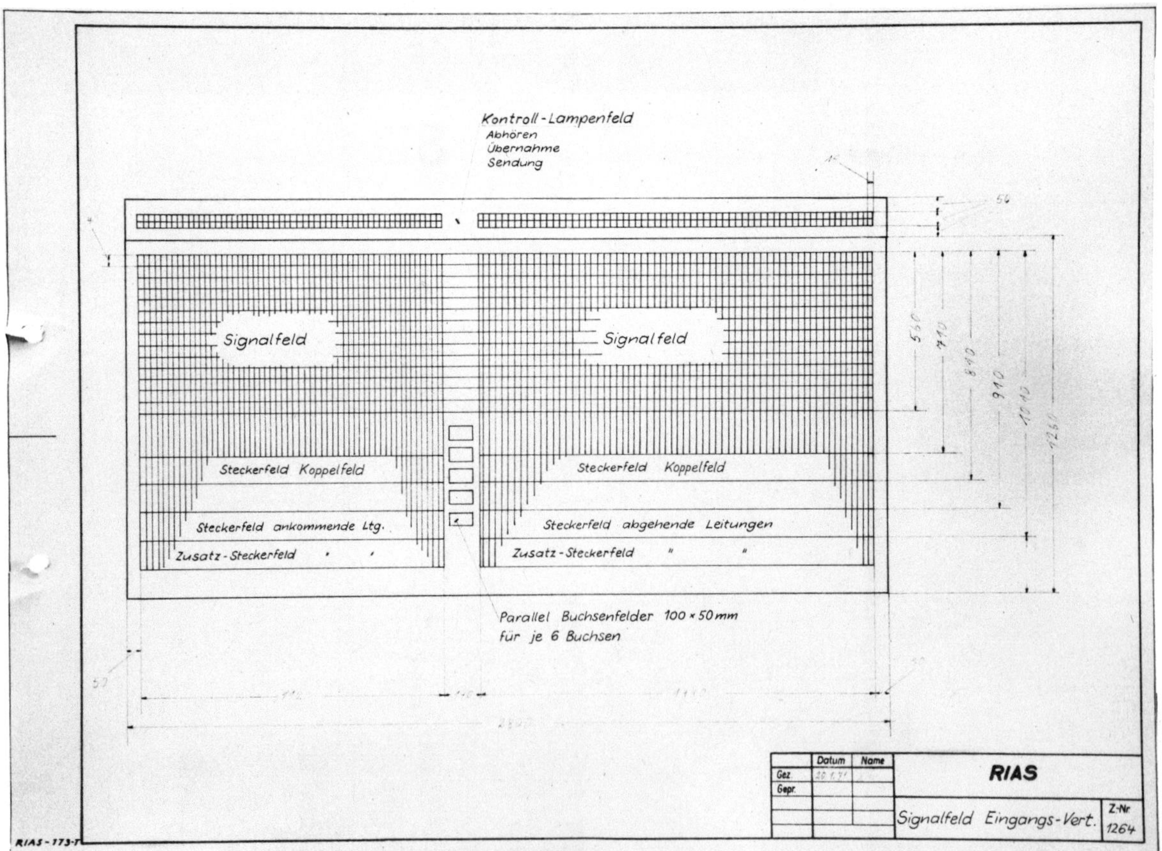
Kontroll - Lampenfeld
Abhören
Übernahme
Sendung
Signalfeld
Signalfeld
Steckerfeld Koppelfeld
Steckerfeld Koppelfeld
Steckerfeld ankommende Ltg.
Steckerfeld abgehende Leitungen
Zusatz - Steckerfeld
Zusatz - Steckerfeld
Parallel Buchsenfelder 100 × 50 mm
für je 6 Buchsen
Datum
Name
Gez
Gepr.
RIAS
Signalfeld Eingangs - Vert.
Z-Nr.
1264
RIAS - 173 - 1

85

Fig. 69 State Radio Committee map showing the coverage and frequency range of different transmission stations in East German territory. Source: Bundesarchiv. DR 6/232.

Fig. 70 State Radio Committee map showing the coverage and frequency range of different transmission stations in East German territory. Source: Bundesarchiv. DR 6/232.

Fig. 71 State Radio Committee map showing the coverage and frequency range of different transmission stations in East German territory. Source: Bundesarchiv. DR 6/232.

Fig. 72 State Radio Committee map showing the coverage and frequency range of different transmission stations in East German territory. Source: Bundesarchiv. DR 6/232.

Fig. 73 Map showing the position of the Burg transmission station relative to the East–West German border. Source: Bundesarchiv. DR 6/242.

Fig. 74 Map comparing the day and nighttime coverage of the Burg transmission station. Source: Bundesarchiv. DR 6/232.

Fig. 75 Map of East German territory showing the series of Western transmission stations ranged strategically along its border. Source: Deutsche Rundfunkarchiv Babelsberg. RIAS 16/94/6.

Fig. 76 Map of East German territory showing the relative positions of the Western transmitters and the GDR jamming stations. The map was developed by the Fernmeldetechnisches Zentralamt, FTZ (Central Office for Communication Technologies) of the West German Post Office, who conducted regular tests of Western signals, paying particular attention to RIAS. Source: Deutsche Rundfunkarchiv Babelsberg. RIAS 16/94/6.

Fig. 77 Map of the entire German territory depicting the position and size of radio transmitters on both sides of the Iron Curtain. Source: Deutsche Rundfunkarchiv Babelsberg. RIAS 16/94/6.

Fig. 78 Map of radio jamming in East Berlin. Source: Der Bundesbeauftragte für die Stasi-Unterlagen, Ministeriums für Staatssicherheit, Bezirksverwaltung Berlin, Abt. III, Nr. 691, Bl. 77.

Fig. 79 Extension of RIAS broadcasting house, cross section. Source: Deutsche Rundfunkarchiv Babelsberg. RIAS F 406-00-00/0007 RIAS 16/94/1.

Fig. 80 Extension of RIAS broadcasting house, elevation. Source: Deutsche Rundfunkarchiv Babelsberg. RIAS F 406-00-00/0007 RIAS 16/94/1.

Fig. 81 Extension of RIAS broadcasting house, longitudinal section. Source: Deutsche Rundfunkarchiv Babelsberg. RIAS F 406-00-00/0007 RIAS 16/94/1.

Fig. 82 Extension of RIAS broadcasting house, view of the new control room. Source: RIAS Berlin, *RIAS: A Free Voice of the Free World* (Berlin: Brüder Hartmann, 1962), 5.

Fig. 83 Extension of RIAS broadcasting house, plan showing the new switch room (*Schaltraum*). Source: Deutsche Rundfunkarchiv Babelsberg. RIAS F 406-00-00/0007 RIAS 16/94/1.

Fig. 84 Extension of RIAS broadcasting house, detail drawing of new control table for the switch room. Source: Deutsche Rundfunkarchiv Babelsberg. RIAS F 406-00-00/0007 RIAS 16/94/1.

Fig. 85 Extension of RIAS broadcasting house, detail drawing of the switchboard for the new switch room. Source: Deutsche Rundfunkarchiv Babelsberg. RIAS F 406-00-00/0007 RIAS 16/94/1.

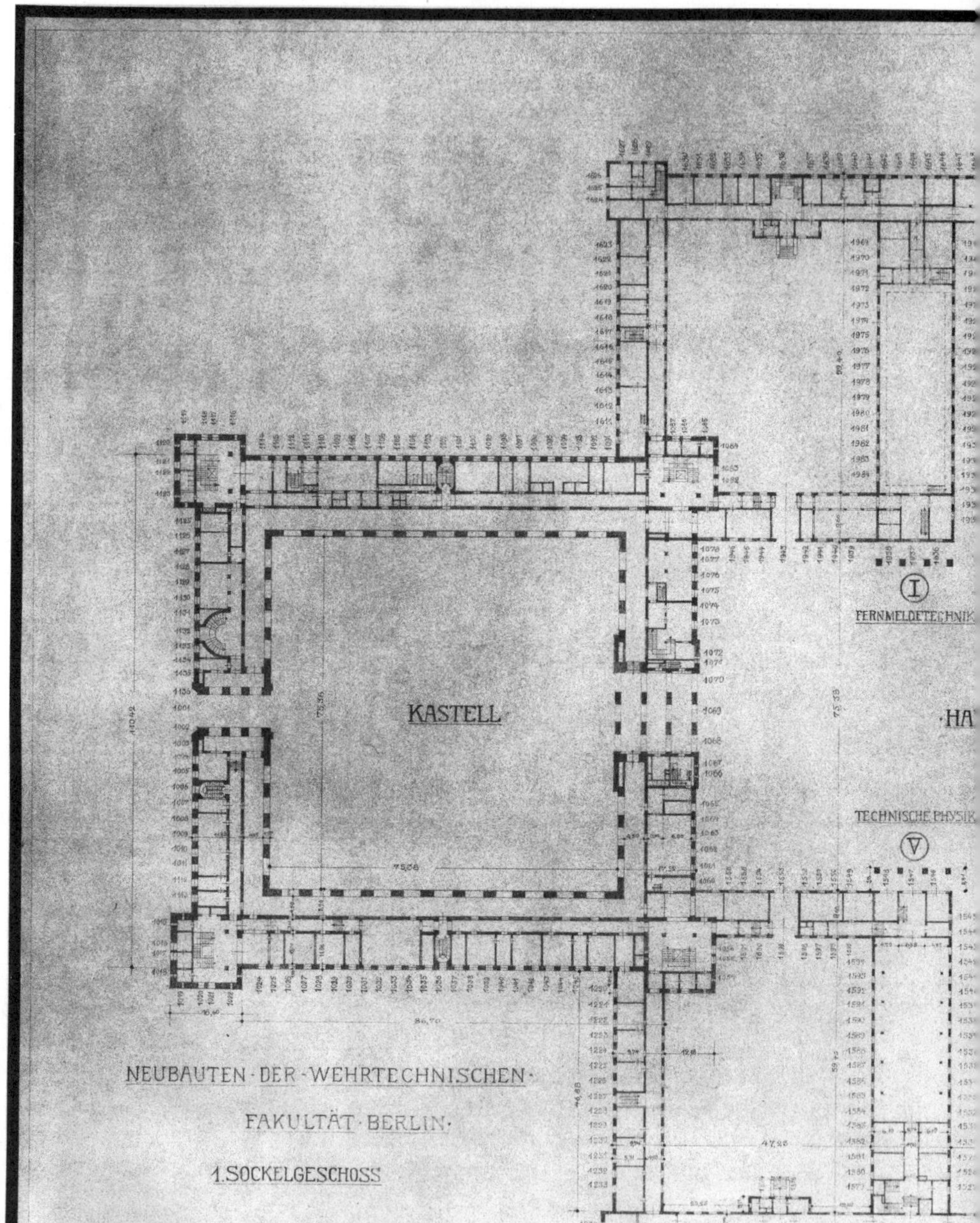
KASTELL
FERNMELDETECHNIK
TECHNISCHE PHYSIK
NEUBAUTEN · DER · WEHRTECHNISCHEN ·
FAKULTÄT · BERLIN ·
1. SOCKELGESCHOSS

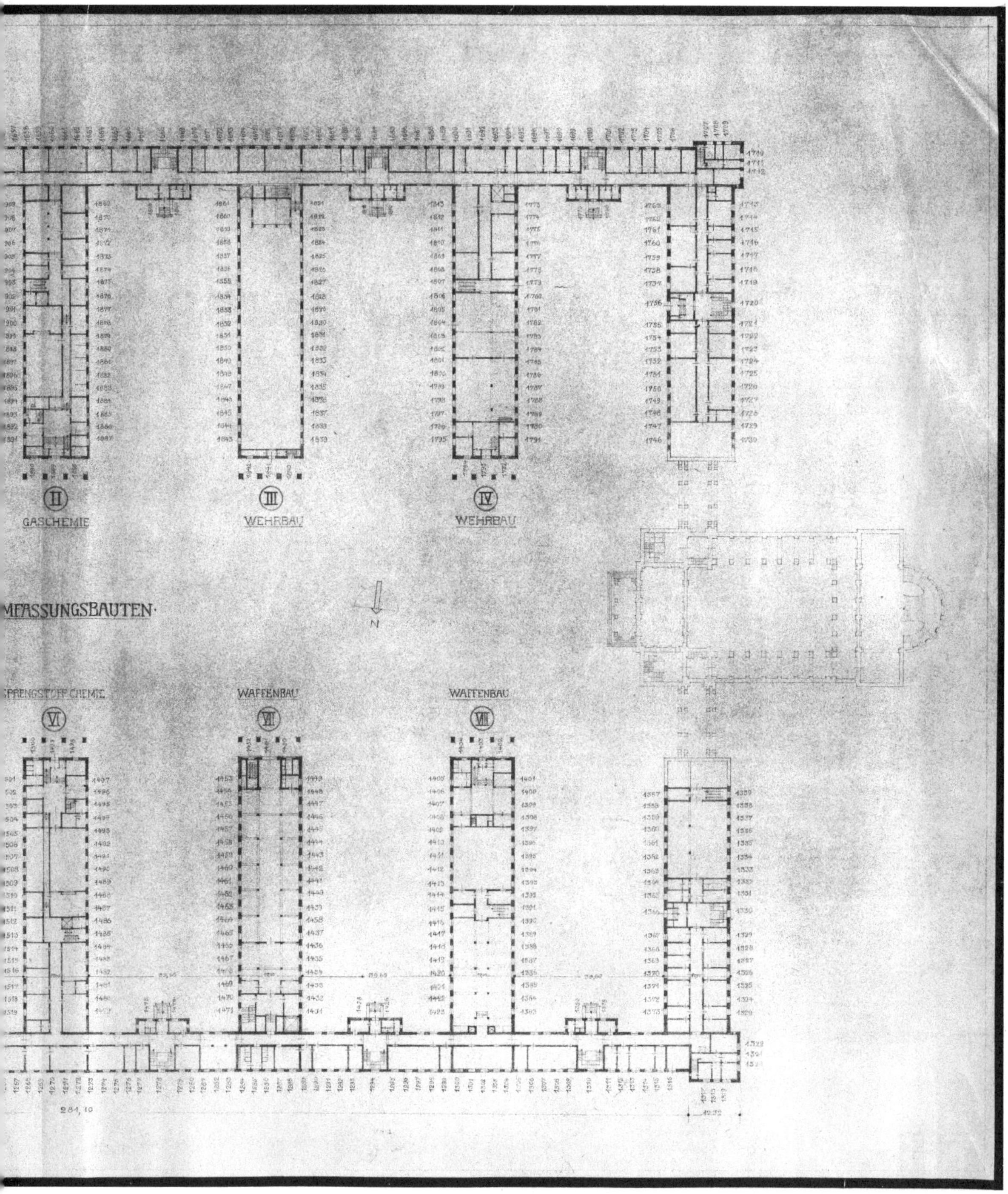
II
GASCHEMIE
III
WEHRBAU
IV
WEHRBAU
MFASSUNGSBAUTEN·
N
SPRENGSTOFFCHEMIE
VI
WAFFENBAU
VII
WAFFENBAU
VIII

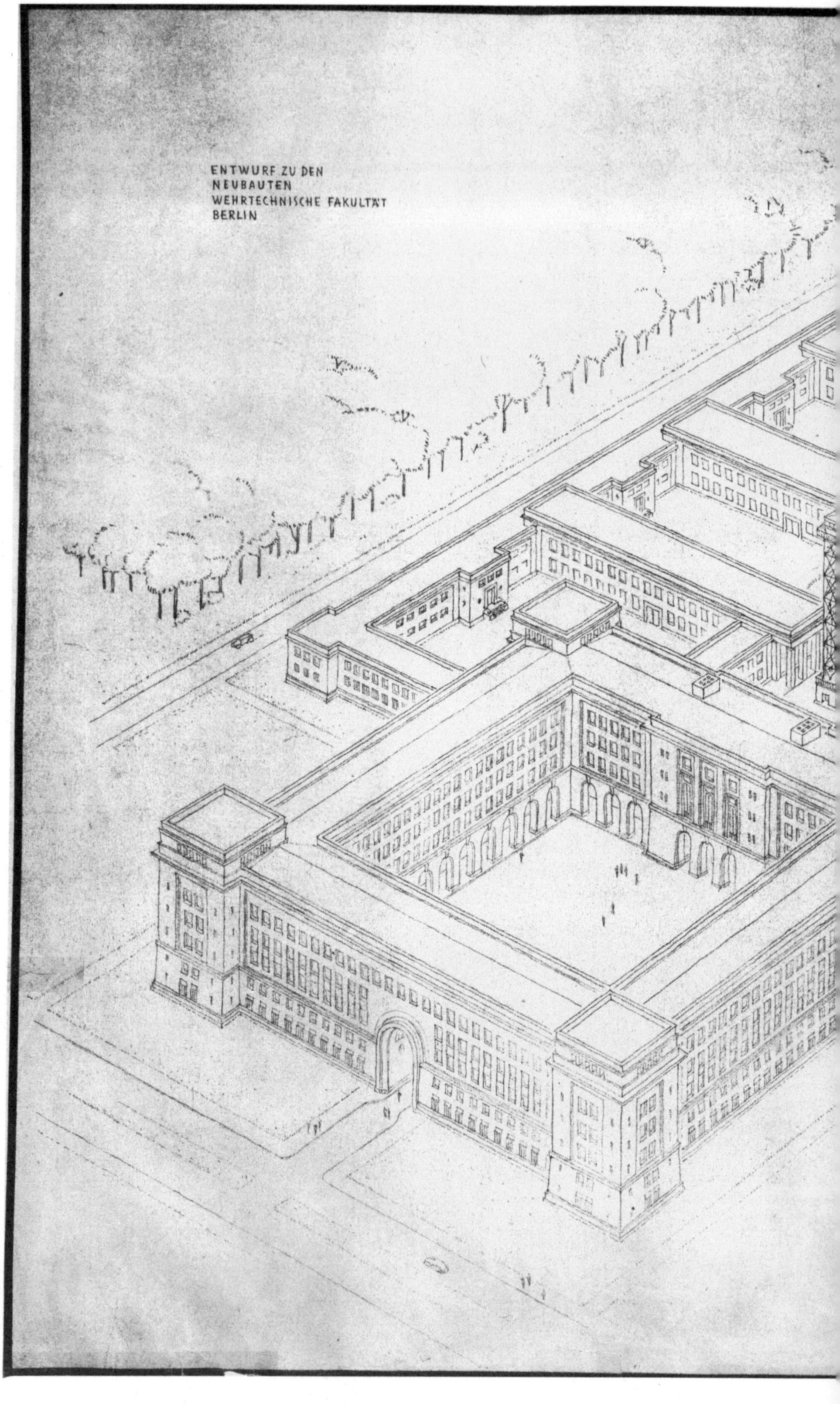
ENTWURF ZU DEN
NEUBAUTEN
WEHRTECHNISCHE FAKULTÄT
BERLIN

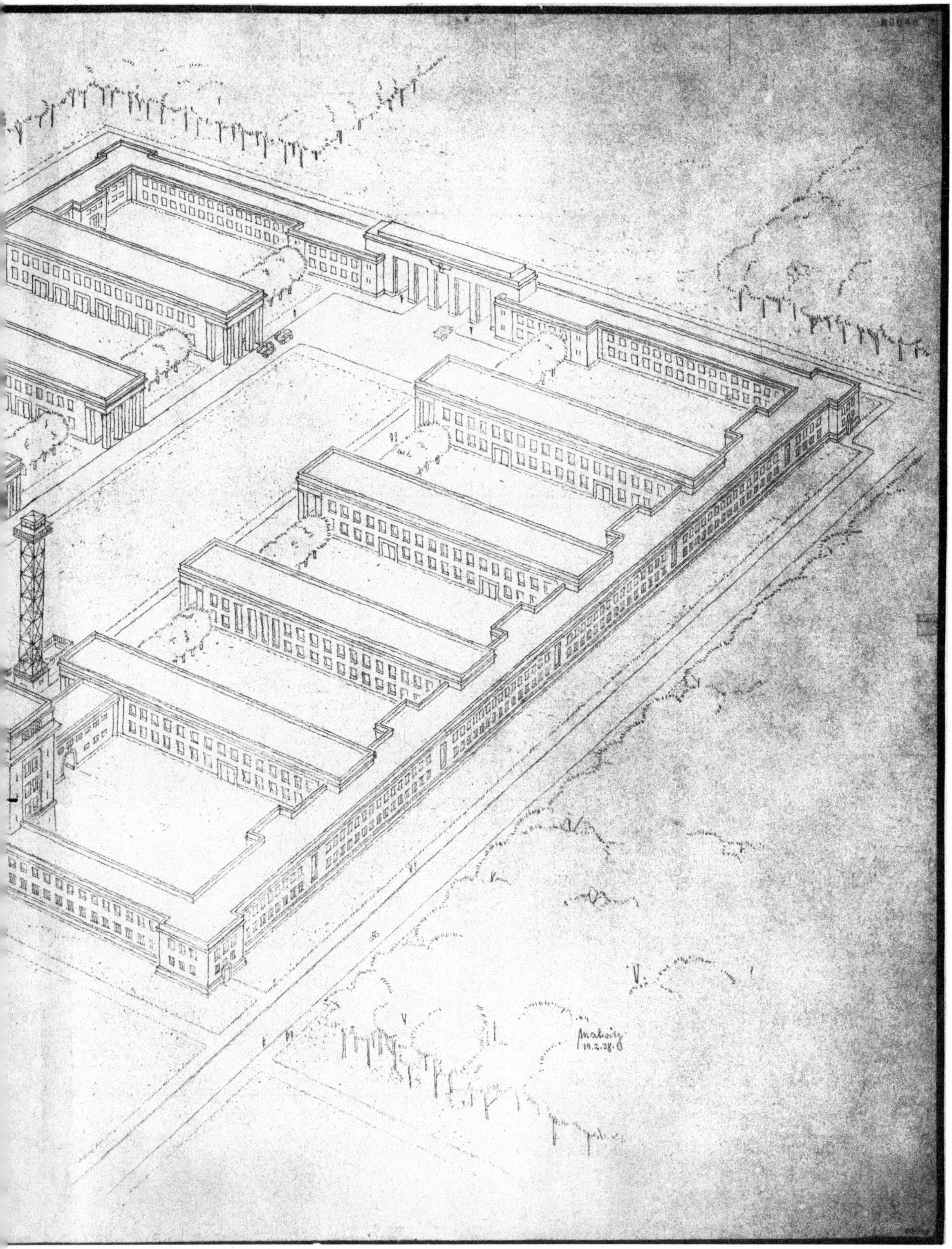

87

-105-

1980 GVS-P-O 026-5/85 Bild 14

Nach Abschluß der Bauvorhaben am
ELOKA-Objekt „Teufelsberg"

Empfangs-u. Auswerte-
einrichtung

Antennenmast

Arctic-Turm mit Radom

ELOKA - Objekt „FIELD STATION BERLIN"

vereinfachte Darstellung der Etagen des ELOKA - Objektes

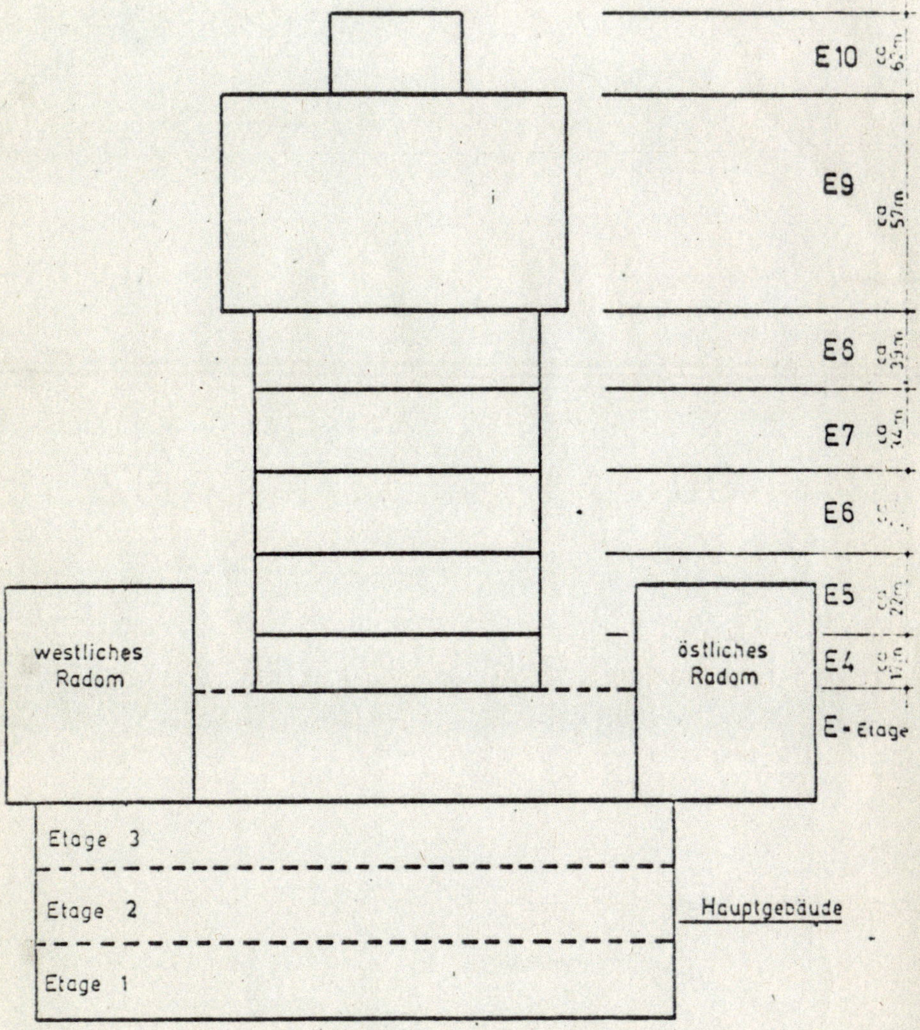

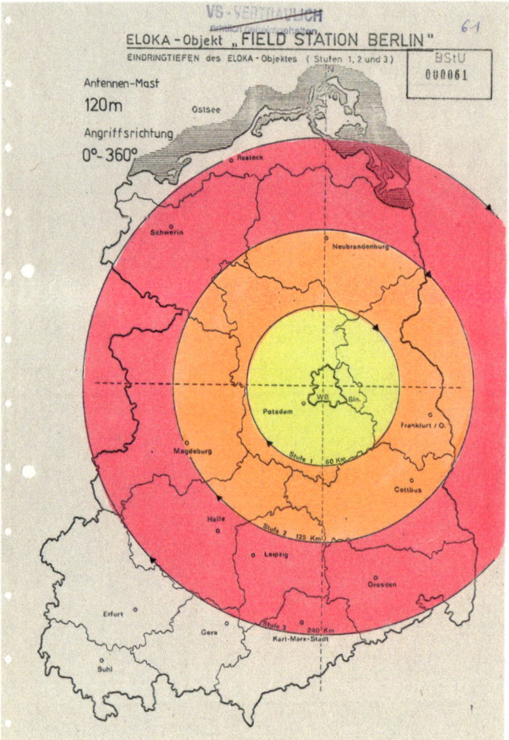
VS-VERTRAULICH
ELOKA-Objekt „FIELD STATION BERLIN"
EINDRINGTIEFEN des ELOKA-Objektes (Stufen 1, 2 und 3)
BStU
000061
Antennen-Mast
120m
Angriffsrichtung
0°-360°
Ostsee
Rostock
Schwerin
Neubrandenburg
Potsdam
WB
Bln.
Frankfurt / O.
Magdeburg
Cottbus
Halle
Leipzig
Dresden
Erfurt
Gera
Karl-Marx-Stadt
Suhl
Stufe 1 60 Km
Stufe 2 125 Km
Stufe 3 240 Km

90

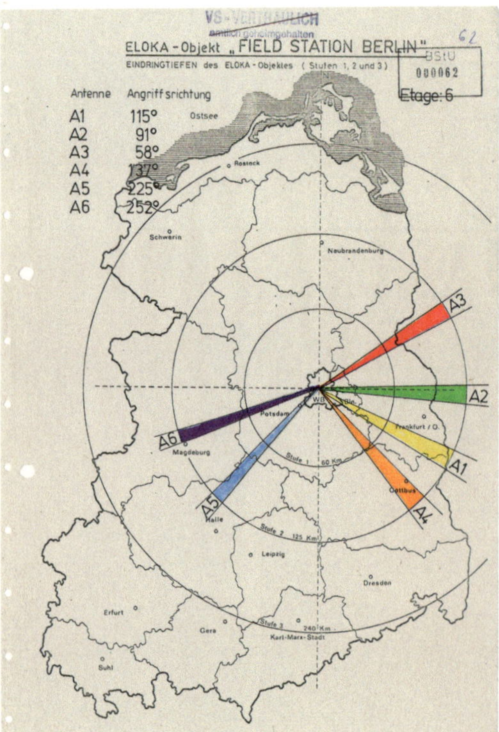
VS-VERTRAULICH
ELOKA-Objekt „FIELD STATION BERLIN"
EINDRINGTIEFEN des ELOKA-Objektes (Stufen 1, 2 und 3)
BStU
000062
Etage: 6
Antenne Angriffsrichtung
A1 115°
A2 91°
A3 58°
A4 137°
A5 225°
A6 252°
Ostsee
Rostock
Schwerin
Neubrandenburg
Potsdam
Frankfurt / O.
Magdeburg
Cottbus
Halle
Leipzig
Dresden
Erfurt
Gera
Karl-Marx-Stadt
Suhl
A1
A2
A3
A4
A5
A6
Stufe 1 60 Km
Stufe 2 125 Km
Stufe 3 240 Km

91

VS-VERTRAULICH
ELOKA-Objekt „FIELD STATION BERLIN"
EINDRINGTIEFEN des ELOKA-Objektes (Stufen 1, 2 und 3)
BStU
Antenne Angriffsrichtung
A1 284°-94°
Etage: 7
Ostsee
Rostock
Schwerin
Neubrandenburg
A1
Potsdam
WB
Bln.
Frankfurt / O.
Magdeburg
Cottbus
Halle
Leipzig
Dresden
Erfurt
Gera
Karl-Marx-Stadt
Suhl
Stufe 1 60 Km
Stufe 2 125 Km
Stufe 3 240 Km

92

VS-VERTRAULICH
000064
ELOKA-Objekt „FIELD STATION BERLIN"
EINDRINGTIEFEN des ELOKA-Objektes (Stufen 1, 2 und 3)
Antenne Angriffsrichtung
A2 14°-184°
Etage: 7
Ostsee
Rostock
Schwerin
Neubrandenburg
A2
Potsdam
Frankfurt / O.
Magdeburg
Cottbus
Halle
Leipzig
Dresden
Erfurt
Gera
Karl-Marx-Stadt
Suhl
Stufe 1
Stufe 2 125 Km
Stufe 3 240 Km

93

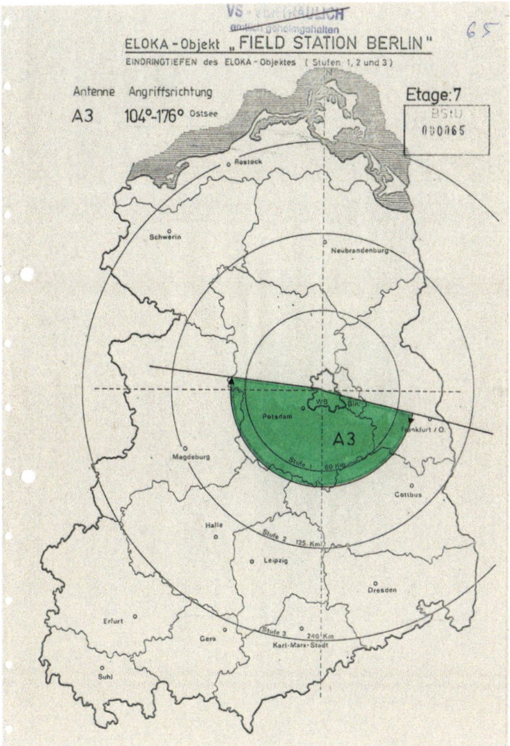
VS-VERTRAULICH
65
ELOKA-Objekt „FIELD STATION BERLIN"
EINDRINGTIEFEN des ELOKA-Objektes (Stufen 1, 2 und 3)
Antenne Angriffsrichtung
A3 104°-176°
Ostsee
Etage: 7
BStU
000065
Rostock
Schwerin
Neubrandenburg
Potsdam
WB
Bln.
A3
Frankfurt / O.
Magdeburg
Stufe 1 60 Km
Cottbus
Halle
Stufe 2 125 Km
Leipzig
Dresden
Erfurt
Gera
Stufe 3 240 Km
Karl-Marx-Stadt
Suhl

94

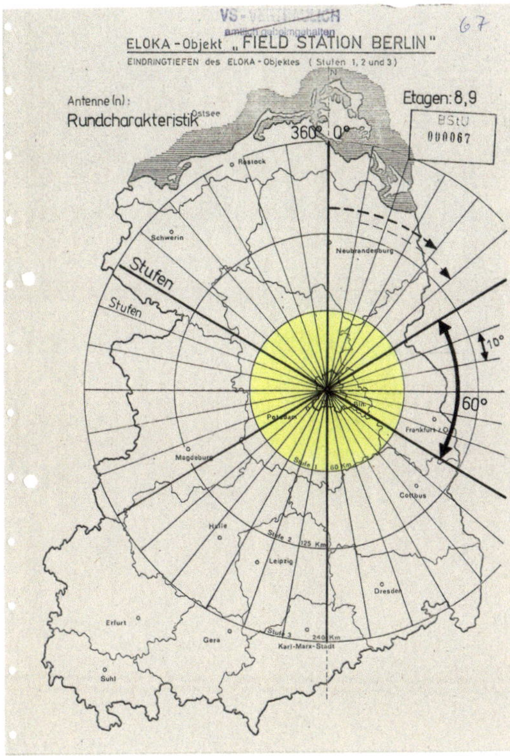
VS-VERTRAULICH
67
ELOKA-Objekt „FIELD STATION BERLIN"
EINDRINGTIEFEN des ELOKA-Objektes (Stufen 1, 2 und 3)
Antenne (n):
Rundcharakteristik
Ostsee
Etagen: 8,9
BStU
000067
360° 0°
Rostock
Schwerin
Neubrandenburg
Stufen
Stufen
10°
60°
Potsdam
Frankfurt / O.
Magdeburg
Stufe 1 60 Km
Cottbus
Halle
Stufe 2 125 Km
Leipzig
Dresden
Erfurt
Gera
Stufe 3 240 Km
Karl-Marx-Stadt
Suhl

96

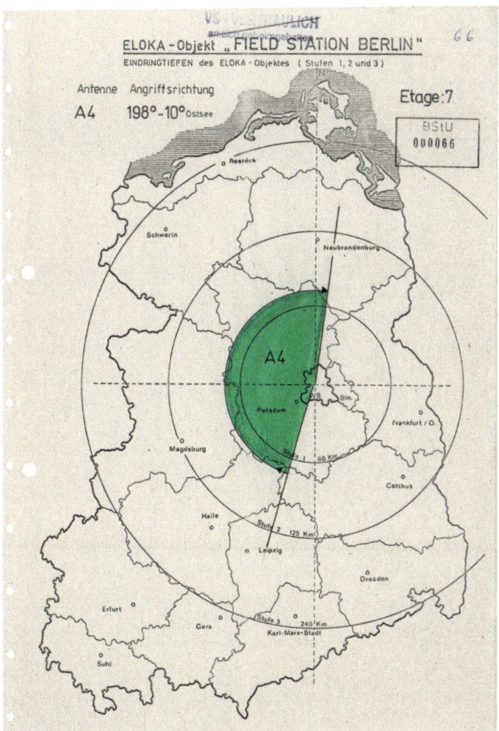
VS-VERTRAULICH
66
ELOKA-Objekt „FIELD STATION BERLIN"
EINDRINGTIEFEN des ELOKA-Objektes (Stufen 1, 2 und 3)
Antenne Angriffsrichtung
A4 198°-10°
Ostsee
Etage: 7
BStU
000066
Rostock
Schwerin
Neubrandenburg
A4
WB
Bln.
Potsdam
Frankfurt / O.
Magdeburg
Stufe 1 60 Km
Cottbus
Halle
Stufe 2 125 Km
Leipzig
Dresden
Erfurt
Gera
Stufe 3 240 Km
Karl-Marx-Stadt
Suhl

95

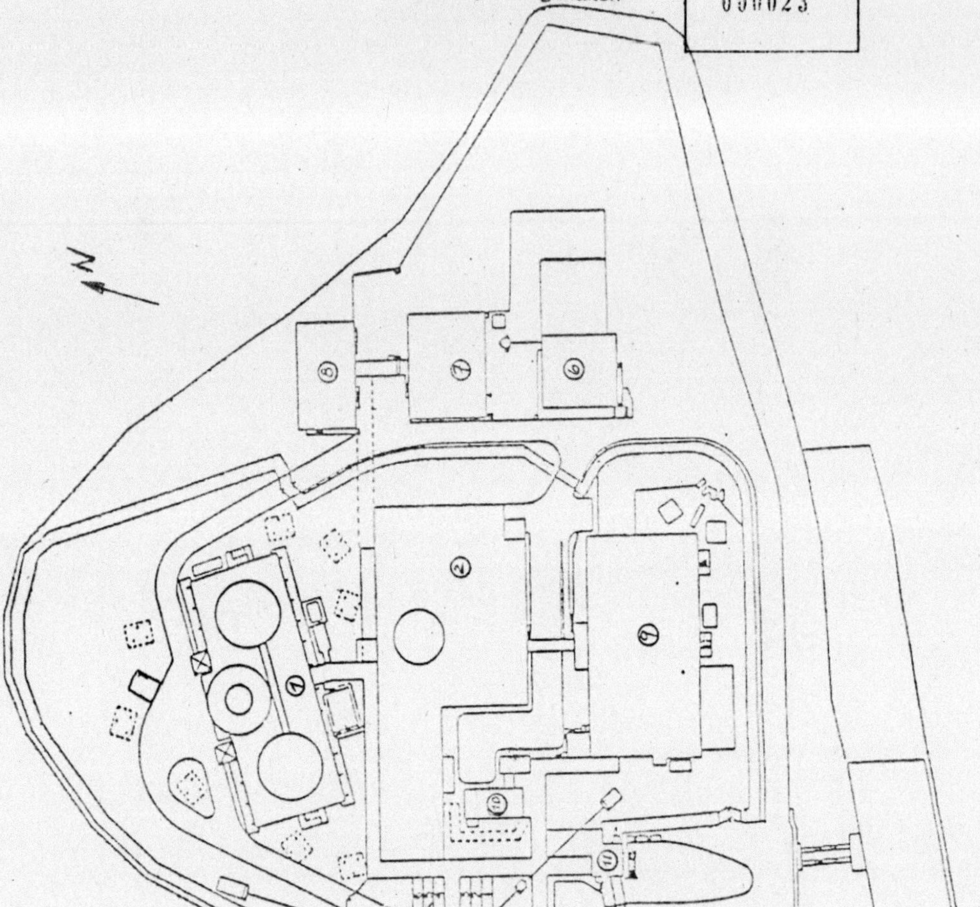

Skizze 2
Hauptoperationsbasis der "US Army Field Station Berlin-USAFSB"
auf dem Westberliner Teufelsberg

-106-

GVS-P-O 026-5/85

1984

Bild 15

Fig. 86 Military academy in Berlin (Wehrtechnische Fakultät der Technischen Hochschule), ground floor. Source: Bundesarchiv. R 4606/3699.

Fig. 87 Bird's-eye view of the military academy in Berlin. Source: Bundesarchiv. R 4606/3699.

Fig. 88 View of the different elements of Field Station Berlin Teufelsberg. Source: Der Bundesbeauftragte für die Unterlagen des Staatssicherheitsdienstes der ehemaligen Deutschen Demokratischen Republik (BStU), MfS, HA III, Nr. 14455, Bl. 94-109, Blatt 105.

Fig. 89 Schematic section of the Field Station Berlin Teufelsberg. Source: Der Bundesbeauftragte für die Unterlagen des Staatssicherheitsdienstes der ehemaligen Deutschen Demokratischen Republik (BStU), MfS, HV A, Nr. 883, Bl. 1-68, Blatt 60.

Fig. 90 Map showing the reception range of the Field Station Berlin Teufelsberg. Source: Der Bundesbeauftragte für die Unterlagen des Staatssicherheitsdienstes der ehemaligen Deutschen Demokratischen Republik (BStU), MfS, HV A, Nr. 883, Bl. 1-68, Blatt 61.

Fig. 91 Map showing the reception range of the Field Station Berlin Teufelsberg. Source: Der Bundesbeauftragte für die Unterlagen des Staatssicherheitsdienstes der ehemaligen Deutschen Demokratischen Republik (BStU), MfS, HV A, Nr. 883, Bl. 1-68, Blatt 62.

Fig. 92 Map showing the reception range of the Field Station Berlin Teufelsberg. Source: Der Bundesbeauftragte für die Unterlagen des Staatssicherheitsdienstes der ehemaligen Deutschen Demokratischen Republik (BStU), MfS, HV A, Nr. 883, Bl. 1-68, Blatt 63.

Fig. 93 Map showing the reception range of the Field Station Berlin Teufelsberg. Source: Der Bundesbeauftragte für die Unterlagen des Staatssicherheitsdienstes der ehemaligen Deutschen Demokratischen Republik (BStU), MfS, HV A, Nr. 883, Bl. 1-68, Blatt 64.

Fig. 94 Map showing the reception range of the Field Station Berlin Teufelsberg. Source: Der Bundesbeauftragte für die Unterlagen des Staatssicherheitsdienstes der ehemaligen Deutschen Demokratischen Republik (BStU), MfS, HV A, Nr. 883, Bl. 1-68, Blatt 65.

Fig. 95 Map showing the reception range of the Field Station Berlin Teufelsberg. Source: Der Bundesbeauftragte für die Unterlagen des Staatssicherheitsdienstes der ehemaligen Deutschen Demokratischen Republik (BStU), MfS, HV A, Nr. 883, Bl. 1-68, Blatt 66.

Fig. 96 Map showing the reception range of the Field Station Berlin Teufelsberg. Source: Der Bundesbeauftragte für die Unterlagen des Staatssicherheitsdienstes der ehemaligen Deutschen Demokratischen Republik (BStU), MfS, HV A, Nr. 883, Bl. 1-68, Blatt 67.

Fig. 97 Site plan of the Field Station Berlin Teufelsberg. Source: Der Bundesbeauftragte für die Unterlagen des Staatssicherheitsdienstes der ehemaligen Deutschen Demokratischen Republik (BStU), MfS, HV A, Nr. 883, Bl. 1-68, Blatt 68.

Fig. 98 View of the different elements that made up the Field Station Berlin Teufelsberg. Source: Der Bundesbeauftragte für die Unterlagen des Staatssicherheitsdienstes der ehemaligen Deutschen Demokratischen Republik (BStU), MfS, HA III, Nr. 14455, Bl. 94-109, Blatt 106.

PART 1
THE CITY

In an essay published in 1971, a decade after the building of the Berlin Wall, Robin Evans proposed a reading of the fundamental element of the wall through the lens of an "environmental history of the war against information." "Walls," he writes, "are the armoury that preserves our personal integrity against the inroads of the rest of humanity and nature."[1] At a time of agitated debate on the relation of cybernetics to architecture, Evans suggests that "information is not such an isotropically good and wholesome commodity," adding that "immersion in a milieu of indiscriminate emblems, images, messages and ideas might just as easily discompose and confuse individuals and communities as enlighten them."[2] Evans's text appears as a bold, yet unspecific, critique of the contemporary architectural discourse around the question of communications and information. By focusing his analysis solely on the element of the wall as a medium, or filter, he avoids speaking about *space*—the word is not even mentioned once in the text. Rather than thinking about architecture and information in terms of enhancing connections, exchange, free flow, and interaction, Evans shifts attention to their flipside, proposing instead to think about architecture as a device for division, disconnection, and fragmentation. Informed by a Cold War ethos, Evans's insightful conceptualization of the wall is especially pertinent to an analysis of the spatial construction of Berlin during those postwar decades, precisely because it puts aside the conventional understanding of space, looking instead at the wall in relation to the "moral, aesthetic, and social" impact of information.[3] The transmission and reception of information had become the new locus of political power, and walls had accordingly to be redefined.

I

From the earliest times, walls signaled power and security to those inside them, and a challenge to those outside.[4] In the first of his *Ten Books on Architecture*, Vitruvius instructs his readers that "the next thing to do," after securing the future city's food supply and

1 Robin Evans, "The Rights of Retreat and the Rites of Exclusion: Notes Towards the Definition of Wall," *Architectural Design* 41, no. 6 (1971): 335–339, here 337. The essay is collected in Robin Evans, *Translations from Drawing to Building* (London: AA Publications, 1997).

2 Evans, "The Rights of Retreat," 335.

3 Evans, "The Rights of Retreat," 339.

4 For a recent study on the cultural history of the wall, see Tobias Prüwer, *Welt aus Mauern: Eine Kulturgeschichte* (Berlin: Klaus Wagenbach, 2018).

means of transport, "is to lay the foundation of towers and walls."[5] Towers and walls together form a system of fortification, delineating a portion of the territory, their geometry and their substance shaped by the primary consideration of defense. "Dimension stone, flint, rubble, burnt or unburnt brick—use them as you find them," he writes. "A faultless wall may be built to last forever."[6] As for the towers, they must be "set at intervals of not more than a bowshot apart,"[7] and be "either round or polygonal. Square towers are sooner shattered by military engines, for battering rams pound their angles to pieces."[8]

It was Gottfried Semper, in the mid-nineteenth century, who first articulated a distinction between the idea of solid walls, "needed for security, for supporting a load, for their permanence," and walls as a vertical element, originally formed of textiles, which serve to visually delimit or enclose space.[9] For the first kind of load-bearing wall, he uses the word *Mauer*, for the latter, space-creating element, *Wand*. Since Semper, so from the second half of the nineteenth century on, the element of the wall has appeared in nearly every relevant theory associated with the concept of "space"—a newer fixture in the argot of architecture, and one whose meaning has proved quite fluid, constantly changing to accommodate different theoretical interpretations and, more importantly here, different built forms.[10]

According to Reyner Banham, it was through the mediation of H. P. Berlage that the relationship between walls and space, initially addressed by Semper, made its way into the twentieth century.[11] "The art of the master-builder lies in this: the creation of space, not the sketching of facades. A spatial envelope is established by means of walls, whereby a space or a series of spaces is manifested, according to the complexity of the walling," Berlage writes in his 1908 "Foundations and Development of Architecture."[12] Notwithstanding the emphasis on the word space in Berlage's writing, the term is still essentially being used to denote that which is enclosed and delimited by walls.

5 Vitruvius, *The Ten Books on Architecture*, ed. Morris Hicky Morgan (Cambridge, MA: Harvard University Press, 1914), 1.5.1.

6 Vitruvius, *Ten Books*, 1.5.8.

7 Vitruvius, *Ten Books*, 1.5.4.

8 Vitruvius, *Ten Books*, 1.5.5.

9 Gottfried Semper, *The Four Elements of Architecture and Other Writings*, trans. Harry F. Mallgrave and Wolfgang Herrmann (Cambridge: Cambridge University Press, 1989), 104. Originally published in German as *Die vier Elemente der Baukunst: ein Beitrag zur vergleichenden Baukunde* (Braunschweig: Friedrich Vieweg, 1851).

10 Adrian Forty traces the genealogy and ever-changing meaning of the idea of "space" in modern architecture in his *Words and Buildings: A Vocabulary of Modern Architecture* (London: Thames & Hudson, 2004), 256–275.

11 Reyner Banham, *Theory and Design in the First Machine Age* (New York: Praeger, 1967), 137–143.

12 Hendrik Petrus Berlage, "Grundlagen und Entwicklung der Architektur" (1908), quoted after Banham, *Theory and Design*, 140. An English version of the text is available in *Hendrik Petrus Berlage, Thoughts on Style, 1886–1909*, trans. Iain Boyd Whyte and Wim de Wit, Getty Publications Virtual Library, https://www.getty.edu/publications/virtuallibrary/0892363347.html.

In 1926, Siegfried Ebeling brought new meaning to the notion of space and, with it, a new idea of what a wall actually constitutes. Rather than seeing the wall as the element separating inside from outside, he reframes it as a membrane. In Ebeling's theory, *Raum als Membran* (Space as membrane), space and wall become fused, and it is precisely the "membrane" that selectively mediates between exterior and interior, bearing in mind that "the proportion of a space, the materials of a wall, the pressure and composition of the air, temperature and degree of luminosity, are all values that are interdependent, with radiation (*Strahlung*) being the common denominator."[13] Here, Ebeling effectively breaches the accepted distinction between interior and exterior by focusing on the emerging problems of radiation and electromagnetism.

Forty-five years after Ebeling's work, Robin Evans's theory of the wall adds yet another degree of complexity to the problem, being arguably the first to include information as a formative agent in the configuration of space.[14] According to Evans, architecture can be understood in terms of artificially manufactured terrains delineated by the material qualities and geometric disposition of walls as they relate to the protection against information. Throughout Evans's analysis, walls are understood as mediators in "two distinct, but not mutually exclusive ways: retreat and exclusion."[15] On the one hand, exclusion refers to the act of keeping a small number of individuals away from the "world." The walls of asylums, clinics, and prisons materially construct this form of exclusion. On the other hand, in a retreat, an individual or small group voluntarily seclude themselves from the "world" using walls, "circumscribing and forgetting about those parts of it that offend them."[16]

To illustrate the "rights of retreat," Evans calls attention to the aesthetic enclosure designed by the protagonist of the canonical novel *À rebours* (1884), by Joris-Karl Huysmans. We read how a jaded aristocrat, Jean des Esseintes, decides to retreat forever from

13 Siegfried Ebeling, *Der Raum als Membran* (Dessau: C. Dünnhaupt, 1926), 32. Author's translation.

14 For a more recent interpretation of the wall as it relates to the politics of separation, see Eyal Weizman, *Hollow Land: Israel's Architecture of Occupation* (London: Verso, 2007), 161–183.

15 Evans, "The Rights of Retreat," 336.

16 Evans, "The Rights of Retreat," 335.

urban life, embracing the idea of "hiding away far from human society, of shutting himself up in some snug retreat, of deadening the thunderous din of life's inexorable activity, just as people deadened the noise of traffic by laying down straw outside a sick person's house."[17] In self-imposed seclusion on the outskirts of Paris, his "project" entails a relentless exercise in spatial construction and reconstruction. For the decadent and secular aristocrat, walls perform the double action of sealing off his subjectivity, blocking an undesired exteriority, while serving at the same time as a projective screen facilitating aesthetic gratification. Aiming for an absolute disconnection from the outside world, Des Esseintes constructs a precisely controlled environment full of interesting architectural tropes.

To celebrate his final public appearance, for example, he orchestrates "a funeral banquet in memory of the host's virility." "Draped in black," the dining room that "opened out to a garden metamorphosed for the occasion." The "dinner itself was served on a black cloth adorned with baskets of violets and scabious; candelabra shed an eerie green light over the table and tapers flickered in the chandeliers."[18] The desire for obsessive control rules over every detail of the house, and life is focused as much on the act of constructing his retreat as it is on experiencing it to the full. To while away the interminable hours of boredom, Des Esseintes would turn "to his portfolios of prints and beg[in] sorting out his Goyas."[19] To dim the colors of a carpet that seems to him too new, too bright, and thus too disturbing to his eyes, he comes up with the novel idea of buying a huge tortoise and having its "buckler glazed with gold": "Back from the workshop..., the reptile blazed as brightly as any sun, throwing out its rays over the carpet, whose tints turned pale and weak" by comparison.[20]

The obverse of "the personal right of retreat" is, as Evans notes, "the corporate rite of exclusion," the isolation of a selected group of individuals from the world.[21] In *The Fabrication of Virtue: English Prison Architecture, 1750–1840* (1982), Evans elaborates

17 Joris-Karl Huysmans, *Against Nature (À rebours)*, trans. Robert Baldick (New York: Penguin, 2004), 9. First published in French in 1884.

18 Huysmans, *Against Nature*, 13.

19 Huysmans, *Against Nature*, 94.

20 Huysmans, *Against Nature*, 54.

21 Evans, "The Rights of Retreat," 336.

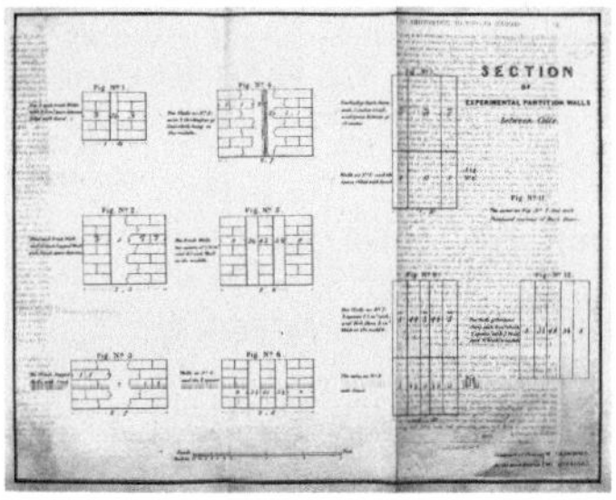

1 → P. 40

on the typology of the prison, which exemplifies the condition of exclusion. He describes a particular set of experimental cell walls designed to increase the distance between the prisoners themselves, constructing an "architecture against communication."[22] Prisoners in solitary confinement were supposed to be isolated from any source of information, including their neighbors.[23] The wall sections show experiments conducted during the development of the Millbank Penitentiary in 1836 by the Inspectors of Prisons, William Crawford and Whitworth Russell, in collaboration with the scientist—and, not by coincidence, future radio pioneer—Michael Faraday.[24] (Fig. 1) The aim was not to dampen the sonic impulse, reducing the noise level generally, but rather to artificially induce reverberation as the sound waves passed from air to brick in the double walls, so diminishing the intelligibility of speech.[25] In concrete terms, a voice would still be detectable, but its meaning would be lost.[26] Evans describes this concerted effort to reduce the transmission of "significant message" as the birth of the "science of information destruction."[27]

The architectural mechanisms for destroying communication were brought nearly to perfection with the Pentonville Model Prison, completed in 1842. According to Joshua Jebb, the engineer responsible for its design, the whole purpose of the building was "that no additional facilities of communication between prisoners in adjoining cells should be afforded ... and, therefore, that the transmission of sound should be carefully guarded against."[28] The precision of the elements tested earlier by Faraday would be amplified to become an entire system of "environmental control."[29] The effect was impeccable: "the silence within Pentonville was so complete—so uncannily still—that [it] felt necessary to look into several cell spy-holes to make sure that prisoners were really there."[30] The hermetic and solitary cells in Pentonville were certainly defined by their physical walls, but the stillness was also augmented by other means—by the felt overshoes of the prison wardens, worn to make their presence imperceptible to the prisoners; by a sophisticated system of mechanical ventilation allowing

22 Robin Evans, *The Fabrication of Virtue: English Prison Architecture, 1750–1840* (Cambridge: Cambridge University Press, 1982), 318.

23 Solitary confinement was a system of imprisonment developed in the United States that European jurists and engineers would learn lessons from in the first half of the nineteenth century. For more, see Evans, *The Fabrication of Virtue*, 318–387.

24 Michael Faraday's experiments with electromagnetism served as a basis for James Clerk Maxwell's theory of the electromagnetic field, which provides the theoretical basis for radio transmission. See Wolfgang Hagen, *Das Radio: Zur Geschichte und Theorie des Hörfunks—Deutschland/USA* (Munich: Wilhelm Fink, 2005). See also Salvo D'Agostino, *A History of the Ideas of Theoretical Physics: Essays on the Nineteenth and Twentieth Century Physics* (Dordrecht: Kluwer, 2000), 111–134. The contribution of the scientific experiments of Michael Faraday to the development of architectural knowledge deserves further investigation. In 1837, Faraday gave a report to RIBA about his studies on the chemical composition of color found in Greek excavations. The report is referenced by Gottfried Semper in *Die vier Elemente der Baukunst* (p. 45) as scientific proof of his theory of polychromatic architecture. For a recent study on Pentonville Model Prison and Millbank Prison, see Moritz Gleich, *Inhabited Machines: Genealogy of an Architectural Concept* (Basel: Birkhäuser, 2023), 212–235.

25 See Bernhard Siegert's analysis of Michel Serres's theory of the parasite in Bernhard Siegert and Geoffrey Winthrop-Young, "Cacography or Communication? Cultural Techniques in German Media Studies," *Grey Room* 29 (Fall 2007): 26–47.

26 Evans, *The Fabrication of Virtue*, 338.

27 Evans, "The Rights of Retreat," 338.

28 Joshua Jebb, *Modern Prisons: Their Construction and Ventilation* (London: John Weale, 1857), 16.

29 Evans, *The Fabrication of Virtue*, 354.

30 Evans, *The Fabrication of Virtue*, 349.

a silent supply of fresh air; by the design of doors and windows mediating access to natural light and food; by a water system designed for individual toilets placed within the cells; and by a militarized division of labor that orchestrated the machine-like operation of the entire facility.[31] What becomes clear in this example is that the effect of a wall is hardly ever determined by a single element; rather, a wall should be seen as an ensemble of artifacts articulated in precise forms to filter and control the agency of different forces. For all the attentive and original interpretation of walls that characterizes Evans's essay, however, and with all due respect to his political and theoretical astuteness, his failure to mention perhaps the most relevant wall of the time, the Berlin Wall, remains a mystery.[32] Whether intentional or not, this omission leaves his argument partially incomplete.

My interest lies in understanding the overlooked agency that built artifacts have relative to the production and transmission of information. As an epicenter of the Cold War, Berlin stands as a synecdoche for a global condition, a focus for analyzing the urgent problem of walls, limits, and information. What follows is an environmental history of information of the city of Berlin that reconstructs the architecture of radio in the specific context of a divided world, when the construction of walls served not just to create or physically demarcate space, but to direct and control the flow of information—an issue that has become more relevant than ever today, when the supple effects of the remote transmission of information are radically challenging the historical solidity of architecture.

II

The advent of an architecture of radio with global aspirations—motivated by ideas of domination over vast swaths of the world, much more than by a striving for technological innovation—can be traced back to the incredibly numerous, but still widely overlooked, structures built for radiotelegraphy and radio broadcasting from the early years of the

31 For more on the construction and ventilation system implemented in the Pentonville Model Prison, including original illustration and drawings, see, besides Joshua Jebb's *Modern Prisons*, his *Report to the Surveyor-General of Prisons on the Construction, Ventilation, and Details of Pentonville Prison* (London: William Clowes and Sons, 1844), and *Observations on the Separate System of Discipline Submitted to the Congress Assembled at Brussels, on the Subject of Prison Reform* (London: William Clowes, 1847).

32 In 1971, a Dutch student at the Architectural Association—where Robin Evans was then teaching—traveled to Berlin as part of a summer field trip that coincided with the publication of Evans's essay on the definition of the wall. Rem Koolhaas decided to do his diploma about the Berlin Wall, and an account of that early experience can be found in his "The Berlin Wall as Architecture" in *S,M,L,XL: Office for Metropolitan Architecture* (New York: Monacelli Press, 1995), 217–231.

2 → P. 40

twentieth century on. In Berlin, this globally ubiquitous form of architecture became fully established with the construction of the Nauen transmitter station (*Grossfunkstelle*) designed by one of the intellectual leaders of the Deutscher Werkbund, Hermann Muthesius. Completed in 1920, in the early years of the Weimar Republic, the large brick facility was conceived as a transmitter station for radiotelegraphy, rather than radio broadcasting. Its inauguration on September 29, 1920 marked both the dawn of a new era of radio in Germany and the decline of a communications empire that had been built up over a period of nearly two decades. The Nauen station, some 40 km northwest of Berlin, was the new base of Telefunken, a telecommunications subsidiary founded in 1903 by two leading German electrical companies, the Allgemeine Elektricitäts-Gesellschaft (AEG) and Siemens & Halske. The aim of the joint venture was to make compatible two existing, independently developed, technical systems in order to compete in the emerging market of wireless telegraphy, which at the time was dominated by the British Wireless Telegraph and Signal Company based on patents of Guglielmo Marconi.[33] Already within the first decade of its existence, wireless telegraphy had become essential for the functioning of commercial, civil, and military enterprises, boosting large sectors of the economy and creating lucrative markets with its capital- and labor-intensive ethos.[34] Telefunken's first director, the radio pioneer Georg von Arco, began his experiments in an improvised laboratory right next to AEG's cable factory and electric power plant in Oberschöneweide, in the east part of Berlin.[35] Using the four chimneys of the AEG cable factory as support for an improvised antenna, and electricity from the adjacent generating station, von Arco successfully transmitted wireless signals to a marine station 275 km away, on Fehmarn Island in the Baltic Sea.[36] It was clear, however, that taking the experiments further would require new and expanded facilities, so Siemens & Halske and AEG once again pooled their capital and their technical expertise to purchase a 40,000 m² site near the city of Nauen. (Fig. 2) By 1906, a 100 m tower had already made its

33 "10 Jahre Telefunken-Technik," in *Telefunken-Zeitung* 2, no. 12 (1913): 142.

34 For an early account on the economic relevance and potential of wireless telegraphy see Gustav Partheil, *Die drahtlose Telegraphie und Telephonie nach Geschichte, Wesen und Bedeutung für Militär und Marine, Verkehr und Schule* (Berlin: Gerdes & Hödel, 1907), 106–114.

35 Michael Bollé, *Die Grossfunkstation Nauen* (Berlin: Willmuth Arenhövel, 1996). For more on Georg von Arco's biography as it relates to early experiments with radio transmission and with his engagement with Telefunken, see Margot Fuchs, *Georg von Arco 1869–1940* (Berlin: Diepholz, 2004), 36–40; 262–289.

36 Partheil, *Die drahtlose Telegraphie und Telephonie*, 92.

mark on an otherwise extremely flat landscape, and the subsequent transformation of the site over the following decades would change forever how the world communicated. From Nauen, German companies set out to conquer the world electrically.[37]

5 → P. 44

10 → P. 49

(Fig. 5) Construction of the new transmitter station at Nauen began in 1916, two years after the start of World War I, and was completed in 1920, two years after the war's end. The sheer scale of the facility speaks of the ever-greater role that Telefunken's operations assumed during the armed conflict. Rather than inaugurating the typology of the transmitter station, the structure designed by Muthesius represented the scaling up of a building type that Telefunken had been busy disseminating around the world: by 1916, they already numbered in their hundreds. After its founding in 1903, the scale of Telefunken's operations expanded almost as quickly as the radio waves, though not without the friction and resistance that material infrastructure usually encounters. Just ten years after it was founded, Telefunken had subsidiaries in New York (Atlantic Communications), in Sydney (Australian Wireless Co.), two in Berlin (Deutsche Betriebsgesellschaft für drahtlose Telegraphie mbH and Deutsche-Südsee-Gesellschaft für drahtlose Telegraphie AG), in Brussels (Société Anonyme Internationale de Télégraphie), and in Shanghai (Telefunken East Asiatic Wireless Telegraph Co.) There were also technical offices—many of them with production capacities—in Buenos Aires, Istanbul, Christiania, London, Madrid, Beijing, Rio de Janeiro, St. Petersburg, Stockholm, Mexico City, and Vienna. This huge corporate, bureaucratic, and technical capacity could be quickly deployed to serve civilian and military enterprises even in the most far-flung corners of the planet. (Fig. 10) In 1912, the journal *Telefunken-Zeitung* proudly announced that the "wireless telegraphy stations built by the Chinese government so far use almost exclusively the Telefunken system. In the period up to 1911 Telefunken has supplied the Chinese government with ten land stations, nine warship stations, one school ship station,

37 Artur Fürst, *Im Bannkreis von Nauen: Die Eroberung der Erde durch drahtlose Telegraphie* (Stuttgart: Deutsche Verlags-Anstalt, 1923), 195. For more on the history of German radiotelegraphy see Artur Fürst, *Ätherwellen: Die Eroberung der Erde durch die drahtlose Telegraphie* (Berlin: Comeniu, 1925), and Hermann Thurn, *Die Funkentelegraphie* (Leipzig; Berlin: Teubner, 1918).

7, 8 → P. 46

9 → P. 48

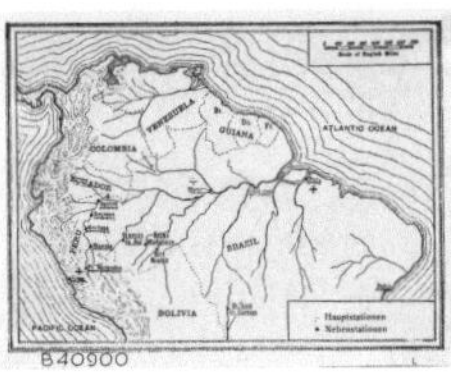

6 → P. 45

11 → P. 49

and eight transportable military stations."[38] (Fig. 7, 8) The reach of Telefunken's radiotelegraphy was measured against the most formidable geographical barriers. Also in 1912, the station in Iquitos, Peru, received the first wireless telegram from Lima, 1,000 km away, establishing an instant connection between the country's capital and the remote places of rubber extraction in the Amazon rainforest. (Fig. 9) The titanic challenge of crossing the Andes, which rose to 6,000 m, was technically surmounted by the transmitter station in Lima, with "its 100 m tower on top of the steep 400 m-high San Cristobal Mountain."[39] (Fig. 6) With these infrastructural additions, a transnational network of stations linked the political centers more closely with the ports in the Pacific (Lima and Puerto Bermúdez), the zones of material extraction in the Amazon rainforest (Iquitos and Manaus), and the Brazil states oriented toward Atlantic trade (Pará). (Fig. 11) The strategic alliance of Telefunken with both young nation-states and German commercial enterprises in Latin America also influenced the construction of several stations in the Bolivian *altiplano*. The journal published a photograph of a German general, Hans von Kundt, on a mission through the Andes, serving as a go-between for Telefunken and the Bolivian Army, building on the "extraordinarily brilliant results" achieved by the "German officers and twenty non-commissioned officers [who] took over the reorganization of the Bolivian army some three years ago."[40] Supported by Telefunken's portable stations, that army, it reported, was "now one of the best in the South American republics."[41] Over the course of 1911, significant improvements in the transmission capacity of Nauen's mother station allowed Telefunken to reach the southernmost tip of the continent, with the construction of a station in Tierra del Fuego.[42] That same year, Telefunken's US subsidiary, Atlantic Communications, bought land in Long Island to build the most powerful station in the Americas, making possible two-way transatlantic communications.[43] Telefunken's construction of transmitting and receiving stations along global north–south axes was of course not limited to the Americas: dozens of

38 Hans Bredow, "Telefunken in China," *Telefunken-Zeitung* 1, no. 4 (1912): 41. Author's translation.

39 "Triumph des Telefunkensystem. Überlandverbindung auf 1000 km über die 6000m hohen Anden," *Telefunken-Zeitung* 1, no. 6 (1912): 86. Author's translation.

40 "Bolivien," *Telefunken-Zeitung* 3, no. 13 (1914): 30. Author's translation.

41 "Bolivien," 30.

42 "Kleine Mitteilungen," *Telefunken-Zeitung* 2, no. 11 (1913): 118.

43 Fürst, *Im Bannkreis von Nauen*, 209.

facilities were built in Africa and across Southeast Asia too. Colonial settlements were the mainstay of Telefunken's worldwide expansion during the 1910s, with the company keen to promote the strategic advantages provided by its technology. Promoting the construction of Telefunken stations in US colonial settlements designed and partially implemented by Chicago architect Daniel Burnham in the Philippines, *Telefunken-Zeitung* used the case of Manila as an example of how "transportable field stations can gain great importance in future uprisings in the colonies as a substitute for the landlines that natives easily destroy."[44]

3 → P. 41

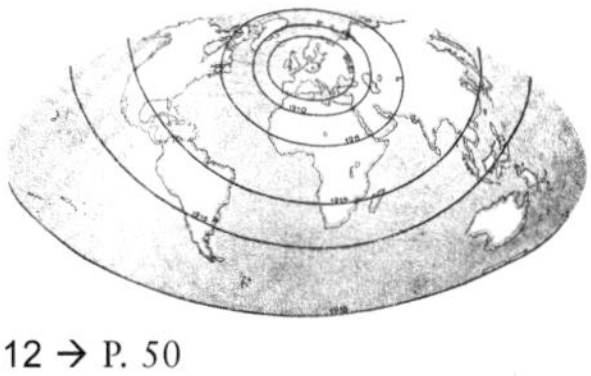

12 → P. 50

(Fig. 3) Indeed, the destruction of the telegraph lines connecting Germany with its colonies in Togo and Cameroon had been one of the central motivations for improving the Nauen station's transmission capacities in 1911.[45] (Fig. 12, 13) Supplied with a new tower and a more powerful engine, the facility would soon be able to reach, without relay stations, a radius of 5,000 km—the distance between Berlin and Kamina in Togo. In 1913, construction began on the new receiving and transmitting station for radiotelegraphy in Kamina, the most powerful one in the entire Global South. (Fig. 14, 15, 16) Huge quantities of material had to be mobilized to connect Togo with Nauen through the apparent immateriality of radio. All the steel elements for the transmission tower were brought from Germany, and Telefunken had to build roads, cross swamps, import cranes, construct a reservoir, and finally lay 5 km of rail track between the port of Lomé and Kamina to transport the heavy material to site.[46] (Fig. 17) By 1914 the station in Kamina was equipped with four 100 m and three 75 m towers and a 100 kW steam electric turbine, enabling it to relay messages received from Nauen to Windhoek in East Africa, to Tabora and Dar es Salaam in Tanzania, and, most importantly, to the station in Monrovia, Liberia, which linked Africa with South America via transoceanic cable.[47] The reach of radio transmission is directly linked to the height of the antenna and the capacity of the machines to amplify current: it is a problem of form and power. In previous

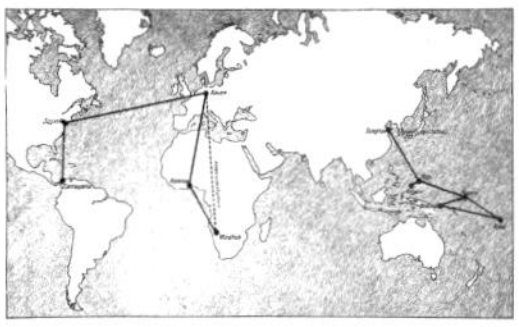

13 → P. 50

14, 15, 16 → P. 52

17 → P. 55

44 "Telefunken auf den Philippinen," *Telefunken-Zeitung* 1, no. 1 (1911): 5. Author's translation. To read more on the US-sponsored masterplan for Manila, designed by Daniel Burnham in 1905, see Daniel Immerwahr, "The Iron Hand of Power: US Architectural Imperialism in the Philippines," in *Architectural History* 64 (2021): 163–186.

45 Fürst, *Im Bannkreis von Nauen*, 202. For a general history of German colonialism, see Karlheinz Graudenz and Hanns-Michael Schindler, *Die deutschen Kolonien: 100 Jahre Geschichte in Wort, Bild und Karte* (Augsburg: Weltbild, 1989). For a more general history of art and architecture related to German

18 → P. 56

19, 20 → P. 57 (–58)

images, we saw the size of the antennas towering over the stations in Nauen and Kamina. There is an intrinsic modernism associated with the form of those towers, with their sheer size, the presence of steel, and the lightness of their structures. (Fig. 18, 19, 20) But there is another thread of modernity connecting these stations, one related to the interiors and the question of power (*Kraft*), and illustrated here by two images that depict the spaces for energy production in the Kamina station and a third image showing one of the most canonical interiors in Berlin, built only a few years before. In 1903, Telefunken's parent company, AEG, had embarked on an ambitious program to manufacture steam turbines for autonomous electricity production. In 1909, work was completed on the new factory that was part of this program—the turbine hall designed by Peter Behrens. A mere two years after the building's completion, the electricity generated by AEG-manufactured steam turbines far outstripped the combined total of electrical energy generated by all other powerplants in Germany.[48] From a functional point of view, arguably the most important effect of the AEG turbine hall was that it allowed vast swaths of Germany's industrial sector to be autonomous from a still underdeveloped electrical grid that depended mostly on access to coal. Like factories in other industrial sectors, the station in Kamina benefited from the remote access to electricity afforded by the steam turbines developed by AEG. Yet what was unique to radiotelegraphy was that it linked the centers of political and economic power (*Macht*) with the periphery where resources were extracted through the power of electricity (*Kraft*). *Macht* through *Kraft* is what allowed Telefunken to define the early years of the architecture of radio.[49]

The station in Kamina and many others worldwide lay within the impressive range of the Nauen transmitter, but by 1916, Telefunken and its new director, Hans Bredow, had set a much more ambitious target radius of 20,000 km, aiming to reach *every* station in the world. Hermann Muthesius was commissioned to design the monumental new facility that would

colonialism, read Itohan Osayimwese, ed., *German Colonialism in Africa and its Legacies: Architecture, Art, Urbanism, and Visual Culture* (London: Bloomsbury Visual Arts, 2023); Michael Falser, *Deutsche-koloniale Baukulturen: Eine globale Architekturgeschichte in 100 visuellen Primärquellen* (Passau: Dietmar Klinger, 2023); and Hollyamber Kennedy, "Infrastructures of 'Legitimate Violence': The Prussian Settlement Commission, Internal Colonization, and the Migrant Remainder," *Grey Room* 76 (2019): 58–97. For a more specific account on the role that radio played for the German colonies, see Reinhard Klein-Arendt, *"Kamina ruft Nauen!": die Funkstellen in den deutschen Kolonien 1904–1918* (Dessau: Herbst, 2010).

46 "Der Bau der Telefunken-Station Kamina (Togo)," *Telefunken-Zeitung* 2, no. 12 (1913): 166–170.

47 For more on the construction of the radio station in Kamina, see "Die Großstation Kamina und der Beginn des Weltkrieges," *Telefunken-Zeitung* 3, no. 16 (1919): 31–36; and Michael Friedewald "Funktelegraphie und deutsche Kolonien," in *2000 Jahre Nachrichten aus dem Krieg: Technik—Medien—Kunst*, ed. Hermann Nörig et al. (Göttingen: Vandenhoeck und Ruprecht, 2009), 51–62.

embody the company's institutional and technical expansion during a time of war. As mentioned earlier, the reach of radio transmission is a problem of form and power. Regarding the latter—and in opposition to the stations in Africa that produced their own energy with steam turbines provided by AEG—the station in Nauen never generated its own electricity but was supplied with a continuous 15,000-volt current by the nearby power plant in Spandau. In the case of Nauen, the close proximity of *Macht* and *Kraft* meant there was no need for a steam turbine. That enormous electrical power, however, had to be transformed into electromagnetic signals. The building was therefore conceived as a device for electric alternation and amplification, housing the largest alternators ever constructed, along with their associated oil cooling systems. The huge rotating machines, which turned the electric current into high-frequency alternating current that could be channeled through the antenna encoded with telegraphic signals, can be seen on the left and right of the photograph of the interior. (Fig. 4) They operated together to supply the larger antenna for transoceanic transmissions, while the smaller alternator between them was used for internal transmissions within Europe. These machines were accommodated—domesticated, as it were—within a carefully proportioned interior, clad in brick. Similarly, the symmetrical and cruciform plan of the main building imposed an institutional and monumental order on an unruly and entropic set of electric components. (Fig. 21, 22) But not a total order: more machinery spilled out of the roof. In the perspective and section, the massive brick building designed to house huge machines is reduced to a miniature, seemingly suspended from the stressed cable structure. The section establishes a direct comparison between this form of architecture and the city, showing the masts of the larger antenna in relation to Berlin Cathedral on the right, and the Brandenburg Gate on the left. The distance spanned by the masts is equivalent to the length of Unter den Linden. What the diagram suggests is effectively a reversal of the design principles behind Muthesius's proposal. Instead of using classical proportions to house and domesticate

4 → P. 43

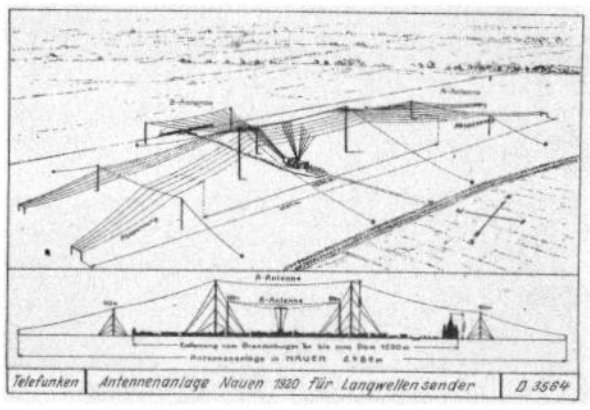

21 → P. 59

22 → P. 60

48 For more information about the design of steam turbines as they related to the larger energetic commercial project of AEG, see the book designed by De Stijl artist César Domela-Nieuwenhuis: *25 Jahre AEG-Dampfturbinen*, ed. Allgemeinen Elektricitäts-Gesellschaft (Berlin: VDI-Verlag GmbH., 1928), 21.

49 I refer here to the semantic distinction between the two words used in German to define power, *Macht* and *Kraft*, in a literal sense. This distinction is logically related but not entirely linked to the distinction articulated in Friedrich Nietzsche, *Der Wille zur Macht: Eine Auslegung alles Geschehens* (Leipzig: Alfred Kröner, 1921).

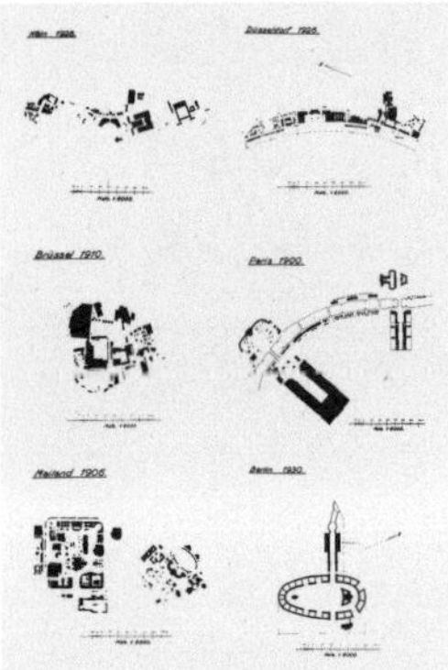

23 → P. 62

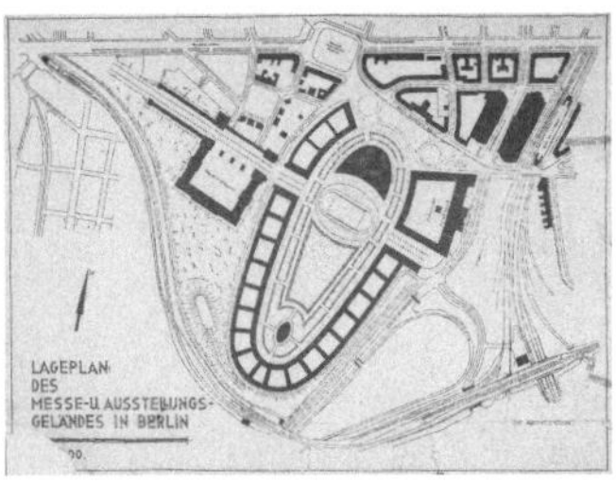

25 → P. 63

an unruly technology, it depicts four hundred years of Prussian civic architecture sheltered beneath the new architecture of radio.

By the time the facility was completed in 1920, however, the central motivation for its construction was long gone. Germany had lost the war and the overseas corporate empire built by Telefunken was dissolved. There was no call any more for point-to-point radiotelegraphy to connect the dispersed colonies with the metropole. But by then a new era for radio was beginning—the era of broadcasting. The interests of German radio institutions turned inward as their global project collapsed, at least temporarily. We will turn inward, too, to analyze the impact of this process on the city of Berlin.

The Westend district of Berlin would become the site of a transformative urban intervention intended to boost the industries of the Weimar Republic: the new exhibition grounds designed by the architects Hans Poelzig and Martin Wagner. In an article in *Das Neue Berlin* in 1929, Wagner theorized the relevance of the "exhibition type" as an engine of economic and cultural progress for the modern city.[50] (Fig. 23) To illustrate his point, he established a formal comparison between the exhibition grounds of Cologne, Düsseldorf, Brussels, Paris, Milan, and Berlin. Not unlike the strategy deployed by Muthesius in Nauen, Wagner's use of the notion of type gives formal cohesion to a seemingly formless program that would normally be governed by the forces of capital or technology, or both.

(Fig. 25) In the midst of the Berlin exhibition grounds stood a radio tower, described by Siegfried Kracauer as a "radiant tree."[51] Wagner and Poelzig's vision was to incorporate the production and transmission of information, through radio, as a fundamental part of the new program of the modern trade fair. They assigned the radio industry a central role in the revival of Germany's fortunes, placing it side by side with construction and the emerging automobile industry.

50 Martin Wagner, "Die Reichsausstellungsstadt Berlin," *Das Neue Berlin* 1 (1929): 7–17.

51 Siegfried Kracauer, "Aus dem Fenster gesehen," in *Straßen in Berlin und anderswo* (Frankfurt am Main: Suhrkamp 2009), 53–55. Author's translation.

In the plan of the exhibition grounds, the Haus des Rundfunks (broadcasting house), designed by Poelzig, terminates the longitudinal axis of the central egg-shaped figure. Two important courtyards sit on either side of the transversal axis, one dedicated to the construction exhibition (*Bauausstellung*), the other centered around two existing structures: the radio tower and adjacent hall (Funkhalle/Haus der deutschen Funkindustrie), which hosted an annual radio exhibition promoting the products of the industry, with Telefunken, Siemens & Halske, and AEG prominently represented. (Fig. 26, 27) These existing structures, completed in 1924–1926 to the designs of Heinrich Straumer, are absorbed into the new master plan for the exhibition grounds by Poelzig and Wagner. In short, the production, transmission, and exhibition of the emerging radio industry was conceived as integral to this new *type*. For Wagner, "the new type of exhibition in Berlin will not be an exhibition in which everyone can exhibit everything."[52] In keeping with his militant social democratic stance (he was an active member of the SPD), he believed "the city of Berlin cannot afford the boundless and shapeless individualistic exhibition, one that only cares for the business of individuals."[53] He had a different aim in mind: "Berlin must become the exhibition city of the main economic and cultural organizations and associations, those who represent German work in the areas of business, administration, politics, science, art, and culture."[54]

26 → P. 65

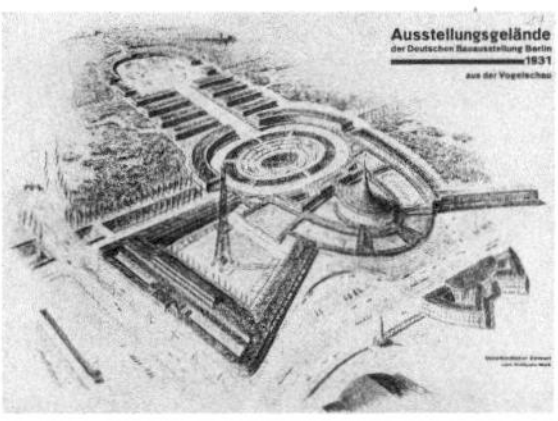

27 → P. 66

28, 29, 31 → P. 67–69

32, 33 → P. 70

(Fig. 28, 29, 31) Straumer's five-story Funkhalle, 131 m long and 41 m wide, was built mainly of wood, since steel would have interfered with the signal reception of the radio devices on display. One of the earliest examples of large-scale timber construction in Germany, it contained a 325 m^2 auditorium, a 400 m^2 restaurant, and housing units for the staff.[55] Straumer had won the commission in a competition in which Poelzig also participated. In contrast to the axial arrangement devised by Straumer, Poelzig had proposed to organize the program beneath a tent. (Fig. 32, 33) The interior rendering of the tent shows twisted lines of iron and steel: the masts holding the tent seem to

52 Wagner, "Die Reichsausstellungsstadt Berlin," here 8. This and subsequent quotes from the document are the author's translation.

53 Wagner, "Die Reichsausstellungsstadt Berlin," 8.

54 Wagner, "Die Reichsausstellungsstadt Berlin," 8. For a broader debate on the notion of *Typ* and *Typisierung* in Germany, see Frederic J. Schwartz, "The Type," in *The Werkbund: Design Theory and Mass Culture Before the First World War* (New Haven: Yale University Press, 1996), 121–146.

55 For more on the political and technological implications of the Funkausstellung, see Eva Susanne Bressler, *Von der Experimentierbühne zum Propagandainstrument: Die Geschichte der Funkausstellung von 1924 bis 1939* (Cologne: Böhlau, 2009), 43–49. See also Erich Blunck, "Das Haus der deutschen Funkindustrie in Berlin-Charlottenburg," *Deutsche Bauzeitung* 59, no. 14 (February 1925): 105–112.

34, 35 → P. 72

36 → P. 75

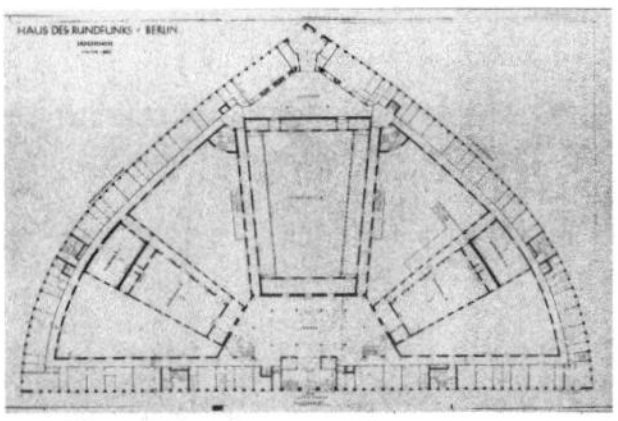

37 → P. 76

38 → P. 78

be electrified, as if they were radio transmitting masts, and the colorful fabric enclosing the space projects the image of internal activity to the space outside.

In a famous László Moholy-Nagy photograph from 1928, Straumer's radio tower appears as an emblem of all the qualities that attracted contemporary modernists: a structure that was lightweight, wall-less, transparent, and ephemeral. (Fig. 34, 35) That lightweight structure would soon be tethered, connected by wires, to another, much more solid building across the street, Poelzig's Haus des Rundfunks, completed in 1931.[56] Commissioned by the Reichs-Rundfunk-Gesellschaft (RRG, Reich Broadcasting Corporation), the Haus des Rundfunks was the largest broadcasting house of its time.[57] (Fig. 36, 37) By absorbing existing buildings into the formal cohesion of the new exhibition type, and by proposing new buildings for the industry to expand into, the exhibition grounds designed by Wagner and Poelzig incorporated radio into the urban structure of Berlin—and not just as technical medium, but as a set of practices and institutions. The same phenomenon can be seen only a few kilometers to the east, in the Heinrich-Hertz-Institut für Schwingungsforschung (HHI, Heinrich Hertz Institute for Oscillation Research), designed by Otto Weissgerber and inaugurated in 1930. (Fig. 38) A pioneering research laboratory dedicated to electroacoustics, the institute was supported by a joint venture between state radio broadcasters like the RRG, electrical industry leaders like Siemens & Halske, AEG, and Telefunken; and technical universities like the Technische Hochschule Berlin-Charlottenburg.[58]

With these projects—the radio tower, the Funkhalle, the Haus des Rundfunks, the master plan of the exhibition grounds, and the Heinrich Hertz Institute—the emerging technology of radio began to generate new building typologies and its own form of urbanism, assuming a powerful visual presence in Berlin during the 1920s and 1930s. The invisible airwaves, combined with the material existence of these buildings within

56 For more on the design of the Haus des Rundfunks, see chapter 2.

57 For more on the formation of the Reichs-Rundfunk-Gesellschaft and its impact on the creation and dissolution of institutions, see chapter 2.

58 Roland Wittje, *The Age of Electroacoustics: Transforming Science and Sound* (Cambridge, MA: MIT Press, 2016), 157–163. For more on the HHI within a general discussion of the space of the laboratory, see chapter 3 of this book.

the city structure, began to reshape the everyday lives of thousands of citizens. The reverberations produced inside the Haus des Rundfunks and transmitted via the radio tower were received instantaneously in people's homes. But the nature of those transmissions would change drastically after January 30, 1933, when Adolf Hitler was sworn in as Chancellor and the Nazis assumed control not only of the state but also of the emergent architecture of radio. Ideas traveled fast via the architecture of radio, which would reframe and expand its capacities as a *medium* towards an *end*. Accentuating the symbolic presence of radio at an urban scale, the master plan for Germania—Hitler's vision of Berlin, remade as the new world capital—located Adolf-Hitler-Platz (today Theodor-Heuss-Platz) adjacent to the Haus des Rundfunks and the exhibition grounds, as one of the central elements of the city's east–west axis.[59] (Fig. 39) At the same time, radio broadcasting seeded a series of built institutions dedicated to the field of electroacoustics and the scientific investigation of the relationship between the transmission of information, sound, and space. The work of these institutions would become an invaluable resource for the Nazis, supporting and accelerating the dissemination of their political ideology.

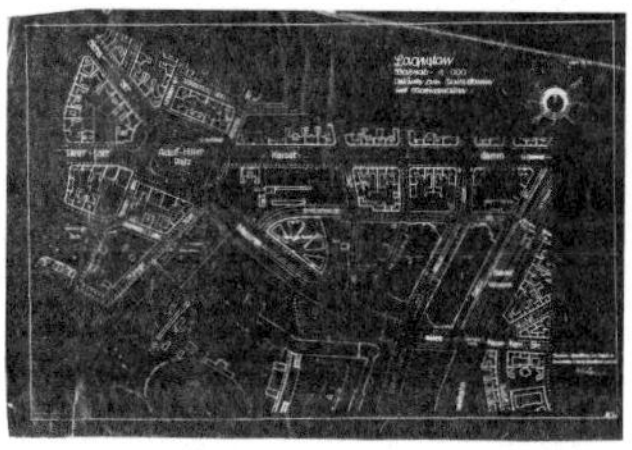

39 → P. 78

40 → P. 80

41 → P. 80

Electroacoustic sound could be disseminated through space either electromechanically (through loudspeakers) or electromagnetically (via radio). With regard to the former, the Nazis first used mass amplification systems during Hitler's presidential campaign of 1932, establishing a material basis for the large political rallies that would follow. Hitler's speeches were amplified through the loudspeaker cart developed by Siemens & Halske. Placed in public spaces and factories, the loudspeakers made it almost impossible to escape the sound of his voice.[60] However, one of the most sophisticated interactions between dispersed loudspeakers and built space was accomplished at the stadium for the 1936 Olympics, designed by Werner March in collaboration with Telefunken. (Fig. 40, 41) Since loudness alone was not sufficient to ensure the simultaneous reception of sound throughout the vast

59 The square at the intersection of Kaiserdamm and Masurenallee, previously named Reichskanzlerplatz, was an important site during the Weimar years for demonstrating and disputing power at an urban scale. Indeed, Heinrich Mendelsohn thought that the center of the new Berlin was being built precisely in this location. He planned two buildings with Heinrich Straumer, the architect of the radio tower, aiming to add the experience of urban cinema to the new media center. Not by coincidence, just after Hitler rose to power, the Reichskanzlerplatz was renamed Adolf-Hitler-Platz. For more on the role of the Reichskanzlerplatz, see Werner Hegemann, "Der Reichskanzlerplatz," *Wasmuths Monatshefte für Baukunst* 14, no. 4 (1930): 436–439. See also Wolfgang Schäche, Daniel Ralf Schmitz, and David Pessier, *Building Berlin: Developers Who Shaped the Emerging Metropolis* (Berlin: Jovis, 2019), 183–211.

60 Carolyn Birdsall, *Nazi Soundscape: Sound, Technology, and Urban Space in Germany 1933–1945* (Amsterdam: Amsterdam University Press, 2012), 39.

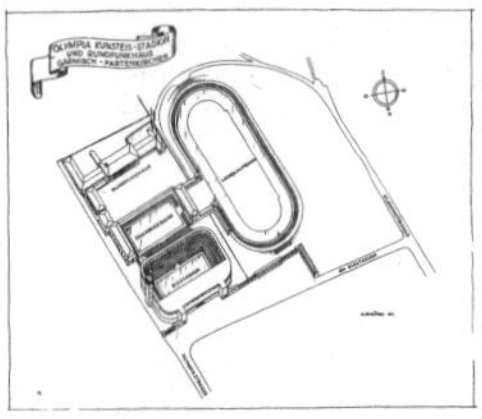
42, 43 → P. 81

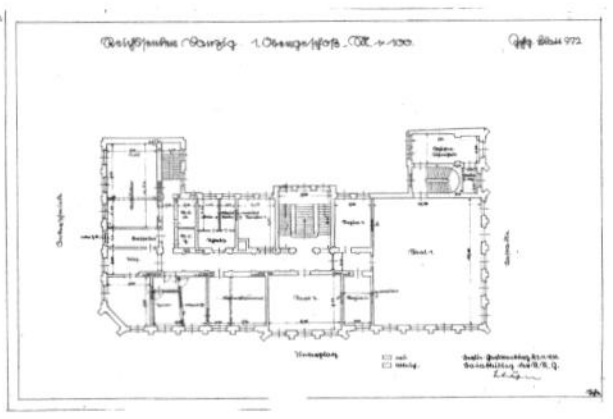
44 → P. 82

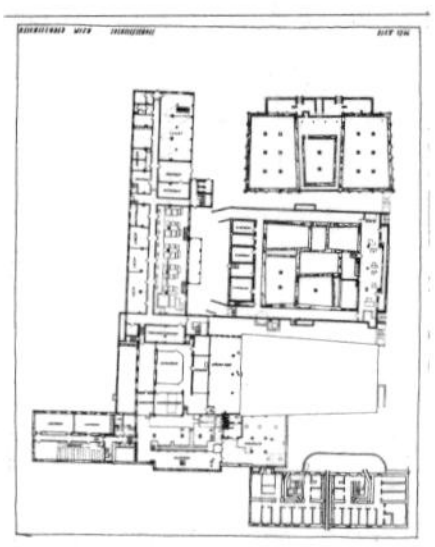
45, 46, 47 → P. 82

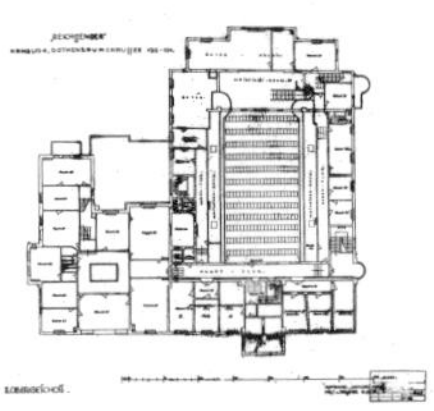
48, 49, 50 → P. 86

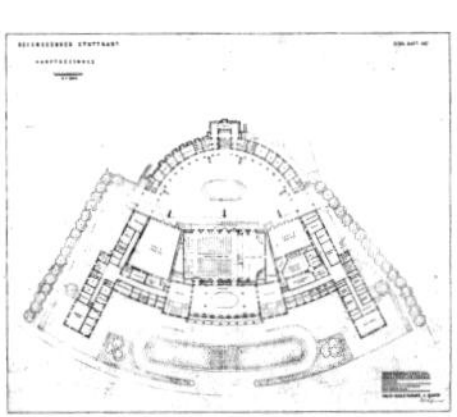
51 → P. 88

venue, synchronicity became an added determinant of the acoustic and architectural design.

Sound, as a mechanical phenomenon, propagates relatively slowly through the air, giving rise to a disparity between sonic and visual perception in large spaces. To compensate for this, Telefunken's first iteration of a sound amplification system for large gatherings, the mushroom speaker (*Pilzlautsprecher*), was designed to bring sound to an audience standing several hundred meters away. The problem with that first design solution, however, was the resulting echo, which reduced the intelligibility of speech.[61] As an alternative to the mushroom speaker, Telefunken created the extinguishing loudspeaker (*Löschstrahler*), which was used for the first time in the Olympic stadium. The acoustician Erwin Meyer described the practical solution, which was

> to mount two series of loudspeakers, one above the other, on a pole at a distance of one or two meters from each other. The essential point is that both wreaths of the loudspeakers, which are of the same kind, are switched on in opposite phases. At greater distances, the low and middle pitch tones vanish, whilst for the higher tones the speakers do not have the same phase of diaphragm velocity. ... At smaller distances near the group of loudspeakers the sound is not extinguished, though selective fading of course occurs and the response curve is irregular. But the effect of this distortion in the sound color is less disturbing than that of the echoes.[62]

Combining the building type of the stadium with the electroacoustic reinvention of the freestanding column, the Olympic stadium would come to embody a process achieved by radio throughout the entire country: a perfectly simultaneous and electronically synchronized acoustic experience.[63] If the design of the stadium gave material expression to the research conducted on the mechanical and electric dissemination of sound in space under the National Socialist regime, there was rapid

61 Wittje, *The Age of Electroacoustics*, 180–182.

62 Erwin Meyer, *Electro-Acoustics* (London: G. Bell and Sons, 1939), 71. It is worth noting that Meyer's book exists in English because he gave a series of lectures on the subject at the University of London in 1938.

63 For more on the work of Werner March during the Third Reich, see Thomas Schmidt, *Werner March: Architekt des Olympia-Stadions 1894–1976* (Basel: Birkhäuser, 1992).

change in the electromagnetic realm as well. On March 14, 1933, Joseph Goebbels was appointed Minister for Public Enlightenment and Propaganda. One of his first moves was to initiate a process of synchronicity (*Gleichschaltung*), placing the dissemination of radio content under central control.[64] The Reich Broadcasting Corporation (RRG)—founded in 1925 to coordinate all the regional radio stations from Berlin—began to operate under the newly created Ministry of Propaganda. Many of the figures behind the diverse institutions related to radio in Berlin during the Weimar years were persecuted and sent to concentration camps. Walter Schäffer, RRG's chief acoustic engineer, committed suicide. Its former director, Kurt Magnus, was persecuted, as were Alfred Braun, theater director and artistic director of Germany's first radio station, Die Funkstunde, and its director, Hans Flesch.

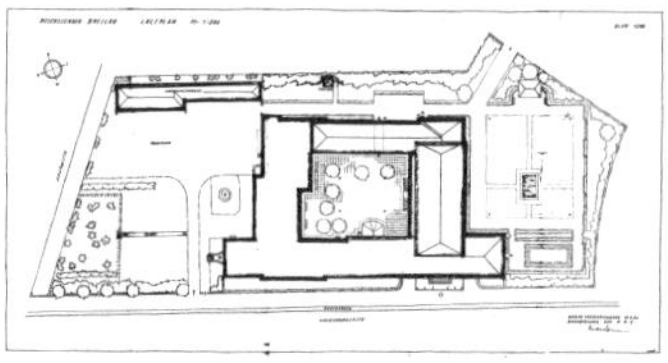

52, 53 → P. 90

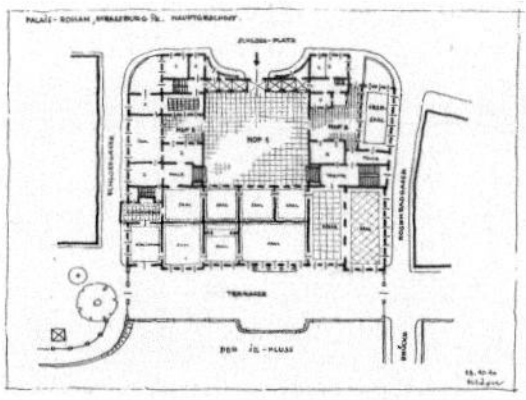

54 → P. 91

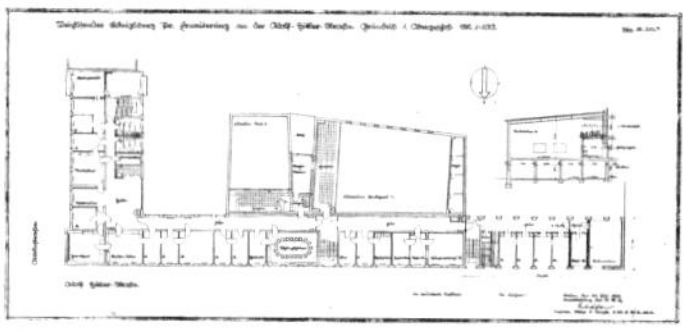

55, 56 → P. 92

57 → P. 96

In 1933, the broadcasting house designed by Hans Poelzig became the center of radio operations for the new regime.[65] From the following year, with Hitler's assumption of supreme power, all the former regional stations were dependent on the centralized institution located in Berlin. All local stations were recast as *Reichssender* (Reich stations), their numbers proliferating in the years up to 1940. The archive of the architecture division of the Reich Broadcasting Corporation reveals the unprecedented speed and scope of the dissemination of the type invented by Poelzig: new Reich stations were built in Gdansk (1939); Vienna (1938); Frankfurt (1937); Cologne (1939); Munich (1939); Hamburg (1938); Leipzig (1938); Stuttgart (1938); Breslau (1939); Strasbourg (1940); Königsberg (now Kaliningrad, 1936); Graz (1940); Krakow (1939); Prague had its Funkhaus (1939); Garmisch-Partenkirchen the Olympia Kunsteisstadion (1936), Herzberg the Deutschlandsender III (1938)—then the tallest structure in Europe.[66] (Fig. 42, 43, 44, 45, 46, 47, 48, 49, 50, 51, 52, 53, 54, 55, 56, 57)

In just one year, from 1933 to 1934, the number of radio receivers in Germany grew from 4.3 million to 8.2 million. By 1940, there were almost 15 million,

64 Bressler, *Von der Experimentierbühne zum Propagandainstrument*, 34.

65 Hans Jürgen Koch and Hermann Glaser, *Ganz Ohr: Eine Kulturgeschichte des Radios in Deutschland* (Cologne: Böhlau, 2005), 140.

66 For more on the building documentation of the transmitter stations and broadcasting houses designed by the RRG, see Deutsches Rundfunkarchiv Frankfurt, A48 RRG Bauzeichnungen, 388–430. For more on the Deutschlandsender III, see Gerd Klawitter, *100 Jahre Funktechnik in Deutschland: Funksendestellen rund um Berlin* (Berlin: Funk Verlag Bernhard Hein, 2004), 127–140. See also Helmut Knuppe, *Ein Riese unter Riesen: Aufstieg und Fall des Deutschlandsenders III in Herzberg (Elster)* (Leipzig: Leipziger Verlagsgesellschaft, 2003).

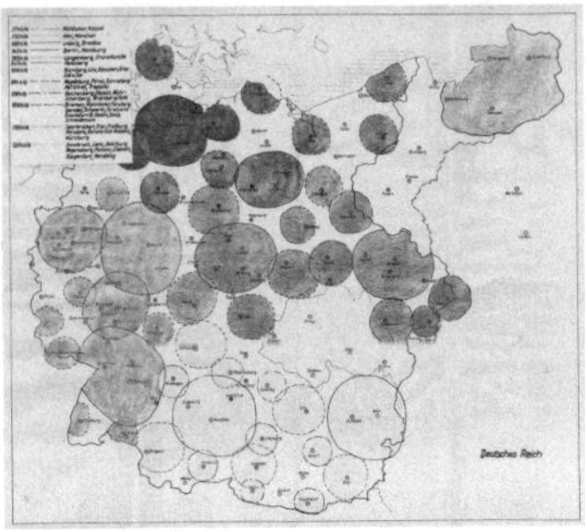

61 → P. 99

with sixty percent of households owning a radio, compared to only twenty-five percent in 1933.[67] During those years, the "people's receiver" (*Volksempfänger*) was welcomed as a new domestic focal point. The DKE 38 model, produced from 1938, and irreverently dubbed "Goebbels' snout" (*Goebbels-Schnauze*), brought not only propaganda but also music and entertainment into the home. Expanding to encompass new territories, both public and in the private circle of family life, the architecture of radio provided the technology for warfare purposes and, soon enough, the actual soundtrack of the war. In this respect, the architecture of radio enacted a new trans-scalar form of producing and occupying space. (Fig. 61)

On January 30, 1945—exactly twelve years after he became German Chancellor—Adolf Hitler gave his last radio address. Weeks later, on March 19, 1945, he issued the so-called Nero Decree, ordering the destruction of "all military transportation, communications, industrial, and supply facilities within the Reich that the enemy might in any way use, immediately or in foreseeable time, to continue fighting."[68] And yet it was not the Nazis who dismantled the Deutschlandsender III, but Soviet troops, following the Fall of Berlin in May 1945. The destruction of Europe's tallest transmission tower was in explicit contravention of an Allied treaty mandating that "No ship, vessel, or aircraft is to be scuttled, or any damage done to their hull, machinery or equipment, and also to machines of all kinds, armament, apparatus, and all the technical means of prosecution of war in general."[69] The Soviets also occupied the Haus des Rundfunks, located in what would become the British sector. Both transgressive acts were as symbolically important as they were technically strategic. They served as the prolegomenon to what has been called "war in the ether" during the Cold War years.[70] The conflict-laden lines of communication across the Iron Curtain are exemplified by the BBC's *East Zone* program targeting East Germany; by Radio Free Europe's broadcasting of "objective" news and entertainment from Munich; by the RIAS (Rundfunk

67 Koch and Glaser, *Ganz Ohr*, 141; Wittje, *The Age of Electroacoustics*, 179.

68 Albert Speer, *Inside the Third Reich* (New York: Simon & Schuster, 1997), 275.

69 For further information, see the act of military surrender by Germany signed in Berlin on May 8 in Charles I. Bevans, ed., *Treaties and Other International Agreements of the United States of America, 1776–1949*, vol. 3 (Washington DC: Department of State, 1969), 1124.

70 Christopher Classen, "Jamming the RIAS: Technical Measures against Western Broadcasting in East Germany," in *Airy Curtains in the European Ether: Broadcasting and the Cold War*, ed. Alexander Badenoch, Andreas Fickers, and Christian Henrich-Franke (Baden-Baden: Nomos, 2013), 321–346, here 323. Classen uses the phrase "war in the airwaves," but he also refers to the German original, "Krieg im Äther," which translates as "war in the ether."

im amerikanischen Sektor) *Aus der Zone für die Zone* program broadcast from West to East Berlin; and, moving in the opposite direction, by the GDR's Berliner Rundfunk and Radio Berlin International, broadcasting music, news, and political propaganda in ten languages from East Berlin to the world.

The idea of a war in the ether is central to understanding the dynamics of postwar Europe. At issue here is how that apparently immaterial exchange of information and entertainment across the Iron Curtain was physically constructed and supported—underpinned by a process of technology transfer. The German media theorist Friedrich Kittler has written about the application of German-designed electronic technology during World War II and its subsequent appropriation by mainstream media, describing, for example, how the magnetic tape recorders discovered by American troops in German stations would make possible the new kind of recording studio that transformed the music industry in postwar Britain and the United States. No Beatles, no Pink Floyd without the electronic technical media developed in Germany for the purposes of the war.[71] Scaling up Kittler's notion to the dimension of buildings, we can see how the strategic relevance of the architecture of radio was recognized early on by both the Western Allies and the Soviets. Within the processes of technology transfer, it was not just tape recorders, mixing boards, and radio amplifiers that were preserved, but a whole set of building typologies including the broadcasting house and the transmission tower. Buildings were protected and transferred as one among many other artifacts, acting as agents mediating between technological gadgets and large infrastructure. Within this process of technology transfer, the bricks (as material), the walls (as elements), and the floor plans (as *dispositifs*) were passed from one system to the other.[72]

Radio either anticipated or largely triggered the effects of the Cold War in Berlin. Attempts by both sides to gain access to the seemingly immaterial technology of radio by taking control of specific—highly material—

71 Friedrich Kittler, "Unconditional Surrender," in *The Truth of the Technological World*, ed. Hans Ulrich Gumbrecht (Stanford: Stanford University Press, 2013), 195–208.

72 On the development of the magnetic tape recorder, see John T. Mullin, "Creating the Craft of Tape Recording," *High Fidelity Magazine* 26, no. 4 (1976): 62–67; Peter Hammar and Don Ososke, "The Birth of the German Magnetophon Tape Recorder 1928–1945," *db Magazine* 16, no. 3 (1982): 47–52.

58 → P. 97

59 → P. 97

60 → P. 97

62 → P. 100

artifacts on the ground became one of the first sources of friction between the Eastern and the Western Blocs. In other words, the definition of the architecture of radio became an urban problem, adding yet another layer of complexity to the structure of the city. Just a few days after occupying the Haus des Rundfunks in May 1945, the Soviets took control of a transmission tower in Tegel, in the French sector. (Fig. 58) With Poelzig's building, the tower, and a group of former Nazi radio technicians now under their command, the Soviets began their own radio transmissions. For their part, the Americans were slower to understand the vital relationship between architecture and the ethereal medium. Denied access to Soviet facilities in 1945, they had to rely on the older wire-based broadcasting (*Drahtfunk*) technology.[73] When DIAS (Drahtfunk im amerikanischen Sektor) began broadcasting in 1945, it did so via telephone, reaching only one percent of the Berlin audience.[74] (Fig. 59) Its range was greatly expanded in 1946, when US forces occupied the telephone exchange building designed by Berlin architects Otto Spalding and Kurt Kuhlow in 1929. Built in the "brick expressionist" style, like Poelzig's Haus des Rundfunks, the exchange's most important feature were the telephone wires that connected it directly with most Berlin homes. (Fig. 60) Using wires or speakers mounted on trucks rather than electromagnetic waves, the Americans provisionally sought to create an alternative architecture of radio.

In 1948, the US radio station converted a former nitrogen factory—the Bayerische Stickstoffwerke designed by Walter Borchard in 1939—into its first official broadcasting house, which was wired to an improvised transmission tower located in Britz. (Fig. 62) DIAS, now able to broadcast using radio waves, became RIAS (Radio in the American Sector).[75] By 1949, and the official splitting of Germany into East and West, into the GDR and the Federal Republic, RIAS was the most popular radio station on both sides of Berlin. Centrally directed by the US government through the Allied High Commission for Germany,[76] it became the vehicle for constructing an alternative

73 Architekten- und Ingenieur-Verein zu Berlin, ed., *Berlin und seine Bauten*, pt. 10, vol. B: *Anlagen und Bauten für den Verkehr (4): Post und Fernmeldewesen* (Berlin: Ernst & Sohn, 1987), 137. See Daniel Gethmann and Florian Sprenger, *Die Enden des Kabels: Kleine Mediengeschichte der Übertragung* (Berlin: Kadmos, 2015); John Harwood, "Wires," in *Architecture / Machine: Programs, Processes, and Performances*, ed. Moritz Gleich and Laurent Stalder (Zurich: gta Verlag, 2018), 82–93.

74 Nicholas J. Schlosser, "The Berlin Radio War: Broadcasting in Cold War Berlin and the Shaping of Political Culture in Divided Germany, 1945–1961" (PhD diss., University of Maryland, 2008), 98.

75 The Bayerische Stickstoffwerke was closely related to IG Farben, the chemical company that served as an important supplier for the German armaments industry during World War II.

76 Schlosser, "The Berlin Radio War," 19.

"atmosphere of objectivity," with its newly renovated building providing the material infrastructure for that mission.[77] Thus, more than a decade before the building of the Berlin Wall, the city was effectively split in two by the divided nature of the architecture of radio. Through this proto-division, there coexisted two different forms of spatial construction—a construction dedicated to the dissemination of sound through electromagnetic waves. But precisely what kind of division was constructed in this respect? For Robin Evans, the birth of the "science of information destruction" began with the design of the Millbank Penitentiary walls in the 1830s. In Berlin, that "science" took on an expanded dimension and suddenly became an urban problem, in the process redefining the meaning and agency of the wall once more.

A number of seemingly unrelated episodes coalesced to become facilitators of the division and fragmentation of the city of Berlin, some of them just as radical and effective as the element of the Berlin Wall itself.[78] The technology of radio, however, does not necessarily conform to such traditional forms of division. There is a photograph showing a white bus passing in front of Poelzig's Haus des Rundfunks and a written sign bluntly stating that what is being broadcast from there was not "west-Berlin radio"—and for sure it was not. (Fig. 63) Even though the building was located in the British sector, it had been used since 1945 as the central broadcasting house of two stations controlled by the East. By 1952, however, the political conflict between East and West had escalated to the point where the subversive power and infrastructural relevance of Poelzig's building could no longer be ignored. In a second image, the perspective is similar, but now we see the perimeter of the building being enclosed with barbed wire.[79] (Fig. 64) On June 3, British military policemen prevented radio employees from entering or leaving. The last East German program was broadcast from the building that afternoon. For five weeks, the Haus des Rundfunks became a barbed-wire island within the British sector, uncannily anticipating an image that would characterize the Western side of

63 → P. 100

64 → P. 100

77 Nicholas J. Schlosser, "Creating an 'Atmosphere of Objectivity': Radio in the American Sector, Objectivity and the United States' Propaganda Campaign against the German Democratic Republic, 1945–1961," *German History* 29, no. 4 (2011): 610–627. For more on the idea of cultural Cold War, see Volker R. Berghahn, *America and the Intellectual Cold Wars in Europe* (Princeton: Princeton University Press, 2001); Frances Stonor Saunders, *The Cultural Cold War: The CIA and the World of Arts and Letters* (New York: New Press, 2000).

78 In the context of the political reorganization of the city after the end of World War II, the Soviets took the lead in appointing people to key positions of power. Hans Mahle, who returned from the USSR together with Walter Ulbricht, was named director of radio, while KPD militant Mattheus Klein was named head of Berlin radio stations. Architect Hans Scharoun was named head of construction and city planning, one of the few to occupy such a position without being an active militant on the political left. For more, see Wolfgang Ribbe, *Geschichte Berlins. Zweite Band: Von der Märzrevolution bis zur Gegenwart* (Munich: C. H. Beck, 1987), 1025–1113.

79 Günter Begoll, "'Eine Nacht, die wir nicht vergessen werden' 1952—Blockade des Funkhauses in der Westberliner Masurenallee—ein Interview," *RFZ-Frequenz* 11 (1952): 2. See also Maral Herbst, *Demokratie und Maulkorb: Der Deutsche Rundfunk in Berlin zwischen Staatsgründung und Mauerbau* (Berlin: Vistas, 2001), 54.

the city after the construction of the Wall.[80] It would subsequently house the Berlin branch of the Cologne-based Nordwestdeutscher Rundfunk (NWDR).[81]

1952 was also the year in which the influence of East German radio stations reached its lowest ebb, losing audiences, on both sides of the divide, to stations in West Berlin.[82] There was one radio receiver for every four inhabitants of East Berlin. Forty-two percent of them reported that they listened to RIAS, while only nine percent said they tuned into NWDR, the other Western presence.[83] Clearly, the manipulative power of "capitalist" media could not be subdued simply through propagandist political speech. In contrast to printed media, radio listeners could move between East and West with the turn of a dial. If they did not want to listen to news or conversations, they could tune their receivers to music instead. Political education had to compete with entertainment. The loss of influence among a wider audience, combined with the escalation of the political conflict that forced the GDR's Berliner Rundfunk out of the enclave of the Haus des Rundfunks, would provoke a radical reaction to the East's architecture of radio.[84]

In 1952, the East German Communist Party (SED, Sozialistische Einheitspartei Deutschlands) announced a concerted effort to construct "the foundations of socialism" at its second party congress.[85] "The work of the democratic radio" was considered to be of key importance in achieving this aim.[86] Following the model established in the Soviet Union, a decision was taken to centralize the "radio activities" (*Rundfunkarbeit*) of the GDR.[87] In a law passed on August 14, 1952, the SED announced the creation of a Staatliches Rundfunkkomitee (State Radio Committee) to oversee all cultural, political, economic, and infrastructural issues related to radio throughout the state.[88] Signed by the GDR Prime Minister, Otto Grotewohl, the announcement baldly stated, "To establish the basis for socialism in the German Democratic Republic, we need to raise the political and cultural level of our people."[89] With the aim of turning

80 For more on the sequence of events triggered by the seizing of the broadcasting house in Masurenalle by British military police and the GDR's reaction, see Letter from General Waronitschew (Administration for questions on Transportation and Communication) to the High Commissioner of the USSR in Germany on September 21, 1953, Bundesarchiv, DR 6, 207. I would like to thank Erica Harrison for this reference.

81 Nordwestdeutscher Rundfunk (NWDR)-Berlin existed from 1949 to 1954, when it changed its name to Sender Freies Berlin (SFB). For more on the institutional dynamics related to radio in postwar Berlin, see Herbst, *Demokratie und Maulkorb*, 88–106.

82 Classen, "Jamming the RIAS," 321–346.

83 International Evaluation Staff (IEV), Country Report on Radio: East Germany, February 1953, in National Archives and Records Administration, RG 306 (USIA), Office of Research, Reports, and Related Studies Maintained or Sponsored by the International Evaluation Staff (IEV) and other Offices of the Department of State, 1948–1953, box 29. Previously cited in Nicholas J. Schlosser, *Cold War on the Airwaves: The Radio Propaganda War against East Germany* (Urbana: University of Illinois Press, 2015), 48.

84 Schlosser, "Creating an 'Atmosphere of Objectivity,'" 610–627. See also Don R. Browne, "RIAS Berlin: A Case Study of a Cold War Broadcast Operation," *Journal of Broadcasting* 10, no. 2 (1966): 119–135.

85 "Stenografisches Protokoll der 2. Parteikonferenz der SED, July 12, 1952," in Bundesarchiv, DY 30/IV 1/1/1 to DY 30/IV 1/1/4. Previously referenced in "DDR-Rundfunk- und Fernsehkomitee: Aufgaben und Geschichte," in *Tele-Visionen: Fernsehgeschichte in Ost und West,* ed. Bundeszentrale für Politische Bildung: 1–15, here 3.

86 "Verordnung über die Bildung des Staatlichen Rundfunkkomitees," *Gesetzblatt der Deutschen Demokratischen Republik*, no. 112 (August 21, 1952): 733–734, here 733. This and subsequent quotes are the author's translation.

87 For a comprehensive history of Russian and Soviet radio, see Stephen Lovell, *Russia in the Microphone Age* (Oxford: Oxford University Press, 2015).

88 The *Gesetzblatt der Deutschen Demokratischen Republik* was the official bulletin for public proclamations published by the GDR from its founding in 1949 until 1968. No law was effective unless it was published in this bulletin.

the tables in the airwaves war against the West, the State Radio Committee was to use radio to "develop the socialist consciousness of the working people, to deeply fill the people with the idea of defending peace, defending our homeland, and hating the imperialist belligerents, militarists, and traitors."[90] To this end, "the population in the German Democratic Republic and in West Germany will be given the opportunity to receive three different, carefully coordinated, high-quality and all-day programs of German Democratic broadcasting."[91] The State Radio Committee would be responsible for designing the three stations. In other words, the document sets out a political and institutional operation with a twofold agenda: securing the internal territory against the transgressive agency of rival signals, and expanding the performative agency of the GDR's own signals in the West.

To ensure the programs could be received throughout the GDR, a series of transmitter stations were designed for strategic locations by the Projektierungsbüro der Deutschen Post (Projects Office of the German Post). Within less than five years, as part of the GDR's first Five-Year Plan, new stations were built in Brocken, Inselberg, Helpterberg, Marlow, Berlin-Köpenick, Görlitz, Schwerin, Petkus, and Burg, and the existing ones in Nauen and Königs Wusterhausen were renovated.[92] A transmitter station is a building type crucial to the agency of the architecture of radio. While having a clearly defined program, its operation requires hardly any human activity. The typology is generally composed of two elements: a "box" and a tower. Any program or content produced in a broadcasting house is transmitted via wire to the box, which houses powerful electric machinery that amplifies the signal and then channels the amplified electric current through a steel mast, usually placed at the center of the tower. As the electric current flows through this vertical element, the constituent atoms begin to vibrate, and the offshoot of that oscillation are the waves—partially electric, partially magnetic, hence the term electromagnetic—radiating from the tower at the speed of light. Electromagnetic waves became a new way of defining

89 "Verordnung über die Bildung," 733.

90 "Verordnung über die Bildung," 733.

91 "Verordnung über die Bildung," 733.

92 For details on the design of the transmitter station in Schwerin, see "Fachgebiet Bauwirtschaft FS und UKW-Sender Schwerin-Zippendorf, 1957," in Bundesarchiv, DM 311 PLAN/100; "FS-Turm Schwerin Strassenbau Übersichtsplan, 1957," in Bundesarchiv, DM 311 PLAN/78; "FS und UKW-Sender Schwerin-Zippendorf, 1956–58," in Bundesarchiv, DM 302/2832. For a general description of transmitter stations built in Germany, see Klawitter, *100 Jahre Funktechnik*.

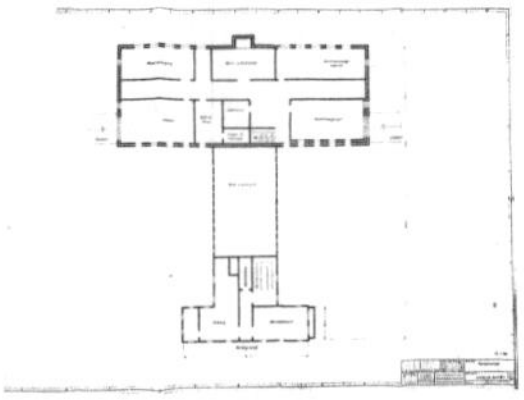

67, 68 → P. 102

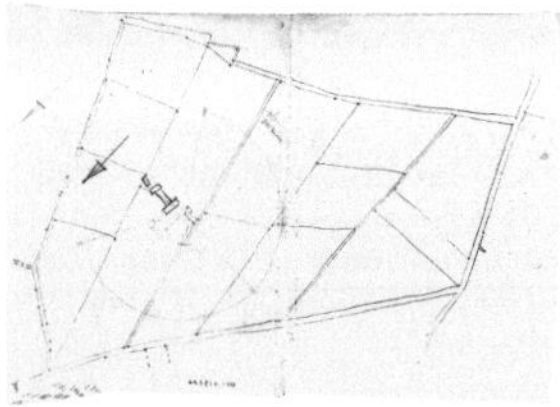

65 → P. 101

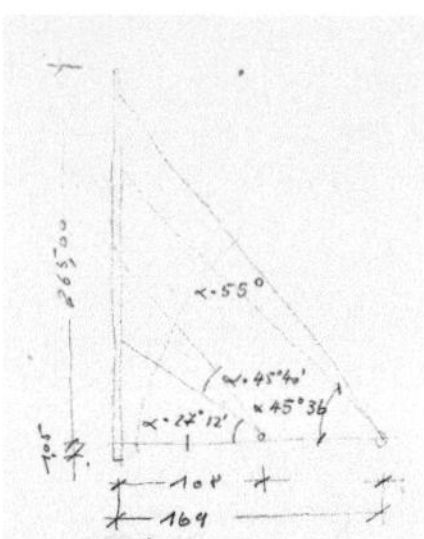

66 → P. 101

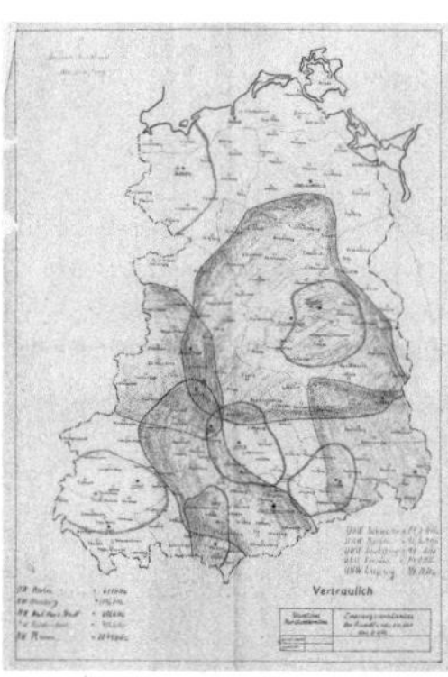

69, 70 → P. 104

71, 72 → P. 106 (–107)

the territory, given form by the size and positioning of transmitter stations. The power and dimension of these stations—in other words, their *design*—would resonate on both sides of the Iron Curtain. Looking closely at the architecture of the transmitter station in Schwerin, for instance, we see how both elements of the type—the tower and the box—relate to one another, but also how their internal size and geometry relate to the larger system. (Fig. 67, 68) In this case, the box is organized along a clearly defined axis of symmetry and has a room for low currents, an electric generator for emergencies, a garage, and another room housing the machines for the water supply. On the level above, at the center of the plan, is the room for the vacuum tubes, flanked on either side by the machinery for ultra-short wave and television transmissions (both technologies closely tied to FM radio transmission). Along the axis are bathrooms, cloakrooms, small offices, and a kitchen, with a recreation room at the end. There is an interaction of activities performed mainly by machines—which explains the various layers defining the exterior walls—combined with a basic program for the few activities that have to be performed by humans.

(Fig. 65) Looking at another sketch drawing of the plan, we see a small auxiliary building that is somehow sheltered by a larger circular shape virtually traced out by the extension of the cables structuring the tower. The circular perimeter is defined by the height of the tower, which in turn is defined by the particular wavelength transmitted by that station. In medium wave frequencies, the height of the tower is between one half and one quarter of the wavelength transmitted by the station, as the sketch in section shows. (Fig. 66) This apparently circumstantial geometrical relationship between the transmission frequency and the height of the tower would leave traces at a territorial scale as well. (Fig. 69, 70, 71, 72) This spatial agency of electromagnetic waves, and the way in which they are registered at a territorial scale, finds a literal expression in a series of maps produced by the State Radio Committee. In the smaller areas marked UKW, denoting the newly developed ultra-shortwave, people could receive high-quality signals through FM radio. MW, for medium wave, afforded broader coverage, while longwave radio, LW, with lower signal-to-noise ratio, could reach audiences far beyond the borders of the GDR. Technical details aside, the definition of how and from where signals were propagated had direct implications for people's everyday lives. Through the series of transmitter stations developed during the first Five-Year Plan, the centrally produced content of the GDR's radio stations could reach the entire population at home. What these maps show is the continuity between the construction of the technical infrastructure—the size and geometry of the new transmitter stations—and the definition of borders, thresholds, and limits at a territorial scale.

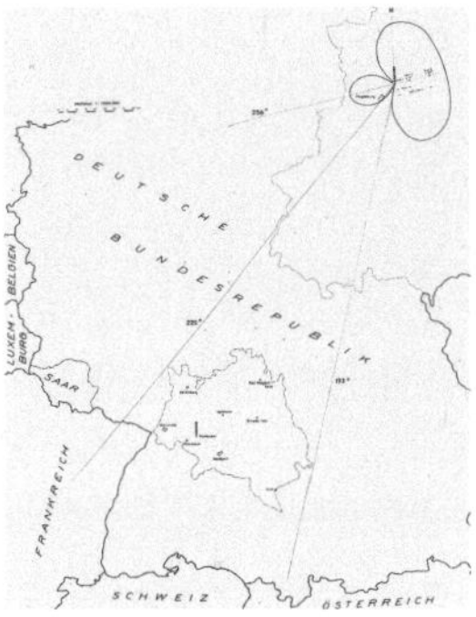

73 → P. 108

74 → P. 109

The patterns defined by those limits, however, indicated not only effective coverage, but also signal strength and quality. Both were crucial factors in a highly contested territory. As the framing of the 1952 law made clear, the motive for centralizing radio activities in the GDR was not only to secure the dissemination of content internally, but also to expand media coverage to the West. These expansionist intentions are graphically illustrated in two maps of the Burg transmitter station built in 1953 near Magdeburg, a few kilometers from the West German border. (Fig. 73, 74) One map shows that the 324 m tower, then the tallest building in Germany, East or West, was directed at a specific swath of Western territory. The other map records variations in the reception of its signals in the West relative to night and day, as well as zones of high interference. Magdeburg, in medieval times a strategic point dividing the Catholic and Slavic regions on either side of the River Elbe, once again became a frontier settlement—albeit one no longer defined by religion or trade routes, but by the geometric patterns and territorial extensions delineated by the logics of radio transmissions.
The Iron Curtain also functioned as a specular axis, triggering symmetrical reactions on either side of the divide. A 1953 report from the US Department of State considered the best strategy for reaching

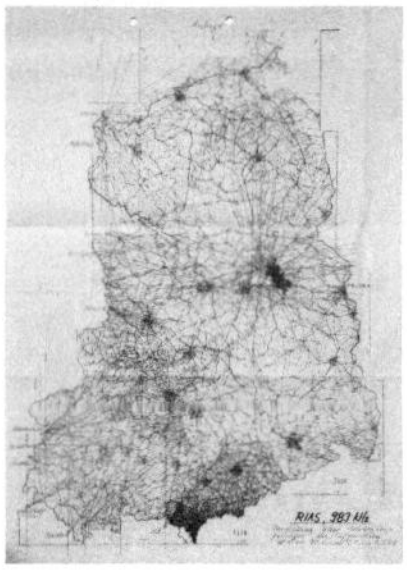

75 → P. 110

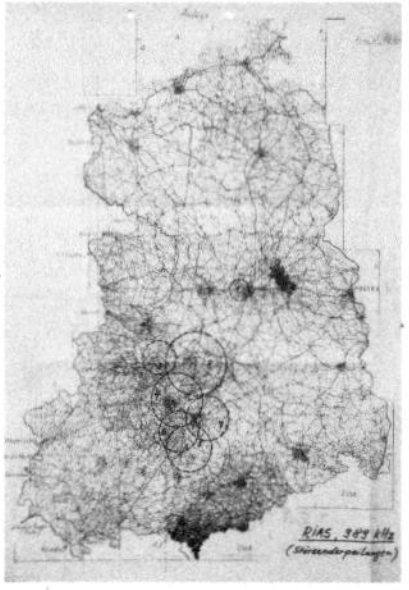

76 → P. 111

77 → P. 112

audiences in East Germany: should it shift to the emerging technology of television, or keep the prime focus on radio, which "for the bulk of the population ... constitutes the primary regular source of reliable information."[93] Its conclusion: "because of the unique position which RIAS, the Radio in the American Sector of Berlin, occupies, and due to its coverage of the entire Soviet Zone through medium wave, shortwave, and FM transmission, the United States has all the opportunities offered by control over a full-scale domestic radio station of great popularity."[94] The range of RIAS airwaves also had territorial implications, which become graphically evident when analyzing a map commissioned by the station. (Fig. 75) Each red dot designates the position of a radio transmitter used for RIAS broadcasting on the medium frequency of 989 kHz. All are located strategically at the border dividing East and West Germany. With its own series of towers ranging from the Baltic port of Lübeck in the north to the Bavarian town of Hof in the south, RIAS defined a border that was solidly constructed—not by a physical wall as such, but by a series of points mapping the information geography within the East. Berlin, however, is located at the center of the map, with red lines radiating out from the city as well—a springboard of Western transmissions placed at the heart of enemy territory. This map clearly shows how the GDR was bombarded on multiple fronts by Western signals. Combating or counterbalancing the invasion of information would require a titanic effort.

(Fig. 76) On a second version of the same map, we see a series of circles marked with thick dotted black lines. At the center of each of those circles was a jamming station (*Störsender*) intended to interfere with the signals emanating from the West. The Fernmeldetechnisches Zentralamt (FTZ, West German Central Telecommunications Office) tested the reception of Western signals on a bi-weekly basis, paying particular attention to RIAS. (Fig. 77) The FTZ's mapping of medium and longwave transmitters depicts the shared landscape of towers, with a particular concentration—almost like the clustering of

93 International Evaluation Staff (IEV), "Country Report on Radio: East Germany, February 1953," in National Archives and Records Administration, RG 306 (USIA), Office of Research, Reports, and Related Studies Maintained or Sponsored by the International Evaluation Staff (IEV) and Other Offices of the Department of State, 1948–1953, box 29, cover page. Quoted in Schlosser, *Cold War on the Airwaves*, 48.

94 "Country Report on Radio: East Germany, February 1953," 8.

towers in the walled town of San Gimignano—around the radio battleground of Berlin.[95] The medium of radio made postwar Germany a contested terrain with overlapping borders—a territory in which both blocs fought to win over their audiences. In this context, the process of centralization initiated by the GDR in 1952 triggered two significant operations defining the East's architecture of radio. One was the redesigning of all media content produced by the GDR's broadcast institutions, a point that will be addressed more fully later on. The other was the creation of an anti-RIAS media campaign. Both strategies were carried out primarily through the medium of architecture, deploying a sophisticated understanding of the agency of walls.[96]

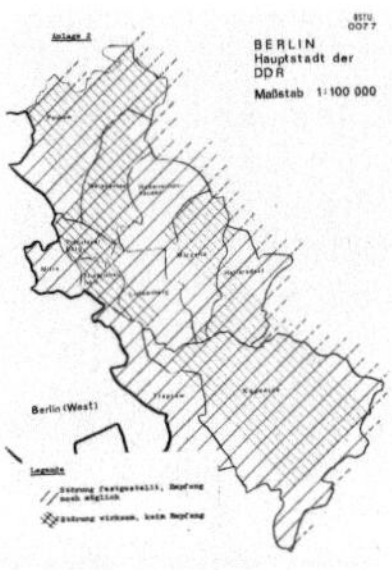

78 → P. 113

95 For more on the full series of reports conducted by the Fernmeldetechnisches Zentralamt der Deutsche Bundespost, see Deutsches Rundfunkarchiv Babelsberg, RIAS, box 16/94/6.

96 Schlosser, *Cold War on the Airwaves*, 108.

97 Schlosser, *Cold War on the Airwaves*, 127.

98 Classen, "Jamming the RIAS," 321–346.

III

At a territorial scale, the GDR's most significant and aggressive action was to jam the undesirable "capitalist frequencies" emitted by RIAS. By 1957, it had built three hundred jammers, which performed as virtual walls protecting its sovereignty of the airwaves.[97] While these were spread across the entire country, Berlin was the main terrain of contention, with the logic of deployment of the stations becoming much more refined. In Berlin, electromagnetic waves were used to draw and define the structure of the city. With great secrecy, the GDR began to grid Berlin via an array of lightweight movable transmitters.[98] A map developed by the Ministerium für Staatssicherheit (Ministry for State Security, better known as the Stasi) shows the impact of these structures, along with the suggestion of what a wall might become according to the logics of radio.

(Fig. 78) The zones demarcated by single diagonal lines show areas of partial jamming. Where those lines are intersected by perpendicular ones, the jamming was conclusive, effectively blocking all unwanted transmissions. In place of the more familiar East/West sectors divided by clearly defined lines, these "radio-activities" transformed the urban fabric of Berlin into an ocean of invisible, overlapping concentric and radiating circles. This map also illustrates the interrelated observations that information creates space, that there is no absolutely impermeable wall with respect to the transgressive agency of radio, and that "information, as such, may be morally neutral but it is certainly not inactive," even at an urban scale.[99] The strategic location of buildings equipped with radio-emitting and radio-receiving capacities turned Berlin into a synthetic geography, a manufactured terrain in which the traditional urban elements of streets, blocks, and squares coexisted with the overlapping agency of different kinds of walls used to turn noises into signals and signals into noise. Even before the construction of the Berlin Wall, this material fabrication of noises and signals twisted and expanded the dialectic conceptualization of the wall in terms of "retreat" and "exclusion." This dialectic distinction, as theorized by Robin Evans, and discussed at the beginning of this chapter, now takes on an expanded meaning. As William Rankin has argued, radio communications and remote signal transmission broke with the historical power of maps, and therefore walls, to give "real-world traction to abstract concepts like jurisdiction and sovereignty," pointing to the fact that maps alone "reinforced a strong split between a space of domestic affairs and a space of foreign relations."[100] Here we see a different kind of map depicting a different kind of wall, and thus a different kind of space. The districts of Prenzlauer Berg, Marzahn, and Köpenick were circumscribed and walled by noises interfering with signals emitted from the secluded and walled West, and, rather than by clearly defined retreats and exclusions, the environment is structured by nested, temporally defined, and simultaneous thresholds.[101]

The GDR's concerted effort to centralize all its stations, expand its infrastructure, and jam RIAS's signals proved to be successful. In 1953, just a year after the creation of the State Radio Committee, RIAS lost its position as the most popular radio station in Berlin for the first time since 1948.[102] A report issued by United States Information Agency in 1957 confirmed that three

99 Evans, "The Rights of Retreat," 335.

100 William Rankin, *After the Map: Cartography, Navigation, and the Transformation of Territory in the Twentieth Century* (Chicago: University of Chicago Press, 2016), 3.

101 Accepting a degree of generalization in the city structure of Berlin, Prenzlauer Berg was a residential district with a fluctuating population, but relevant for the GDR because of its proximity to the newly planned center of Alexanderplatz. Marzahn was a district with a high density of massive *Plattenbau* or prefabricated housing complexes for the working class. Köpenick was an industrial area in which, probably not coincidentally, the GDR's "media industries" for radio and television were located.

102 RIAS's popularity was reduced to thirty-seven percent according to a report by Nordwestdeutscher Rundfunk. See Schlosser, *Cold War on the Airwaves*, 127.

new GDR stations (Radio DDR, Berliner Rundfunk, and Deutschlandsender) attracted seventy-seven percent of the audience in Berlin.[103] The decline in RIAS's popularity became a source of diplomatic tension, as the United States demanded authorization to increase the power of its transmitters, which the West German government was reluctant to grant, fearing an escalation in the war on the airwaves.[104] For RIAS, the waning of its popularity called for a prompt response in the form of improvements to its own architecture of radio.[105] In 1953, the station increased the power of its transmitter station in Britz to 300 kW in the medium wavelength. In 1961, after major diplomatic maneuvers shortly before the construction of the Berlin Wall, RIAS was allowed to transmit via FM radio from Britz as well. By 1962, as the station proudly announced in an information booklet, "All wavelengths except FM are jammed, but RIAS gets its message through by using several different frequencies on all wavelengths, and by virtue of the fact that its main and very powerful transmitters are located within the relatively small area to which it broadcasts. It has been estimated that the Soviet Zone jammers use enough electricity each day to light the city of Leipzig."[106] The RIAS's now "very powerful transmitters" were located a stone's throw from the border.[107]

Improvements in the intensity and quality of RIAS's signals were matched by a transformation of content production. In 1963, work began on extending the RIAS broadcasting house in Fritz-Elsas-Strasse, located in the former administrative building of the Bayerische Stickstoffwerke AG (Bavarian Nitrogen Factory). The five-story extension was to fill the entire adjacent plot, which had remained empty since the end of the war. While it had its own independent access and circulation, it was directly connected to the existing broadcasting house. The basement of the building contained sound archives: the newly developed technology of tape recording allowed for much longer recordings, and all that information needed to be organized and stored. From the ground level, a curved stair led up to a foyer, 4 m above, that

103 Research Staff, Office of Public Affairs, American Embassy, "Radio Listening in the GDR, Report A-18, series 3, December 19, 1957," in National Archives and Records Administration, RG 306 (USIA), Office of Research, Field Research Reports 1953–1978, box 35. Schlosser, *Cold War on the Airwaves*, 128.

104 Schlosser, *Cold War on the Airwaves*, 130.

105 Gerd Klawitter, *100 Jahre Funktechnik in Deutschland: Funksendestellen rund um Berlin* (Berlin: Funk Verlag Bernhard Hein, 2004), 247–256.

106 For more on the original booklet, see RIAS Berlin, *RIAS: A Free Voice of the Free World* (Berlin: Brüder Hartmann, 1962), 25.

107 The territorial implications of radio broadcasting remained a central preoccupation of the United States and influenced its diplomatic relationships with West Germany until the 1980s. In a National Security Decision Directive signed by President Ronald Reagan issued in 1982, the political problem of radio broadcasting and jamming is addressed in several points. The documents declared: "Diplomatic requirements deriving from this modernization and expansion program shall be given high priority by the Department of State. Acquisition of new transmitting sites and facilities should be a priority matter on the political agenda of bilateral relations with appropriate countries. Attention should also be given to renewal of facility agreements which have lapsed, with permission for augmentation as necessary, and to negotiation for the use of out-of-band frequencies." The document also speaks directly about the problem of jamming, stating, "particularly in view of the recent renewal of jamming of VOA by the Soviet Union, it is essential that the US take all possible steps to overcome jamming of US international broadcasts and to ameliorate its effects. Continuing interagency study needs to be given to political as well as technical aspects of jamming and possible countermeasures." And finally, the document refers to Germany as a "jammed environment," suggesting that "further study should be undertaken immediately of the technical parameters of a variety of plausible DBS [Direct Broadcasting by Satellite] systems for international broadcasting, with particular attention to the role of DBS in a jammed environment." See Executive Secretariat, National Security Council: National Security Decision Directives (NSDD): Records, 1981–1987, in Reagan Library Collections, NSDD 45 [United States International Broadcasting], Box 2.

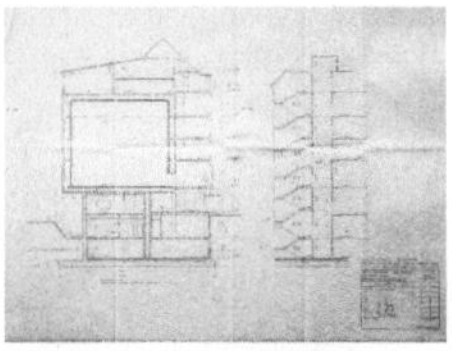
79, 80, 81 → P. 114

82 → P. 117

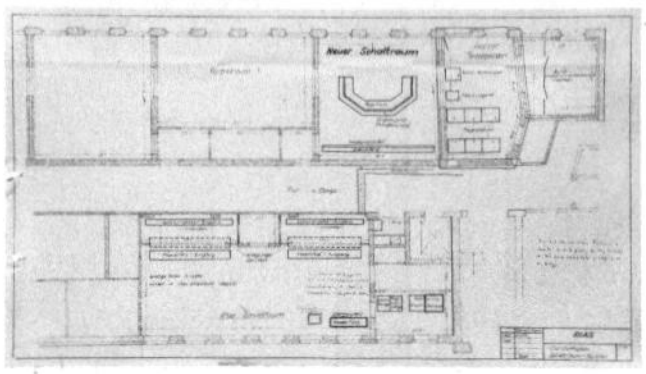
83 → P. 117

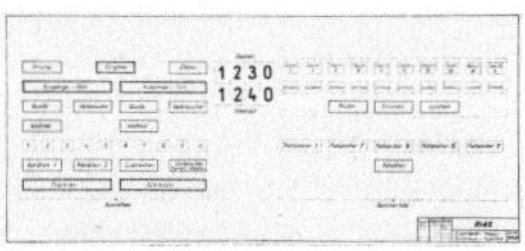
84 → P. 118

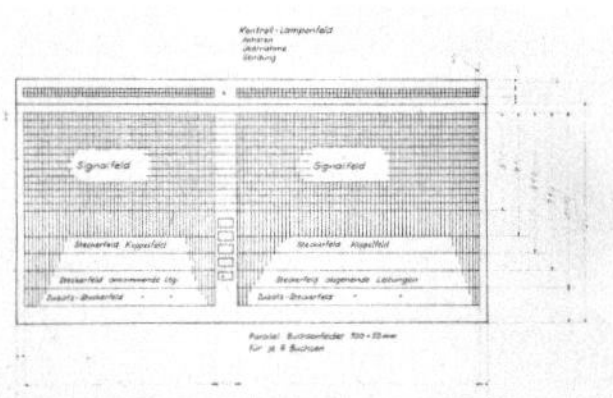
85 → P. 118

concertgoers and employees and offered views out onto the street. At that level, the new building was connected with the older infrastructure. (Fig. 79, 80, 81) A new studio, 25 m by 20 m by 15 m, was designed to increase the quality of recordings of the RIAS symphony orchestra: the new space literally floated, being structurally independent from the rest of the building, a design feature that mimicked predecessors of the same type. Offices were oriented towards the street on three levels, in a sense replicating the typological structure initiated by Hans Poelzig a few decades earlier, though in a denser context.[108]

The most relevant aspect of the renovation of the broadcasting house, however, was the new control room (*Schaltraum*) on the top floor of the extension, the design of which speaks to the new form that the architecture of radio had taken for RIAS and other Western stations.[109] (Fig. 82) A photograph shows the basic organization of the control room, with one technician working at the control table, which regulated output to the transmitter station, while the other attends to the switchboard. (Fig. 83, 84) Both these elements had been redesigned in line with new requirements. (Fig. 85) Through the new switchboard, with input lines (*ankommende Leitungen*) on the left and output lines (*abgehende Leitungen*) on the right, signals were channeled into the room from many different sources and places. Looking at the new list of input lines, we find the RIAS broadcasting house was wired to—and receiving signals from—the Philharmonie, Trade Fair, Congress Hall (today, Haus der Kulturen der Welt), Reichstag, Olympic stadium, and Schöneberg Town Hall.[110] The RIAS's new control room thus served to bind together different cultural and political institutions scattered across the city of Berlin. Looking at the output lines, we see that the broadcasting house was effectively connected to the transmitter station in Hof, Bavaria (both for the medium wave and the ultra-short FM frequencies), to the public radio service, Sender Freies Berlin, and to the facility in Britz, Berlin (on several frequencies on medium wave and ultra-short wave). In precisely orchestrating the timing and content of

108 For more on the renovation of RIAS's broadcasting house including construction documents, see "Alte Baupläne des RIAS-Funkhauses, 1963–1971," in Deutsches Rundfunkarchiv Babelsberg, RIAS, Box 16/94/1.

109 The control room (*Schaltraum*) is where the program output is regulated. Simply stated, all the programs and content produced in a broadcasting house are amplified and delivered, through wire, from the control room to the transmitter stations. Therefore, the control room houses the machines that process the electric signal for the last time before leaving the house, and it is also the room in which the timing of the transmissions is regulated.

110 For the complete renovation of the control room and the complete list of input and output lines, see "Übersichtsplan Schaltraum-Neubau, 1971," in Deutsches Rundfunkarchiv Babelsberg, RIAS, Box 16/94/1.

its broadcasts for people on both sides of the Wall, the control room was just like a spider, managing and manipulating a vast territory by pulling threads from its position in the middle of its web. It is in this sense that the broadcasting house type created a new form of centrality.[111]

Looking at territory from the perspective of radio, another dimension of Berlin's strategic location within a divided Europe becomes apparent. The singular location of the city gave the West a privileged position from which to broadcast to the East and, equally, to intercept enemy signals. In other words, by virtue of the ubiquitous presence of radio, Berlin provided the West with a formidable mouthpiece in the midst of enemy territory, as well as a unique position from which to listen to the enemy with radio-sensitive ears. In 1963, a peculiar construction began to rise from the top of Teufelsberg in Grunewald, West Berlin. The structure, built by the US Army, was named Field Station Berlin.

The highest landform in the city, Teufelsberg is an artificial hill made of war rubble—more than 75 million cubic meters of it, piled up to a height of around 80 m (or 120 m above sea level).[112] A fuel shortage in West Berlin after the blockade of 1948 meant that most of the rubble was left in the Western sector, and in 1950 the area next to a glacial lake, the Teufelssee, became a designated site for its disposal. It was here, on top of this manmade peak, that the US Army and National Security Agency (NSA) chose to build a station, not for the purpose of radio transmission, but rather to receive, analyze, and redirect signals coming from the East. The site, however, had not lain empty before the creation of the rubble hill. In fact, Teufelsberg was built on the ruins of another structure, the unfinished military academy (Wehrtechnische Fakultät) designed by Hitler's architect, Albert Speer. The military academy was strategically located at the western end of the east–west axis of the projected world capital, Germania, not too far from the trade fair designed by Martin Wagner and Hans Poelzig. Its program made ample provision for the study of radio communication technologies: two

111 For a broader discussion on the role of centrality in media houses, see Staffan Ericson and Kristina Regent, eds., *Media Houses: Architecture, Media, and the Production of Centrality* (New York: Peter Lang, 2010).

112 Benedict Anderson, *Buried City, Unearthing Teufelsberg: Berlin and Its Geography of Forgetting* (New York: Routledge, 2017).

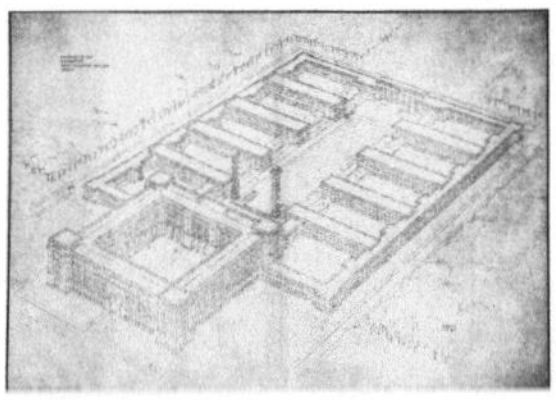

86, 87 → P. 121

88, 89 → P. 124

symmetrically positioned radio towers overlooked the entire complex (Fig. 86, 87).[113] The new field station designed by the NSA was using as its foundation the very school from which the Nazis researched radio communication technologies for their own war purposes.[114]

The NSA began to operate from the top of Teufelsberg in 1961—the same year the Berlin Wall was built—and construction of the station began in 1963. A report issued by the Stasi makes it clear that the authorities in the East were keeping a close eye on its activities.[115] It notes that the station was part of ECHELON, a global network of stations operated by the Intelligence and Security Command (INSCOM) for the US National Security Agency.[116] As a listening station, with antennas and radars located at the top of the hill, its remit was to intercept signals and to interpret and decode any received information. The report concludes that the main objective of the station was to have "a central coordinating function for the activities related to electronic espionage against the GDR, the Soviet Armed Forces located in Germany, and the People's Republic of Poland."[117] It adds that "the high level of development of the technical equipment, as well as the human resources, allow the facility to record, around the clock, radio communications right into the territory of the USSR, and to listen to the satellites and troposcatter communications of the member states of the Warsaw Pact."[118]

(Fig. 88, 89) The field station built by the US Army atop Teufelsberg has a very specific architecture encompassing multiple interactions between people and electroacoustic technologies, articulated at many scales simultaneously. The central building of the facility is shown on the left-hand side of a photograph taken by the Stasi. Its main slab is crowned by two geodesic domes on the east and west corners. A 45 m tower rises from the center of the structure, also crowned by a geodesic dome. Each of these domes housed a radar station, a spherical structure built from light steel triangular frames clad with synthetic

113 For the complete set of construction drawings of the Wehrtechnische Fakultät der Technischen Hochschule, see Bundesarchiv, R 4606/3620; Bundesarchiv, R 4606/3699.

114 As opposed to Allied countries, Germany did not pursue scientific research and development in electroacoustics for war purposes after World War I, mostly due to the agreements established in the Versailles Peace Treaty. Therefore, most of the electroacoustic research developed during the Weimar Republic was conducted for civil purposes and more specifically for radio broadcasting. After 1933, however, there was a remilitarization of German technology. In this context, the construction of the new military academy could be read as the material basis for increasing research in various technical fields directed towards war purposes, an infrastructure that was lacking during the interwar period. For further reading see Wittje, *The Age of Electroacoustics*, 184–188.

115 To read the full report, see "Analyse der Struktur und Tätigkeit der US Army Field Station Berlin (USAFSB) Teufelsberg, 1985," in Der Bundesbeauftragte für die Unterlagen des Staatssicherheitsdienstes der ehemaligen Deutschen Demokratischen Republik (BStU), MfS, HV A, Nr. 883, Bl. 1–68.

116 The project was part of the signal intelligence (SIGINT) cooperation between Australia, Canada, the United Kingdom, New Zealand, and the United States. A very recent study has declassified another signal alliance named Maximator between Denmark, Sweden, Germany, the Netherlands, and France. The collaboration relied on the Swiss producer Crypto AG for cryptographic devices. See Bart Jacobs,

fabric. The eastern radar had a remote-controlled parabolic antenna capable of receiving signals in 360 degrees operating in the frequency range from 250 to 4,000 MHz. The western radar had a similar parabolic antenna that operated from 1 to 5 GHz. The larger 12 m-diameter radar on the top of the tower also operated from 250 to 4,000 MHz but reached a greater distance.[119] The 120 m antenna in the center of the picture, fabricated by Rohde/Schwarz in West Germany, received signals in the frequency range from 20 MHz to 1,118 GHz.[120]

(Fig. 90) From the top of Teufelsberg, this antenna had a 360-degree directionality and, when operating at maximum capacity, a range of 240 km, so it could receive signals from as far away as Poland. An examination of the building section reveals direct connections between the architecture and its territorial implications. Six manually operated parabolic directional antennas were located on the sixth level (marked "E6"), each focusing the reception of signals at 7.5 degrees. These antennas covered specific areas of the territory, each of them operating at a different frequency spectrum (Fig. 91). On the level above, there were another four manually operated parabolic antennas receiving signals up to 8 GHz. Each of these covered a designated range of 90 degrees, allowing the four of them together to cover the entire circumference around the structure (Fig. 92, 93, 94, 95).[121] From the eighth floor, thirty-six directional antennas each covered a range of 10 degrees and, on the level above, another six antennas covered the frequency range from 100 to 1,000 MHz. (Fig. 96) What is especially relevant here is that the construction of the section of the building had a specific relationship to the areas and frequency ranges that the station was meant to survey. By considering how the section of the building defined a specific territory, we grasp how the station on Teufelsberg had an influence at the scale of the entire country.

Rigorously secured by fences and by the US Military Police, the complex was laid out without any

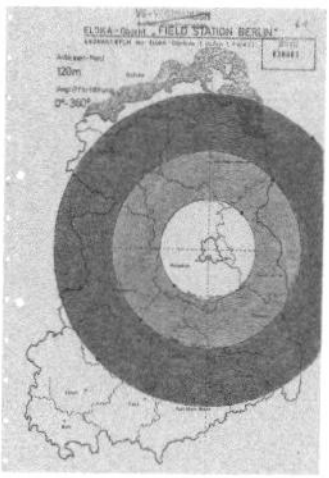

90 → P. 126

91 → P. 126

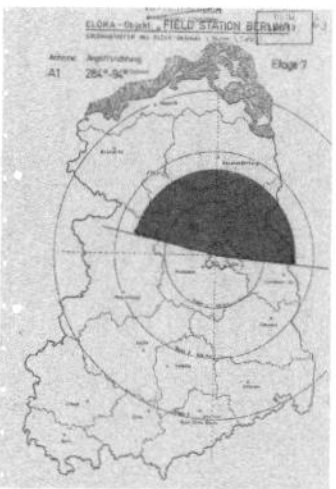

92, 93, 94, 95 → P. 126

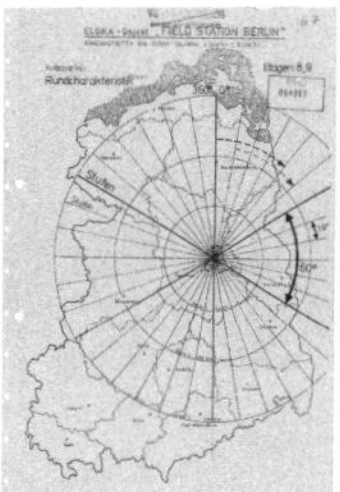

96 → P. 127

"Maximator: European Signals Intelligence Cooperation, from a Dutch Perspective," *Intelligence and National Security* 35, no. 5 (2020): 659–668.

117 "Analyse der Struktur und Tätigkeit," Bl. 1–68, here 4. This and subsequent quotes are translated by the author.

118 "Analyse der Struktur und Tätigkeit," 4.

119 "Analyse der Struktur und Tätigkeit," 55.

120 "Analyse der Struktur und Tätigkeit," 53.

121 "Analyse der Struktur und Tätigkeit," 56.

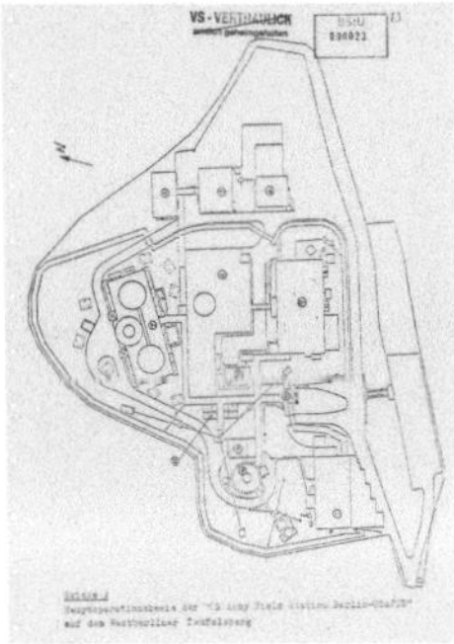

97 → P. 128

122 "Analyse der Struktur und Tätigkeit," 22.

123 "Analyse der Struktur und Tätigkeit," 25.

outwardly discernible logic. Its twelve different buildings operated semi-independently, each performing a specific function. (Fig. 97) On the eastern side of the complex, marked with the numbers 6, 7, and 8, were three smaller buildings housing, respectively, the technical equipment for the electric power station, the heating plant, and the electrical system that secured a constant supply for the electronic equipment throughout the complex. The three largest buildings were located along the north–south axis of the complex and marked with the numbers 1, 2, and 9. Building number 9 was allocated to the British Army (the station was officially in the British sector), while building number 2 housed those responsible for maintaining all of the station's electronic equipment as well as the subsystem named "PAPA," dedicated to detecting and deciphering Morse code. The central building (number 1) contained the subsystem for collecting and storing telecommunication information, as well as the office for filing and controlling classified documents and the department for information processing and analysis. The 45 m tower housing the complex section of antennas described above sat atop the three levels of offices and administrative program. Access to the station was from the western end of the complex: alongside the entrance, building number 5 contained administrative offices along with a conference room, dining room, and the headquarters of the S-3 unit in charge of the field station.[122] The facility functioned almost as a city within the city, employing more than one thousand soldiers—of whom three hundred were staff members, while another three hundred worked in Company A, signal intelligence (SIGINT), and three hundred worked in Company B, radio engineering and electronic intelligence (ELINT).[123] The Stasi identified the station as a nodal point of a global network of listening stations, referring to it by the acronym ELOKA, short for *Elektronische Kampfführung* (Electronic Warfare). With this example we see how the architecture of radio had produced another type of building, which also created its own centrality while organizing large swaths of surrounding territory, mostly by means of a routine architectural configuration.

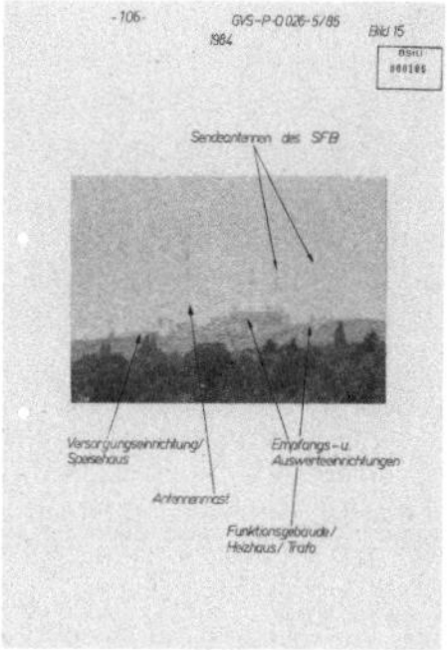

98 → P. 129

However it is not just the field station built by the US that is of interest here. Built on the ruins of the massive project designed by Albert Speer, shaped out of war rubble, piled high with countless fragments of other buildings from all over the city, then crowned with a radio station with antennas, radars, and facilities to house hundreds of soldiers, Teufelsberg, as a whole, is the material outcome of one of modernity's most extreme media-environmental conflicts. Combining trees, birds, and insects with a large signal-emitting and receiving infrastructure, we can understand Teufelsberg as a palimpsest of different layers of natural and technological elements. A constitutive element of Berlin's architecture of radio, the field station atop Teufelsberg was an important element of the city's skyline, its profile complemented by other radio towers, including those of Sender Freies Berlin (SFB). (Fig. 98) The sky above those buildings, far from being empty space, was occupied and delineated by the overlapping agencies of structures dedicated to the transmission or reception of electromagnetic waves.

By analyzing the way in which the territory was understood from the vantage point of radio, we see how the early 1950s were crucial in defining the institutional and infrastructural base of the newly divided architecture of radio in Berlin. It becomes apparent that the Iron Curtain, rather than strictly dividing the city, made possible reciprocal and symmetrical moves. On the one hand, we see how the radio stations of the West organized a territorial strategy to access the Eastern Bloc via their signals, with the specific and complex location of the city of Berlin playing a particularly important role. We see how the West's electromagnetic waves crossed the border without resistance, but also, and more relevantly, how that invisible and apparently immaterial mode of transgression was defined and structured by the construction of physical infrastructure all along the border. On the other hand, we see how the GDR designed its own electromagnetic territory by strategically locating structures to transmit signals within and beyond its borders. What was at stake, then, in this reciprocal battle of symmetrical moves on either side of the Iron Curtain? What was to be conquered and what was to be defended, if there was no territorial invasion, strictly speaking? Can we speak about two different forms of producing space that in fact coexist in the same territory—a fight over sovereignty relative to borders that are no longer clearly delineated, but rather temporally defined by another form of spatial production and dissemination?

"Has state socialism produced a space of its own?": this was the question Henri Lefebvre posed in 1974, in arguably one of the most compelling critiques ever written about space.[124] For Lefebvre, the evolving characteristics of "capitalist space" could be clearly defined: it was "abstract space, which includes the 'world of commodities,' its 'logic' and its worldwide strategies, as well as the power of money and that of the political state. This space is founded on the vast network of banks, business centres, major productive entities, as also on motorways, airports and lattices. Within this space the town—once the forcing-house of accumulation, fountainhead of wealth and centre of historical space—has disintegrated."[125] By comparison, the nature of "socialism's space" proved more elusive, but was nonetheless central for his analysis, as he famously stated that "A revolution that does not produce a new space has not realized its full potential; indeed it has failed in that it has not changed life itself, but has merely changed ideological superstructures, institutions or political apparatuses. A social transformation, to be truly revolutionary in character, must manifest a creative capacity in its effects on daily life, on language and on space—though its impact need not occur at the same rate, or with equal force, in each of these areas."[126]

Beginning his analysis, Lefebvre immediately invokes Berlin, asking: "Where can an architectural production be found today that might be described as 'socialist'—

124 Henri Lefebvre, *The Production of Space* (Cambridge: Blackwell, 1991), 54. The same question has previously been problematized by Philip Ursprung in his essay, "In Search of Socialist Space," in *Historiografía Performativa*, ed. Philip Ursprung (Santiago: Ediciones ARQ, 2018), 102–127.

125 Lefebvre, *The Production of Space*, 53.

126 Lefebvre, *The Production of Space*, 54.

or even as new when contrasted with the corresponding efforts of capitalist planning? In the former Stalinallee, East Berlin—now renamed Karl-Marx-Allee? In Cuba, Moscow or Peking?"[127] Lefebvre acknowledges that large and monumental interventions in the urban realm are not sufficient to counteract the hegemony of the capitalist mode of production, "which weighs down so heavily upon the whole planet, imposing its division of labour on a worldwide scale."[128] "There is no easy affirmative answer to the question of 'socialism's space,'" Lefebvre avers, but later in his analysis, we find hints of an answer framed in the form of a question—one that is extremely relevant to the material examined here. "Ought we to look upon architectural or urbanistic works as a type of mass medium?" Lefebvre asks.[129] In the context of this study, that question could be turned around, to ask whether radio, as both a technical medium and mass media, could be understood as a type of architecture. Reframing the question in this way, we can perhaps begin to determine how and where "socialist space" was constructed.

In previous studies, it has been argued that the construction of a socialist idea of space was at the core of the socialist program—a project in opposition to the overwhelming presence of the rival project of capitalist space. Traditionally, this kind of space is associated with an almost megalomaniac approach towards urbanity and the built environment: large avenues for parades, representational buildings to host new institutions, and the construction of buildings for new programs oriented towards a wider population.[130] Common references from East Berlin include Stalinallee, Kino International, and the Palast der Republik.[131] But what is clear from the reciprocal and symmetrical territorial and institutional reconfiguration of Germany after 1952 is that the architecture of radio was fully committed to shaping subjectivity by accessing domestic spaces on both sides of a divided country.[132] In both East and West, the architecture of radio was devised to enter and access the everyday life of the inhabitants.

127 Lefebvre, *The Production of Space,* 54.

128 Lefebvre, *The Production of Space,* 54.

129 Lefebvre, *The Production of Space,* 131.

130 David Crowley and Susan E. Reid, eds., *Socialist Spaces: Sites of Everyday Life in the Eastern Bloc* (Oxford: Berg, 2002).

131 For a historical account of the built environment in East Berlin after 1945, see Andreas Butter and Ulrich Hartung, *Ostmoderne: Architektur in Berlin 1945–1965* (Berlin: Jovis, 2004). For a recent study on the dialectical aspect of the city of Berlin, see Arno Brandelhuber, Florian Hertweck, and Thomas Mayfried, eds., *The Dialogic City: Berlin wird Berlin* (Cologne: Walther König, 2015). See also Anders Åman, *Architecture and Ideology in Eastern Europe during the Stalin Era: An Aspect of Cold War History* (Cambridge, MA: MIT Press, 1992).

132 Henri Lefebvre, *Critique of Everyday Life* (London: Verso, 2014); Michel de Certeau, *The Practice of Everyday Life* (Berkeley: University of California Press, 1984).

Recent studies have argued that the space of everyday life was just as important in constructing the socialist ideal as it was for the capitalist model. The medium of radio, however, tends to blur the familiar dividing line between domesticity and monumentality, between private and public realms.[133] As we have seen, the dissemination of content at the speed of light across physical and political borders shifts the traditional understanding of centrality and periphery, turning the domestic realm into an urban, infrastructural, and territorial problem.[134] It can be concluded that the West understood very well how radio could be used to penetrate the domestic space of the inhabitants of the Eastern Bloc, and thereby access this extremely fragile, and at the same time fundamental, locus of subjectivity construction and political power. Simultaneously, the East understood that radio was subversively infiltrating its territory, but recognized the potential to reverse the flow and to use those frequencies to expand its "spatial construction" beyond the limits demarcated by the Iron Curtain. By turning the domestic realm and the space of everyday life into a contested terrain, radio destabilized the traditional distinction between private and public, between urban and domestic, between periphery and centrality. The territory sealed in by the Berlin Wall and the Iron Curtain was transgressed in both directions by the medium of radio. Consequently, rather than disappearing altogether, the architectural element of the wall took on a much more relevant and complex agency. The architecture of radio in Berlin became an intrinsically trans-scalar construction: a complex aesthetic, technological, political, and social construction that made monumentality and everyday life fold into each other.

133 In 1960, American political scientist Erich Goldhagen wrote an interesting report on how the idea of *progress* was being conceived in socialist countries. Calling it "the discipline of leisure," Golhagen spoke precisely about the intentions to merge private and public realms in socialist countries and about the use of certain technologies for achieving that purpose. See Erich Goldhagen, "The Glorious Future—Realities and Chimeras," *Problems of Communism* 9, no. 6 (November 1960): 10–18.

134 Important for these reflections has been the study of German philosopher Rahel Jaeggi, *On the Critique of Forms of Life* (Cambridge, MA: Harvard University Press, 2018).

PART 2
THE BROADCASTING HOUSE SOURCE MATERIAL

99

100

37

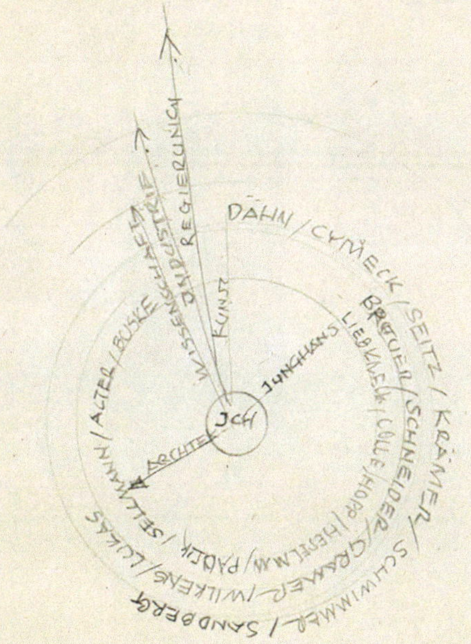

entgegengenommen am 5.VI.57
durch Oblt. Hettwer von GI
Neumann.

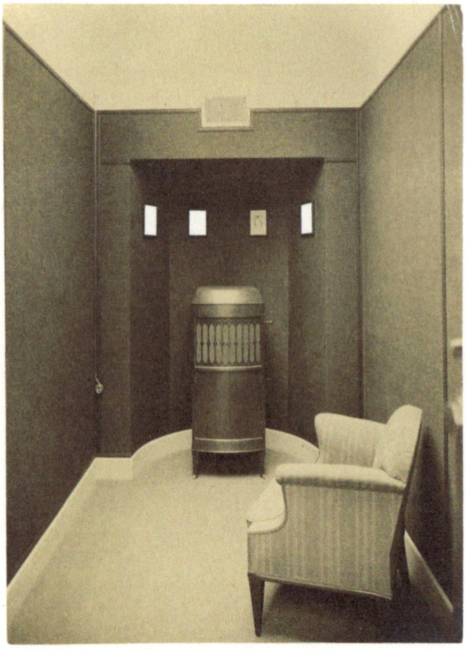

102

103

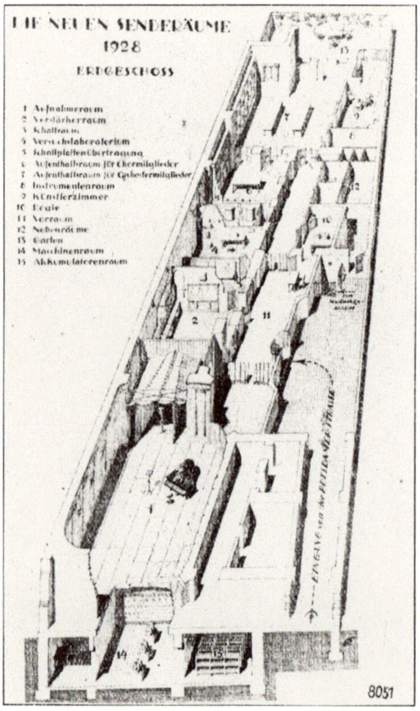

104

105

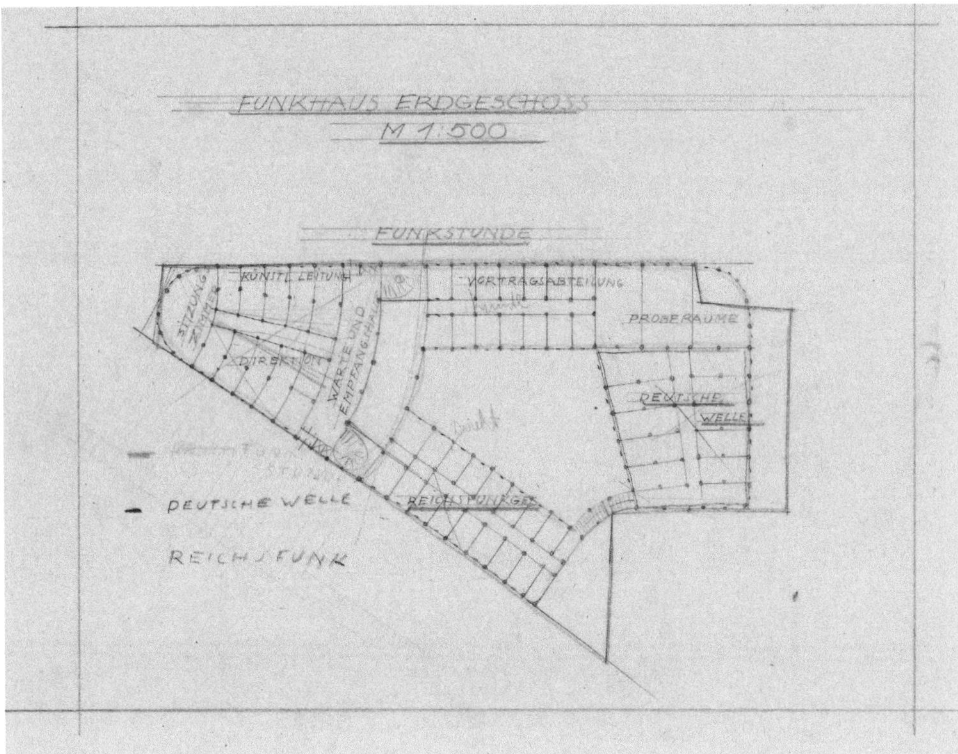
FUNKHAUS ERDGESCHOSS
M 1:500
FUNKSTUNDE
KÜNITL LEITUNG
VORTRAGSABTEILUNG
PROBERÄUME
DIREKTION
WARTE UND EMPFANGSHALLE
DEUTSCHE WELLE
REICHSFUNKGES
DEUTSCHE WELLE
REICHSFUNK

106

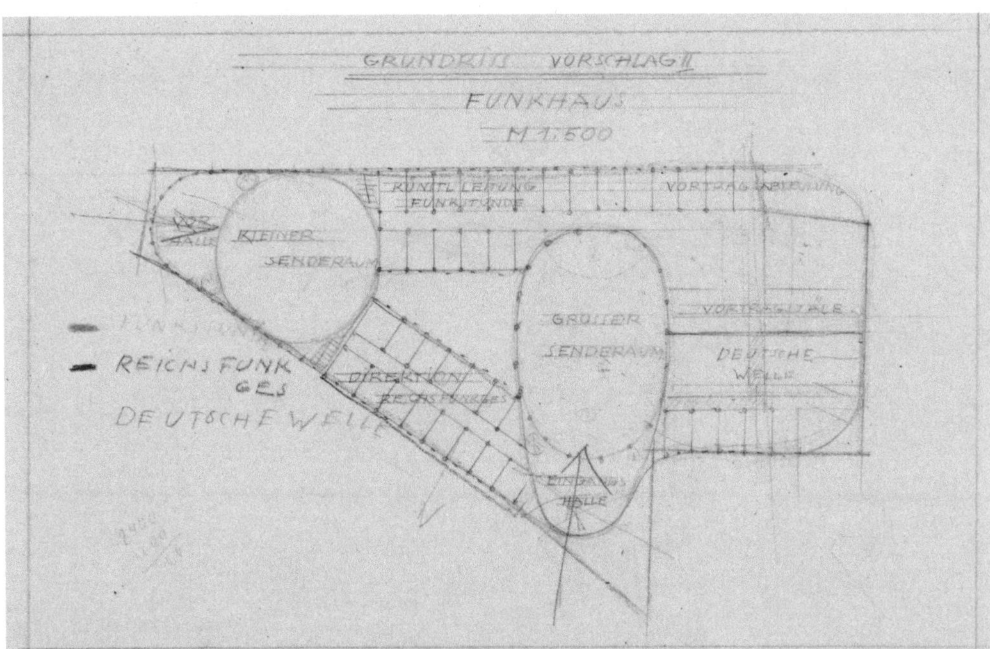
GRUNDRISS VORSCHLAG II
FUNKHAUS
M 1:500
KÜNITL LEITUNG
FUNKSTUNDE
VORTRAGSABTEILUNG
KLEINER SENDERAUM
GROSSER SENDERAUM
VORTRAGSSÄLE
DEUTSCHE WELLE
DIREKTION
REICHSFUNKGES
EINGANGSHALLE
REICHSFUNK GES
DEUTSCHE WELLE

107

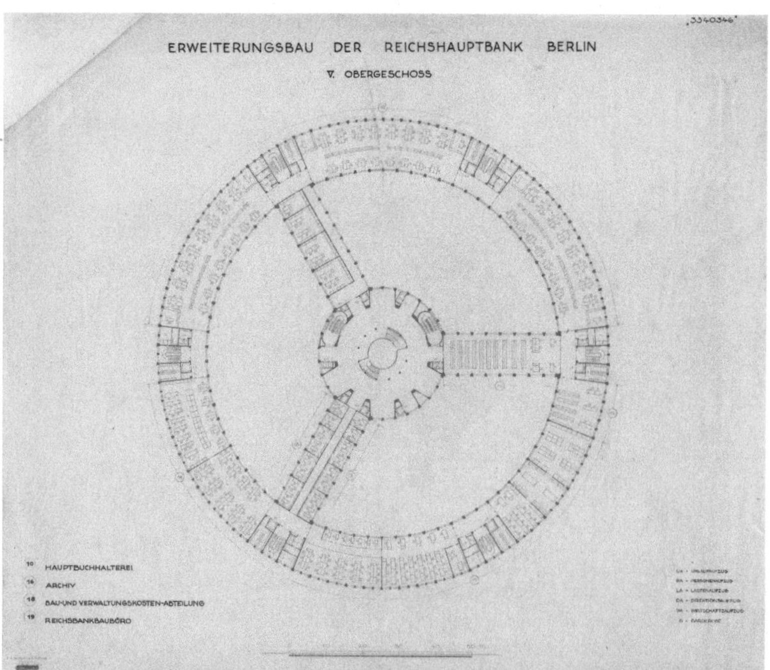
„3340346"
ERWEITERUNGSBAU DER REICHSHAUPTBANK BERLIN
V. OBERGESCHOSS
10 HAUPTBUCHHALTEREI
16 ARCHIV
18 BAU-UND VERWALTUNGSKOSTEN-ABTEILUNG
19 REICHSBANKBAUBÜRO

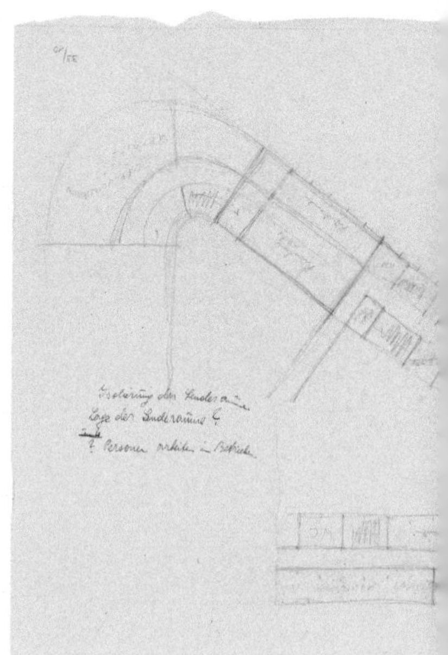

109

BAUATELIER POELZIG
BERLIN-CHARLOTTENBURG 2
HARDENBERGSTRASSE 33

108

110

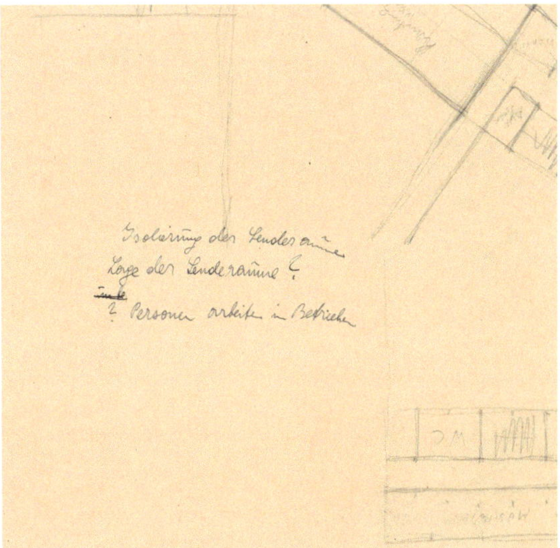
Isolierung der Senderäume
Lage der Senderäume?
? Personen arbeiten in Betrieben

111

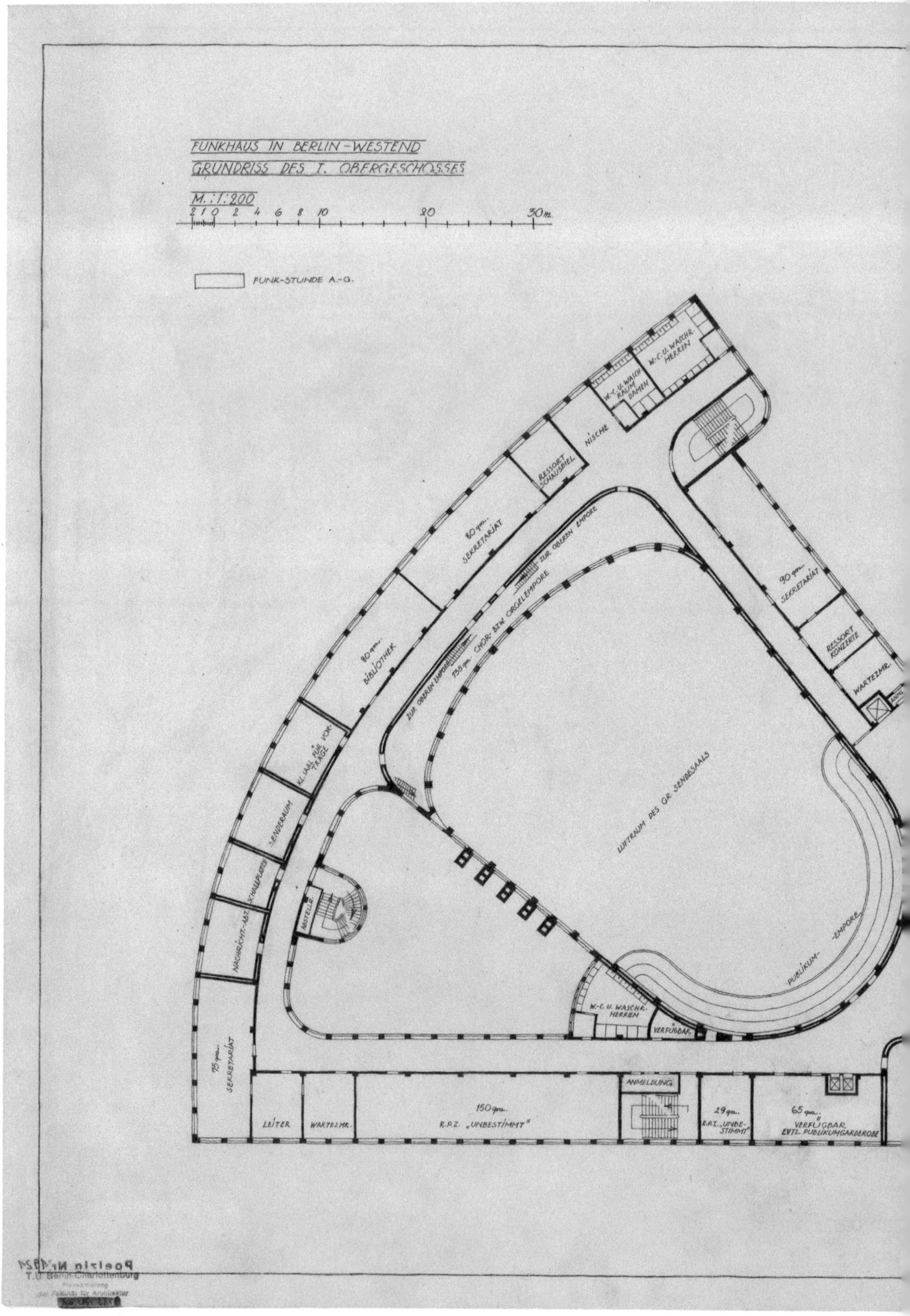
FUNKHAUS IN BERLIN-WESTEND
GRUNDRISS DES I. OBERGESCHOSSES
M.: 1:200
2 1 0 2 4 6 8 10 20 30m.
FUNK-STUNDE A.-G.
W.-C. U. WASCHRAUM DAMEN
W.-C. U. WASCHR. HERREN
NISCHE
RESSORT SCHAUSPIEL
80 qm. SEKRETARIAT
ZUR OBEREN EMPORE
135 qm. CHOR- BZW. ORGELEMPORE
80 qm. BIBLIOTHEK
KLIAAL FÜR VORTRÄGE
SENDERAUM
SCHALLPLATTE
NACHRICHT.-ABT.
BESTELLR.
90 qm. SEKRETARIAT
RESSORT KONZERTE
WARTEZMR.
LUFTRAUM DES GR. SENDESAALS
PUBLIKUM- -EMPORE
W.-C. U. WASCHR. HERREN
VERFÜGBAR
75 qm. SEKRETARIAT
ANMELDUNG
LEITER
WARTEZMR.
150 qm. R.P.Z. „UNBESTIMMT"
29 qm. R.P.Z. „UNBESTIMMT"
65 qm. „VERFÜGBAR EVTL. PUBLIKUMGARDEROBE
Poelzig Nr. 1824
T.U. Berlin-Charlottenburg

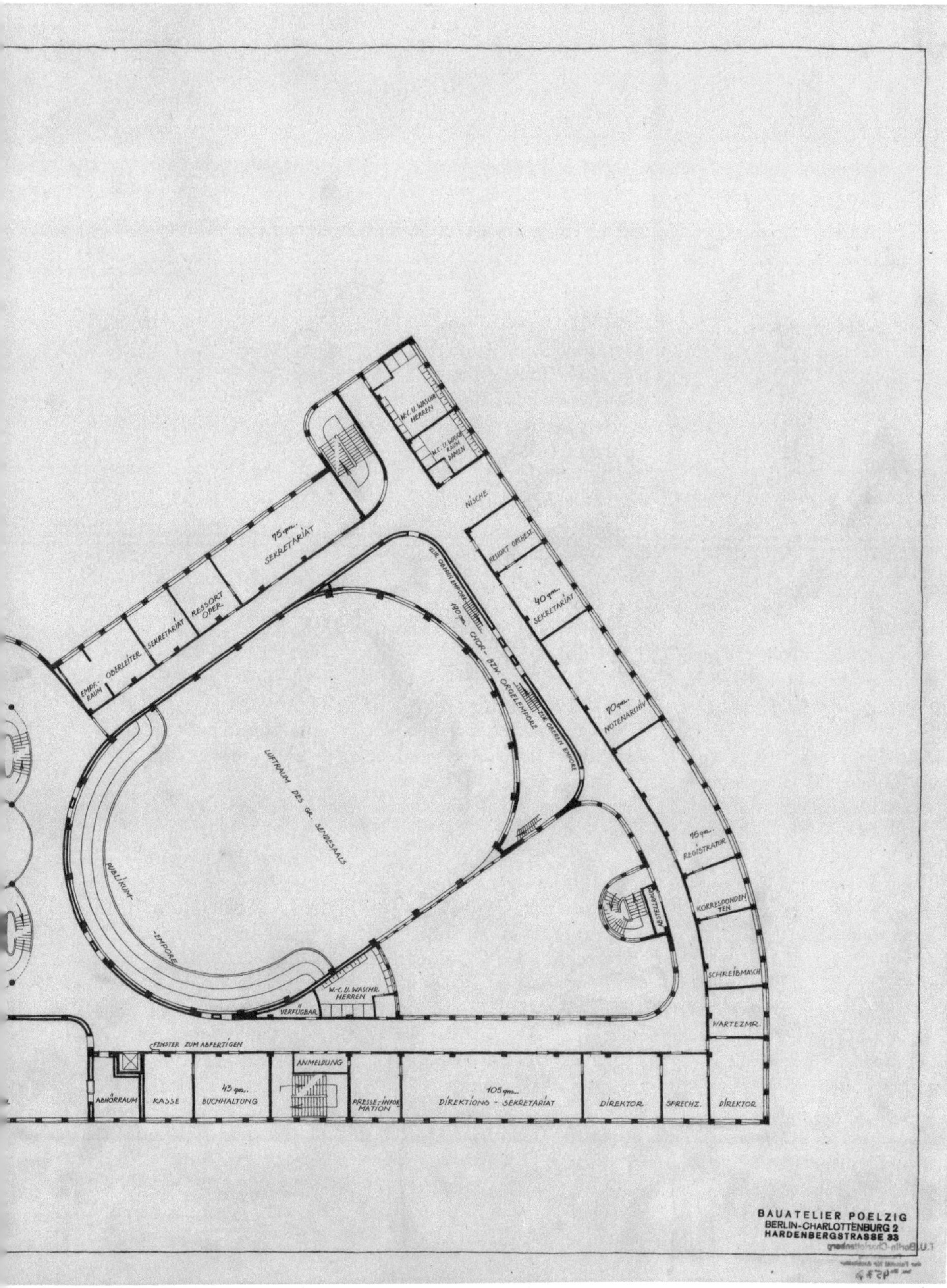
W.-C. U. WASCHR. HERREN
W.-C. U. WASCHR. RAUM DAMEN
NISCHE
75 qm. SEKRETARIAT
RESSORT OPER
SEKRETARIAT
OBERLEITER
EMPF.-RAUM
RESSORT ORCHEST.
40 qm. SEKRETARIAT
190 qm. CHOR- BZW. ORGELEMPORE
ZUR OBEREN EMPORE
90 qm. NOTENARCHIV
LUFTRAUM DES GR. SENDESAALS
PUBLIKUM-EMPORE
15 qm. REGISTRATUR
KORRESPONDENTEN
SCHREIBMASCH.
WARTEZMR.
W.-C. U. WASCHR. HERREN
VERFÜGBAR
FENSTER ZUM ABFERTIGEN
ANMELDUNG
ABHÖRRAUM
KASSE
43 qm. BUCHHALTUNG
PRESSE-INFORMATION
105 qm. DIREKTIONS-SEKRETARIAT
DIREKTOR
SPRECHZ.
DIREKTOR
BAUATELIER POELZIG
BERLIN-CHARLOTTENBURG 2
HARDENBERGSTRASSE 33

112

Fig. 99 A Stasi officer interrogates a political prisoner in a scene from *Das Leben der Anderen* set inside the Funkhaus Berlin-Nalepastrasse. Source: *Das Leben der Anderen / The Lives of Others*. Directed by Florian Henckel von Donnersmarck. Germany: Wiedemann & Berg Film, 2006.

Fig. 100 Funkhaus Berlin-Nalepastrasse, view from across the River Spree. Source: Bundesarchiv. Bild 183-J0701-0301-001.

Fig. 101 Diagram drawn by Franz Ehrlich depicting himself as a central actor within the artistic, architectural, scientific, and political circles of the GDR. Source: Der Bundesbeauftragte für die Unterlagen des Staatssicherheitsdienstes der ehemaligen Deutschen Demokratischen Republik (BStU), MfS AIM 6503/75, Band 1.

Fig. 102 Vox-Haus on Potsdamerstrasse, interior view of renovation. Source: gta Archiv, ETH Zurich. Otto Rudolf Salvisberg: Vox-Haus: Umbau 24-055 F.4.

Fig. 103 Vox-Haus on Potsdamerstrasse, interior view of renovation. Architect: Otto Rudolf Salvisberg. Source: gta Archiv, ETH Zurich. Otto Rudolf Salvisberg: Vox-Haus: Umbau 24-055 F.2.

Fig. 104 Vox-Haus on Potsdamerstrasse, perspective of renovated building. Source: Ludwig Kapeller, "Die neue Senderäume der Berliner Funk-Stunde," *Funk* 38 (1928): 269.

Fig. 105 Vox-Haus on Potsdamerstrasse, interior view of renovation. Source: gta Archiv, ETH Zurich. Otto Rudolf Salvisberg: Vox-Haus: Umbau 24-055 F.3.

Fig. 106 Haus des Rundfunks, schematic plan from competition entry. Source: TU Berlin Architekturmuseum, Inv. Nr. HP 039, 002.

Fig. 107 Haus des Rundfunks, schematic plan from competition entry. Source: TU Berlin Architekturmuseum, Inv. Nr. HP 039, 003.

Fig. 108 Haus des Rundfunks seen from Masurenallee, perspective from competition entry. Source: TU Berlin Architekturmuseum, Inv. Nr. 4586.

Fig. 109 Reichshauptbank in Berlin-Mitte, plan from Hans Poelzig's competition entry. Source: TU Berlin Architekturmuseum, Inv. Nr. 5217.

Fig. 110 Haus des Rundfunks, schematic plan from competition entry. Source: TU Berlin Architekturmuseum, Inv. Nr. HP 039, 010.

Fig. 111 Haus des Rundfunks, detail of a schematic plan from competition entry. The caption reads "Isolierung der Senderäume?" (Isolate the recording studios?). Source: TU Berlin Architekturmuseum, Inv. Nr. HP 039, 010.

Fig. 112 Haus des Rundfunks, level 1 plan from final competition scheme. Source: TU Berlin Architekturmuseum, Inv. Nr. 4577.

113

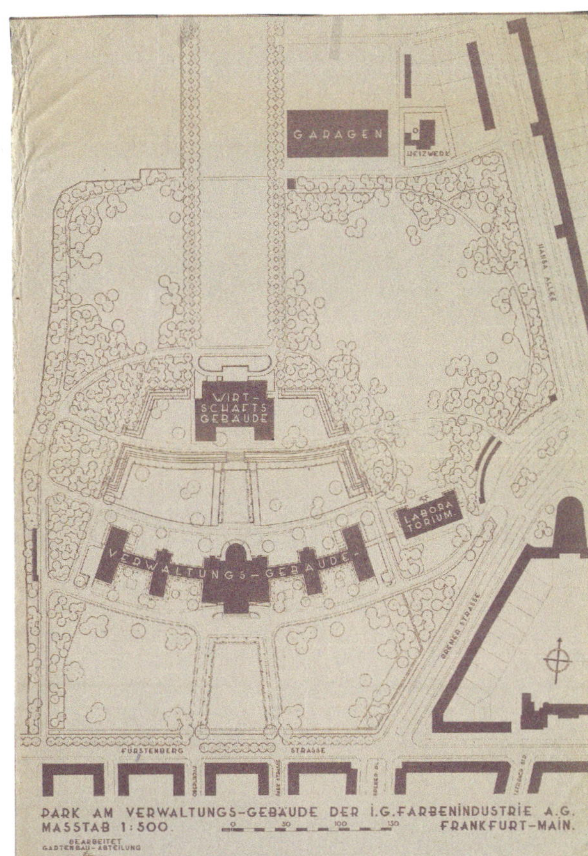
GARAGEN
WIRT-
SCHAFTS
GEBÄUDE
LABORA
TORIUM.
VERWALTUNGS-GEBÄUDE-
FÜRSTENBERG
STRASSE
PARK AM VERWALTUNGS-GEBÄUDE DER I.G.FARBENINDUSTRIE A.G.
MASSTAB 1:500.
0
50
100
150
FRANKFURT-MAIN.
BEARBEITET
GARTENBAU-ABTEILUNG

114

115

116

117

118

119

120

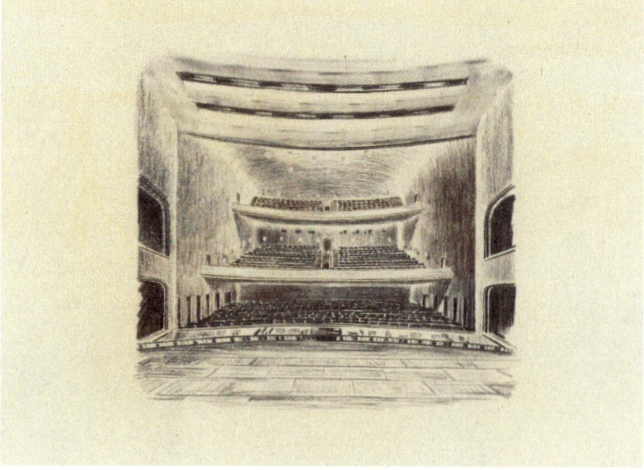

121

122

GROSSES SCHAUSPIELHAUS BERLIN 1919
Kuppel im Zuschauerraum

123

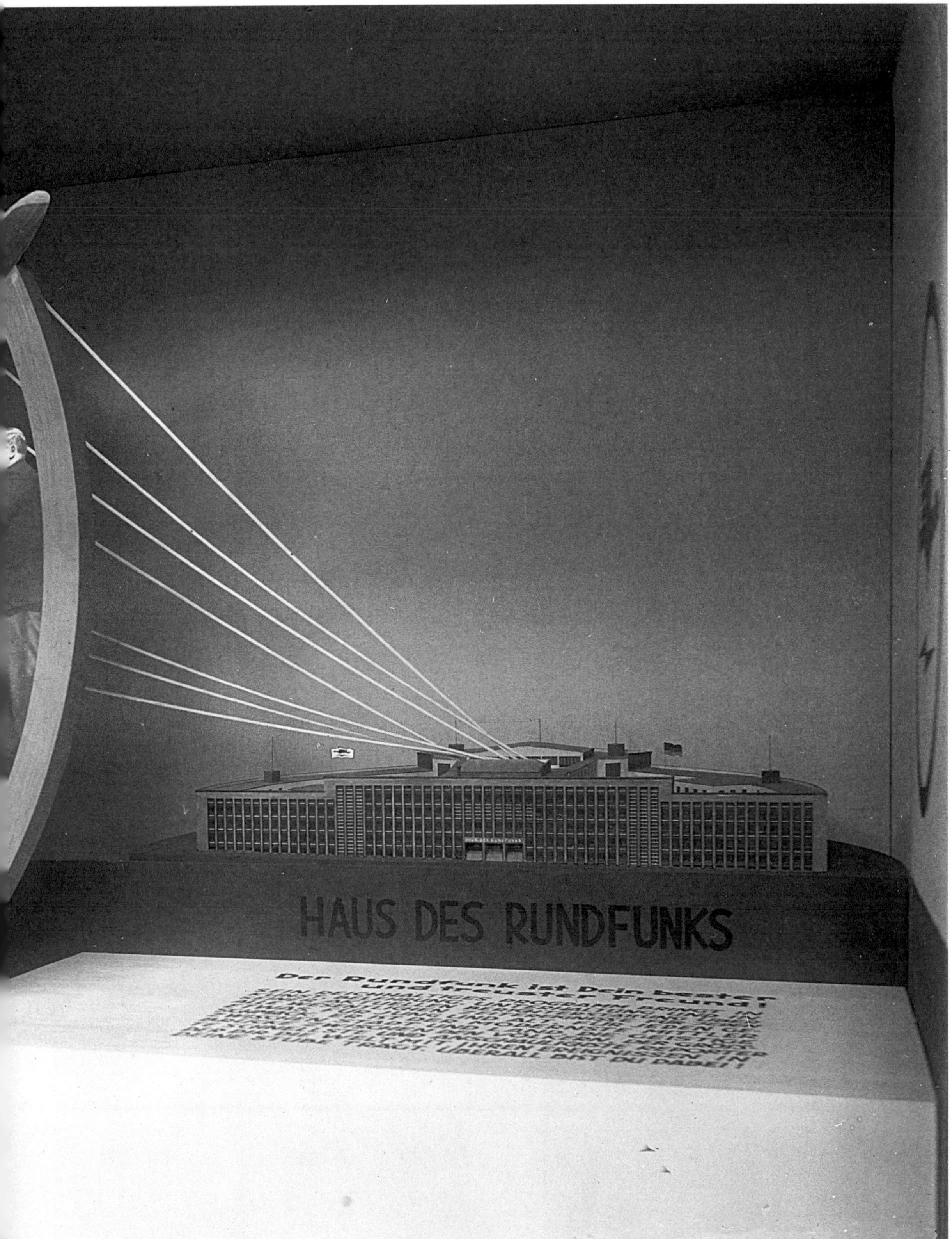
HAUS DES RUNDFUNKS
Der Rundfunk ist Dein bester und treuster Freund!
ÜBERALL BIST DU DABEI!

124

125

126

127

Fig. 113 Milch & Co. chemical factory, Luban, side elevation. Architect: Hans Poelzig. Source: TU Berlin Architekturmuseum, Inv. Nr. 2640.

Fig. 114 IG Farben headquarters, Frankfurt am Main, site plan. Architect: Hans Poelzig. Source: TU Berlin Architekturmuseum, Inv. Nr. 3923.

Fig. 115 Cassirer cable works, Spandau, Berlin, perspective view. Architect: Hans Poelzig. Source: TU Berlin Architekturmuseum, Inv. Nr. 3773.

Fig. 116 Kino Babylon, Berlin, interior view. Architect: Hans Poelzig. The cinema was designed as part of a larger housing complex in Berlin-Mitte. Source: TU Berlin Architekturmuseum, Inv. Nr. 1789.

Fig. 117 Capitol am Zoo, Berlin, interior view. Architect: Hans Poelzig. Source: TU Berlin Architekturmuseum, Inv. Nr. 3188.

Fig. 118 Deli-Kino, Breslau, interior view. Architect: Hans Poelzig. Source: TU Berlin Architekturmuseum, Inv. Nr. 1730.

Fig. 119 View of the watchtower designed by Hans Poelzig as part of the scenography for *The Golem: How He Came into the World*, directed by Paul Wegener and Carl Bosse. Germany: PAGU. 1920. Source: TU Berlin Architekturmuseum, Inv. Nr. 1629.

Fig. 120 View of Hans Poelzig's scenography for *Living Buddhas*, directed by Paul Wegener. Germany: Wegener-Film. 1925. Source: TU Berlin Architekturmuseum, Inv. Nr. 1638.

Fig. 121 Friedrich-Theater, Dessau, interior sketch. Architect: Hans Poelzig. Source: TU Berlin Architekturmuseum, Inv. Nr. HP 047, 002.

Fig. 122 Festspielhaus, Salzburg, interior view. Designed by Hans Poelzig for Max Reinhardt. Source: TU Berlin Architekturmuseum, Inv. Nr. 2777.

Fig. 123 Grosses Schauspielhaus, Berlin, interior view. Renovated by Hans Poelzig for Max Reinhardt. Source: TU Berlin Architekturmuseum, Inv. Nr. 2753.

Fig. 124 Photograph of a model of the recently competed Haus des Rundfunks presented at the Radio Exhibition in Berlin in 1931. Source: Bundesarchiv. Bild 102-12189.

Fig. 125 Sketch of the stage design for Bernhard Reich's adaptation of Shakespeare's *King Lear*. Source: Hans Poelzig, *Der dramatische Raum: Hans Poelzig, Malerei, Theater, Film* (Krefeld: Museum Haus Lange, 1986), 86.

Fig. 126 Sketch of the stage design for Ludwig Hoerth's adaptation of Mozart's opera *Don Giovanni*. Source: Hans Poelzig, *Der dramatische Raum: Hans Poelzig, Malerei, Theater, Film* (Krefeld: Museum Haus Lange, 1986), 81.

Fig. 127 Sketch of the stage design for Gustav Hartung's adaptation of Shakespeare's *Midsummer Night's Dream*. Source: Hans Poelzig, *Der dramatische Raum: Hans Poelzig, Malerei, Theater, Film* (Krefeld: Museum Haus Lange, 1986), 88.

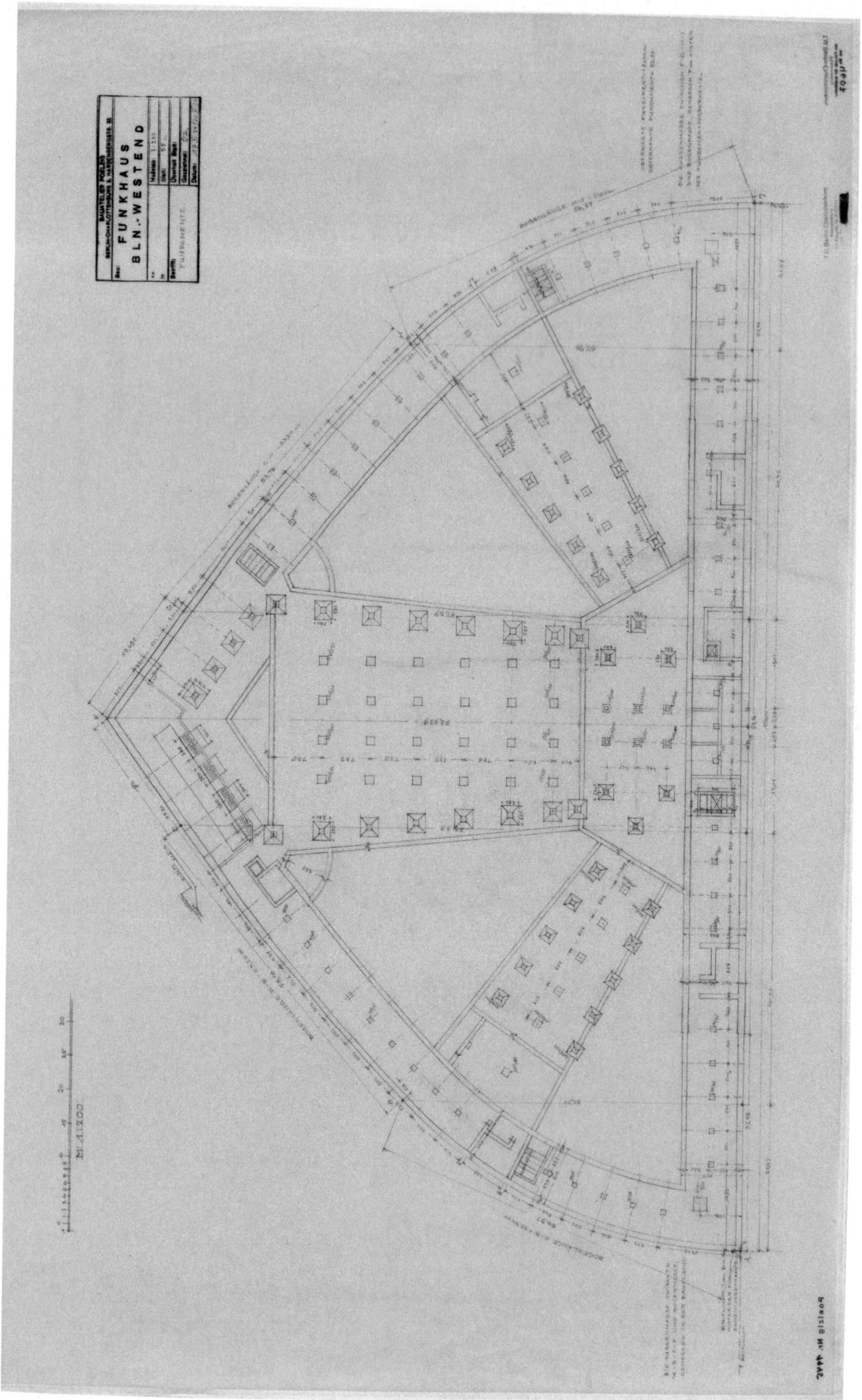

128

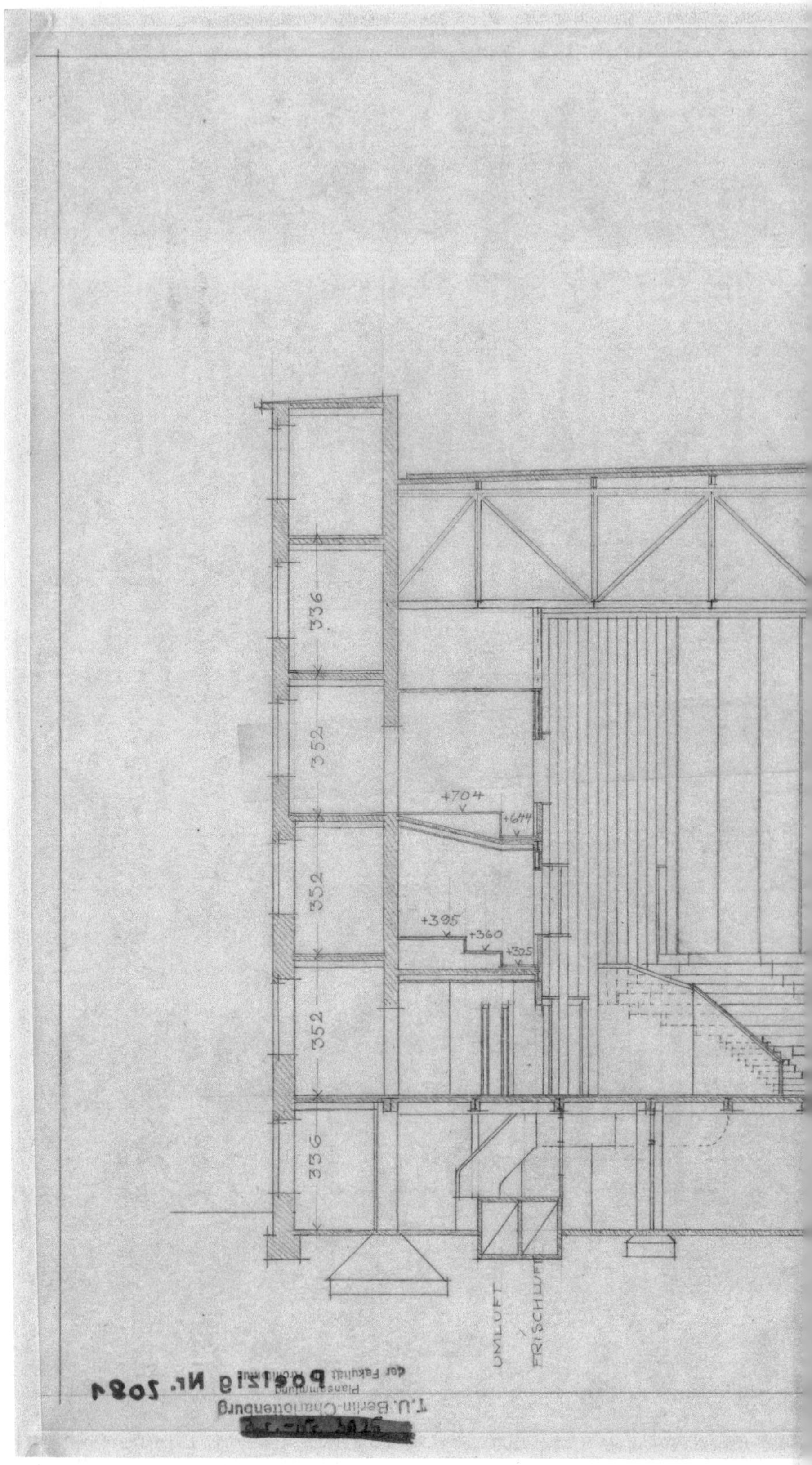
336
352
+704
+644
352
+395
+360
+325
352
336
UMLUFT
FRISCHLUFT
Poelzig Nr. 2084
T.U. Berlin-Charlottenburg

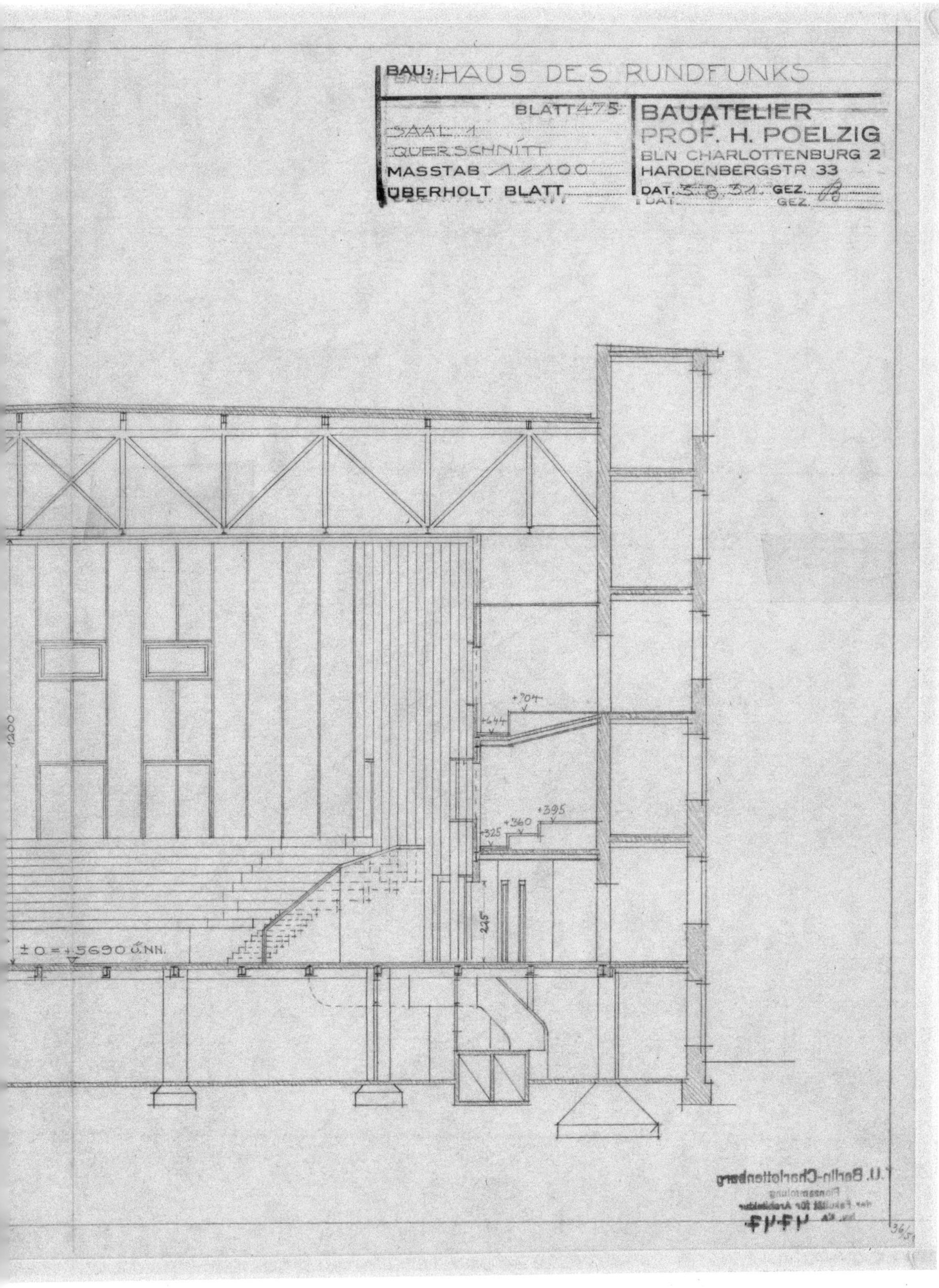
BAU: HAUS DES RUNDFUNKS
BLATT 475
SAAL 1
QUERSCHNITT
MASSTAB 1:100
ÜBERHOLT BLATT
BAUATELIER
PROF. H. POELZIG
BLN CHARLOTTENBURG 2
HARDENBERGSTR 33
DAT. 3.8.31. GEZ.
+704
+644
+395
+360
+325
225
1200
±0=+5690 ü.NN.
T.U. Berlin-Charlottenburg

130

131

132

133

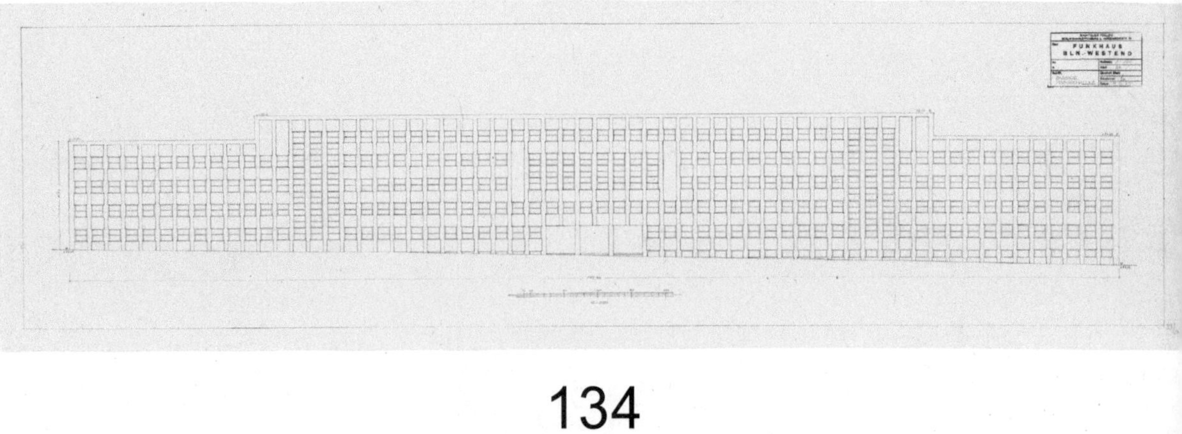

134

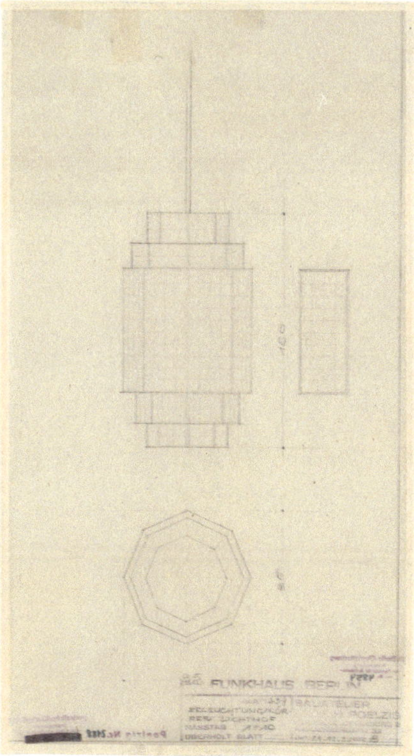

135

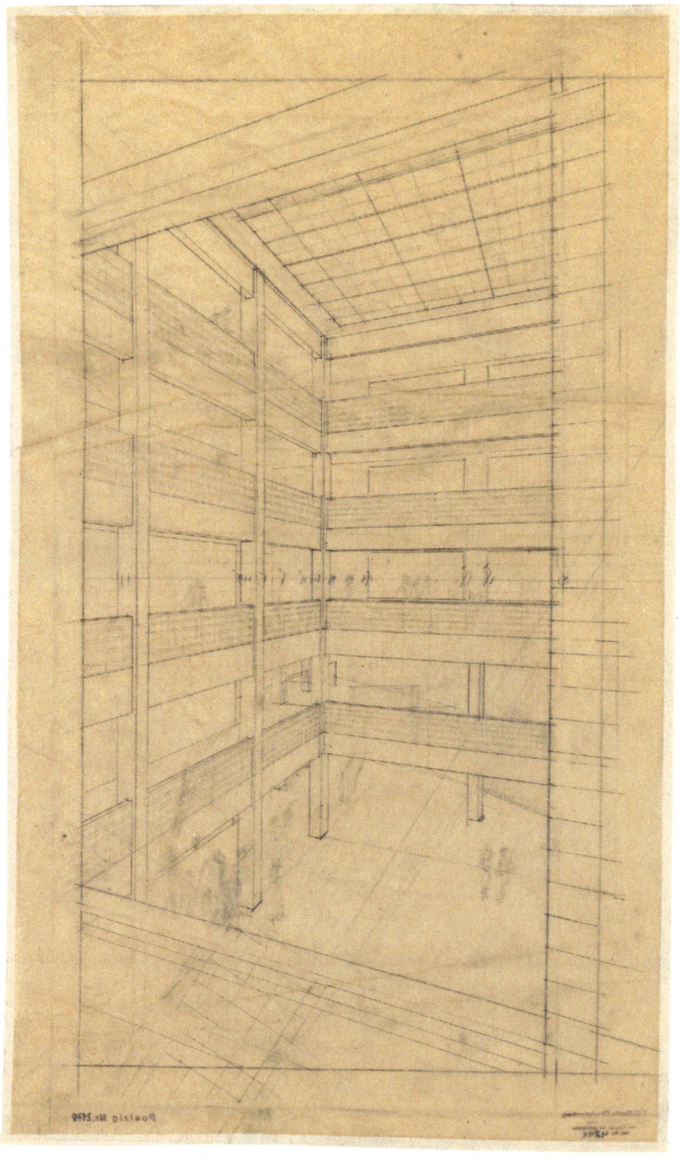

136

137

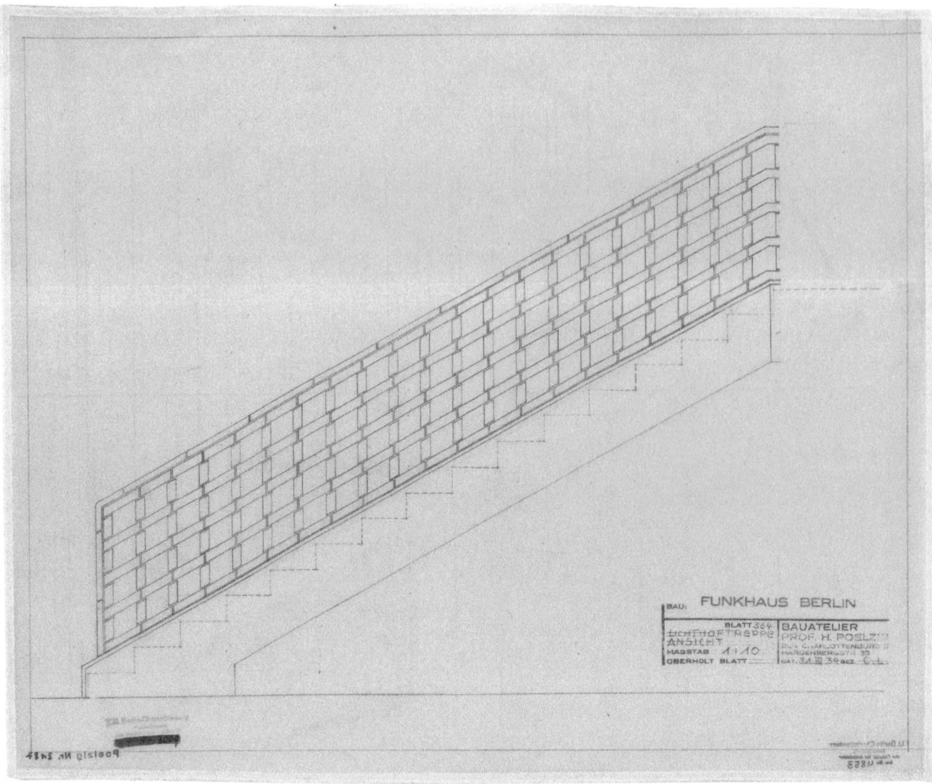
BAU: FUNKHAUS BERLIN
LICHTHOFTREPPE
ANSICHT
MASSTAB 1:10
BAUATELIER
PROF. H. POELZIG

138

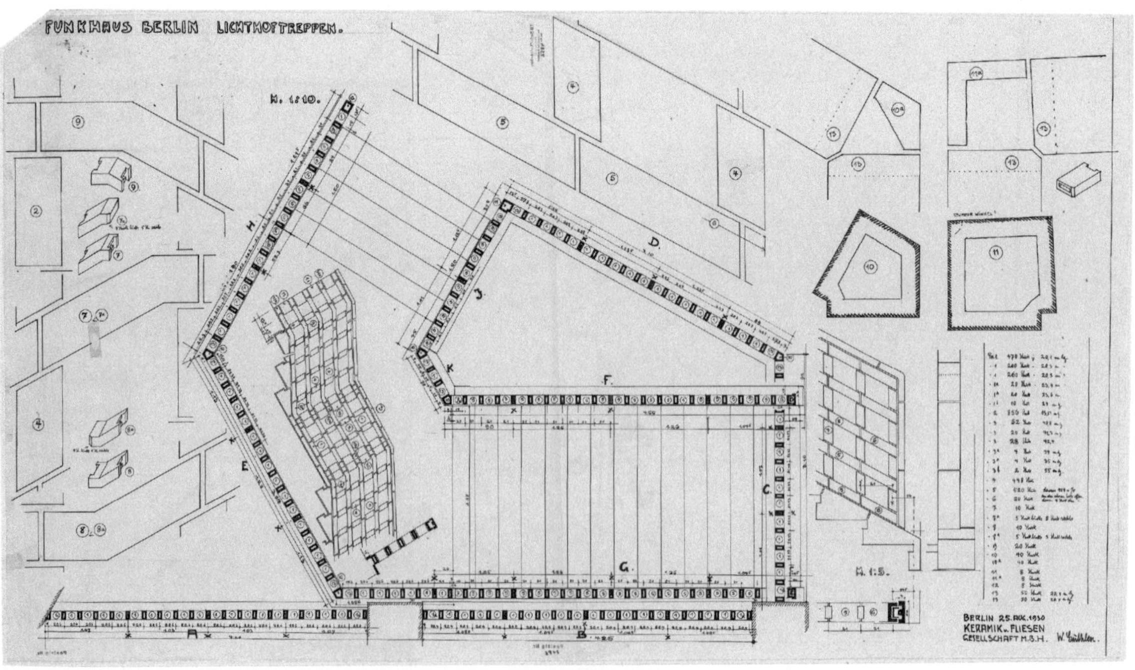
FUNKHAUS BERLIN LICHTHOFTREPPEN.
M. 1:10.
M. 1:5.
BERLIN 25. AUG. 1930
KERAMIK- u. FLIESEN
GESELLSCHAFT M.B.H.

139

140

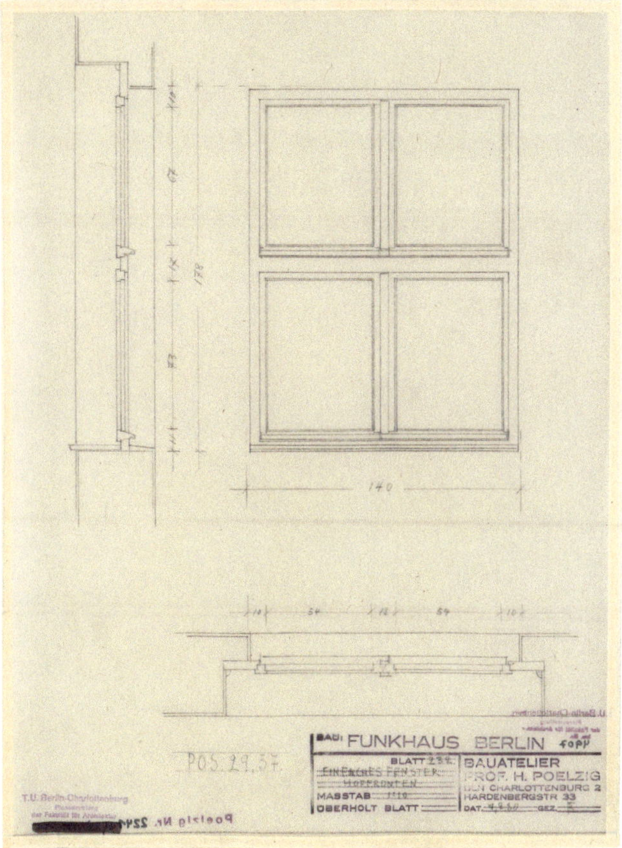
140
POS. 29.57
BAU: FUNKHAUS BERLIN
BAUATELIER
PROF. H. POELZIG
HARDENBERGSTR 33
MASSTAB
ÜBERHOLT BLATT
T.U. Berlin-Charlottenburg

141

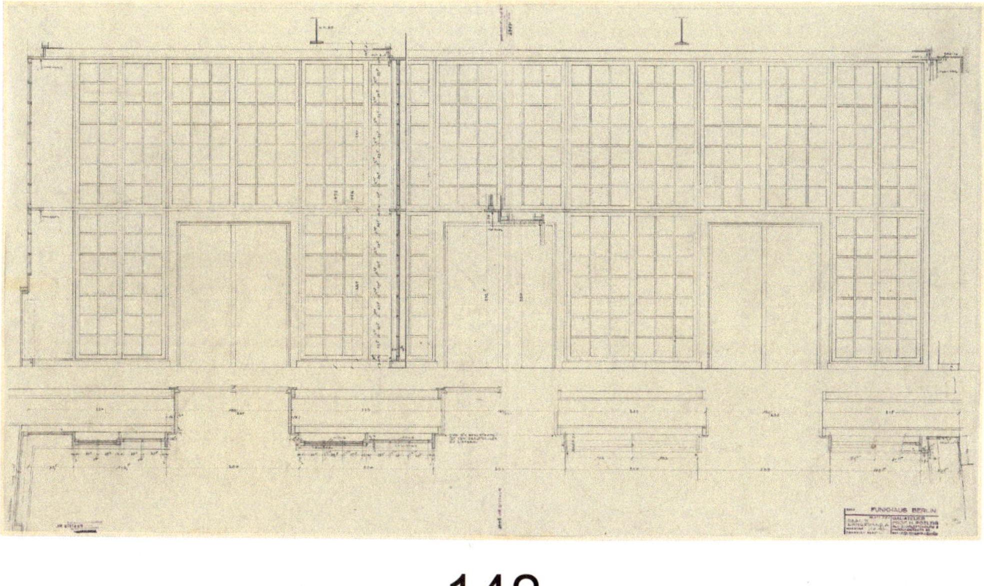
FUNKHAUS BERLIN

142

143

Fig. 128 Haus des Rundfunks, foundation plan. Source: TU Berlin Architekturmuseum, Inv. Nr. 4607.

Fig. 129 Haus des Rundfunks, plan of level 1 during the design development phase. Source: TU Berlin Architekturmuseum, Inv. Nr. 4620.

Fig. 130 Haus des Rundfunks, cross section through main recording studio. Source: TU Berlin Architekturmuseum, Inv. Nr. 4747.

Fig. 131 Haus des Rundfunks, interior view of a recording studio showing the movable and folding panels. Source: Walter Schäffer, "Raumakustik für Rundfunk-Darbietungen," *Funktechnische Monatshefte* 2 (1932): 53.

Fig. 132 Inaugural concert of the Haus des Rundfunks in 1931. Photograph by Georg Pahl. Source: Bundesarchiv. Bild 102-11093.

Fig. 133 Haus des Rundfunks, perspective from Masurenallee. Source: TU Berlin Architekturmuseum, Inv. Nr. 4913.

Fig. 134 Haus des Rundfunks, Masurenallee elevation. Source: TU Berlin Architekturmuseum, Inv. Nr. HP 039, 028.

Fig. 135 Haus des Rundfunks, detail of lamps designed by Hans Poelzig. Source: TU Berlin Architekturmuseum, Inv. Nr. 4854.

Fig. 136 Haus des Rundfunks, interior perspective of central hall. Source: TU Berlin Architekturmuseum, Inv. Nr. 4844.

Fig. 137 Haus des Rundfunks, interior view of central hall. Photograph by the author.

Fig. 138 Haus des Rundfunks, staircase detail. Source: TU Berlin Architekturmuseum, Inv. Nr. 4853.

Fig. 139 Haus des Rundfunks, staircase detail in plan, section, perspective, and elevation. Source: TU Berlin Architekturmuseum, Inv. Nr. 4852.

Fig. 140 Haus des Rundfunks, view of a staircase. Photograph by the author.

Fig. 141 Haus des Rundfunks, window details. Source: TU Berlin Architekturmuseum, Inv. Nr. 4907.

Fig. 142 Haus des Rundfunks, detail of interior elevation. Source: TU Berlin Architekturmuseum, Inv. Nr. 4782.

Fig. 143 Haus des Rundfunks, interior perspective of main recording studio. Source: TU Berlin Architekturmuseum, Inv. Nr. HP 039, 034.

144

145

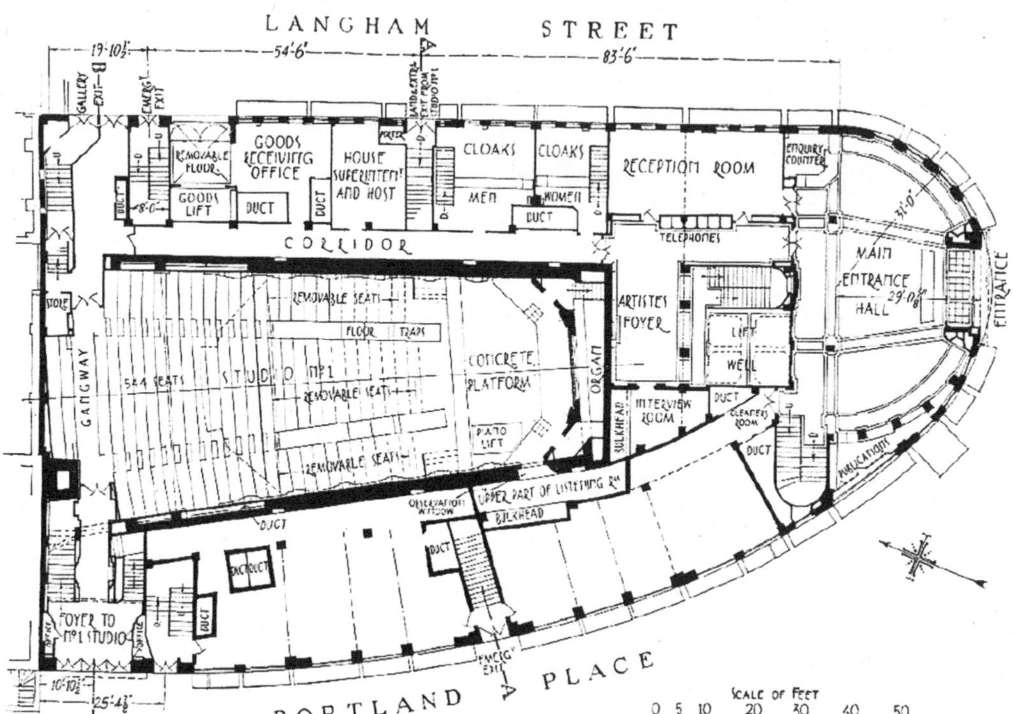
LANGHAM STREET
PORTLAND PLACE
GALLERY EXIT
EMERG^Y EXIT
REMOVABLE FLOOR
GOODS RECEIVING OFFICE
HOUSE SUPERINTENT AND HOST
GOODS LIFT
DUCT
CLOAKS
CLOAKS
MEN
WOMEN
RECEPTION ROOM
ENQUIRY COUNTER
TELEPHONES
CORRIDOR
MAIN ENTRANCE HALL
ENTRANCE
REMOVABLE SEATS
FLOOR TRAPS
544 SEATS
STUDIO Nº 1
CONCRETE PLATFORM
ORGAN
ARTISTES FOYER
LIFT WELL
GANGWAY
STORE
INTERVIEW ROOM
CLEANERS ROOM
BULKHEAD
PIANO LIFT
UPPER PART OF LISTENING R^M
OBSERVATION WINDOW
PUBLICATIONS
FOYER TO Nº1 STUDIO
EMERG^Y EXIT
SCALE OF FEET
0 5 10 20 30 40 50

146

148

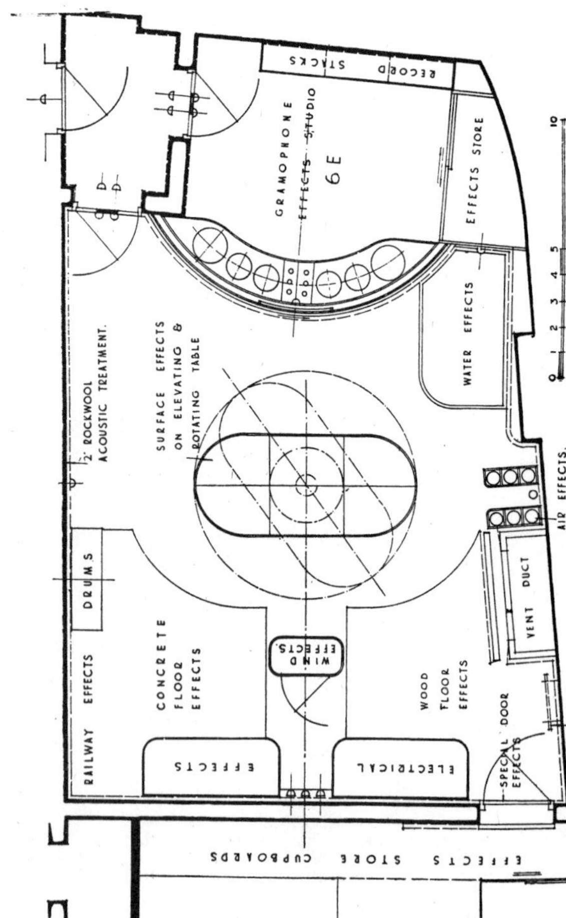
RECORD STACKS
GRAMOPHONE EFFECTS STUDIO 6E
EFFECTS STORE
WATER EFFECTS
2" ROCKWOOL ACOUSTIC TREATMENT.
SURFACE EFFECTS ON ELEVATING & ROTATING TABLE
AIR EFFECTS.
DRUMS
VENT DUCT
WIND EFFECTS.
RAILWAY EFFECTS
CONCRETE FLOOR EFFECTS
WOOD FLOOR EFFECTS
SPECIAL DOOR EFFECTS
THUNDER EFFECTS
EFFECTS
ELECTRICAL
EFFECTS STORE CUPBOARDS
0 1 2 3 4 5 10
SCALE OF FEET

147

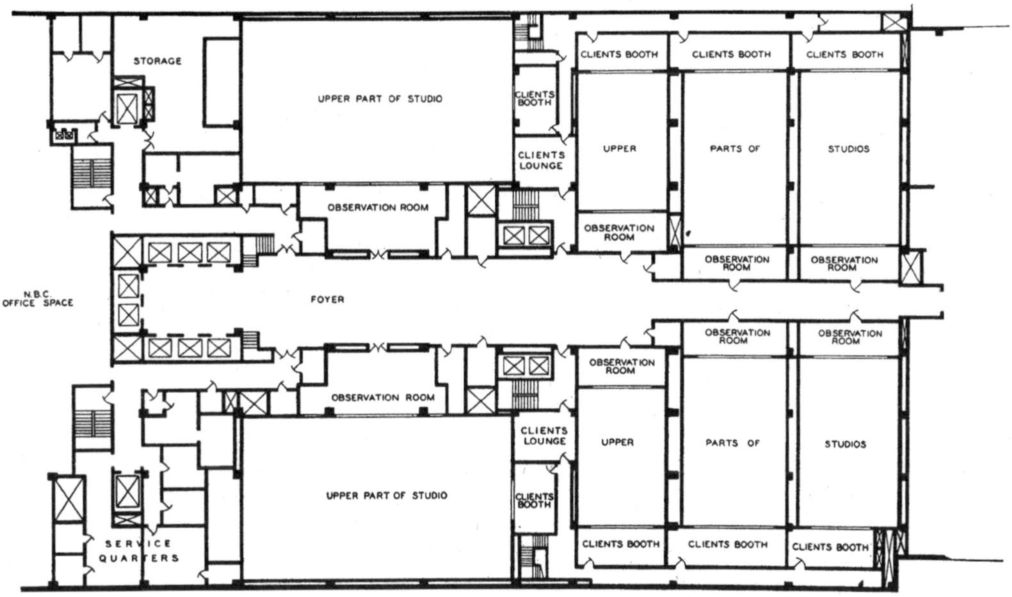
STORAGE
UPPER PART OF STUDIO
CLIENTS BOOTH
CLIENTS LOUNGE
CLIENTS BOOTH
CLIENTS BOOTH
CLIENTS BOOTH
UPPER
PARTS OF
STUDIOS
OBSERVATION ROOM
OBSERVATION ROOM
OBSERVATION ROOM
OBSERVATION ROOM
N.B.C. OFFICE SPACE
FOYER
OBSERVATION ROOM
OBSERVATION ROOM
OBSERVATION ROOM
OBSERVATION ROOM
OBSERVATION ROOM
CLIENTS LOUNGE
UPPER
PARTS OF
STUDIOS
UPPER PART OF STUDIO
CLIENTS BOOTH
SERVICE QUARTERS
CLIENTS BOOTH
CLIENTS BOOTH
CLIENTS BOOTH

149

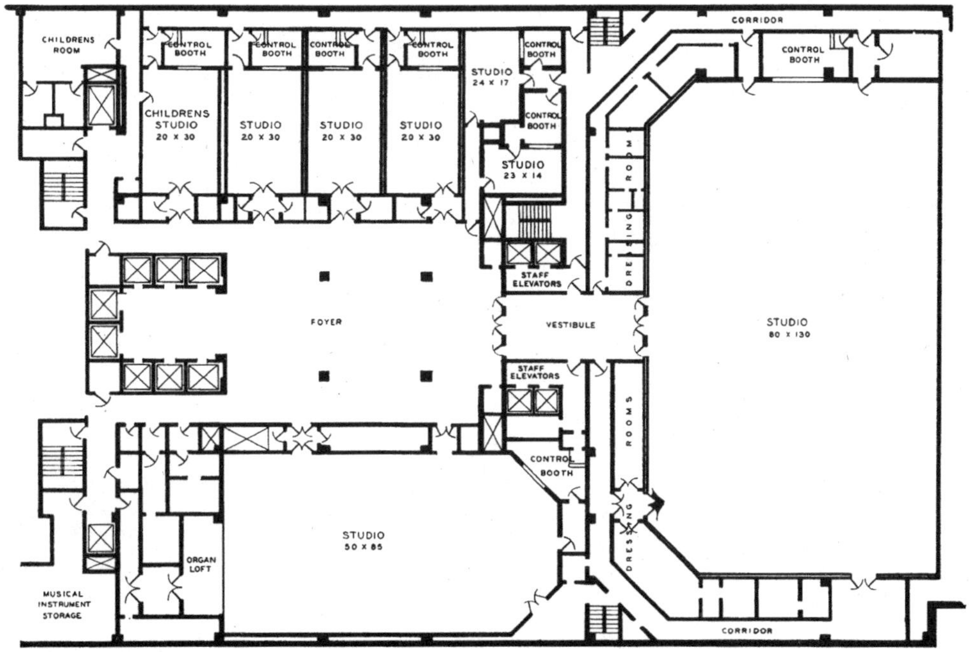
CHILDRENS ROOM
CONTROL BOOTH
CONTROL BOOTH
CONTROL BOOTH
CONTROL BOOTH
STUDIO 24 X 17
CONTROL BOOTH
CORRIDOR
CONTROL BOOTH
CHILDRENS STUDIO 20 X 30
STUDIO 20 X 30
STUDIO 20 X 30
STUDIO 20 X 30
CONTROL BOOTH
STUDIO 23 X 14
DRESSING ROOMS
STAFF ELEVATORS
FOYER
VESTIBULE
STUDIO 80 X 130
STAFF ELEVATORS
ROOMS
CONTROL BOOTH
DRESSING
STUDIO 50 X 85
ORGAN LOFT
MUSICAL INSTRUMENT STORAGE
CORRIDOR

150

151

152

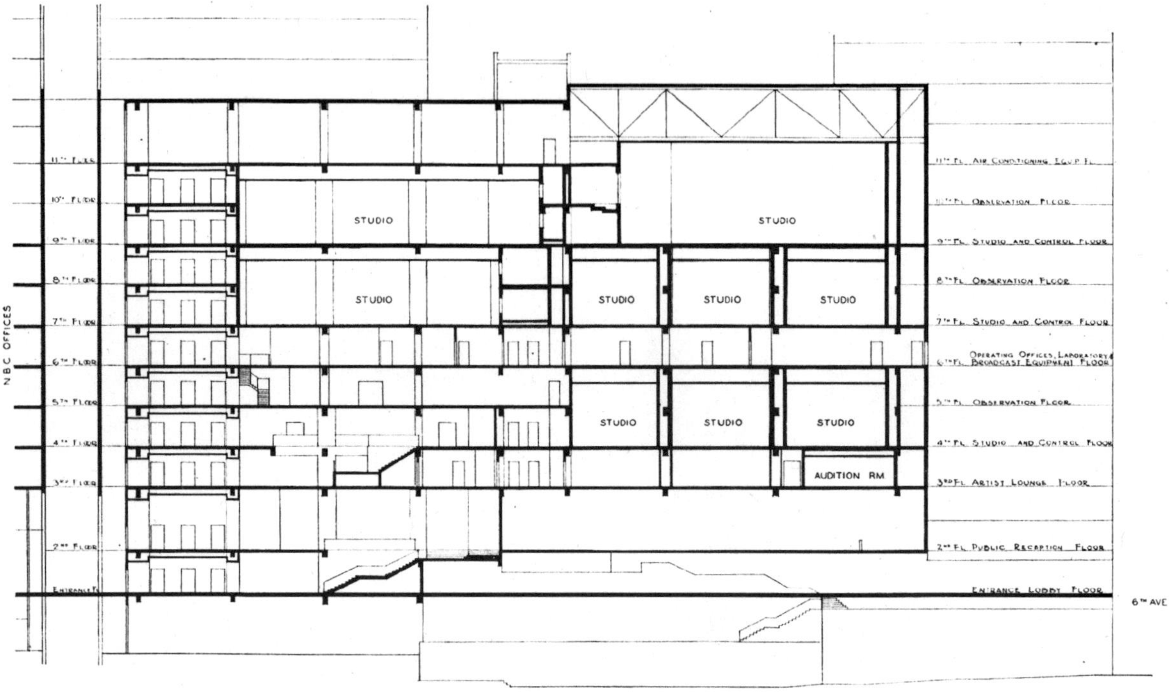
NBC OFFICES
STUDIO
STUDIO
STUDIO
STUDIO
STUDIO
STUDIO
STUDIO
STUDIO
STUDIO
AUDITION RM
11TH FL. AIR CONDITIONING EQUP FL.
10TH FL. OBSERVATION FLOOR
9TH FL. STUDIO AND CONTROL FLOOR
8TH FL. OBSERVATION FLOOR
7TH FL. STUDIO AND CONTROL FLOOR
6TH FL. OPERATING OFFICES, LABORATORY & BROADCAST EQUIPMENT FLOOR
5TH FL. OBSERVATION FLOOR
4TH FL. STUDIO AND CONTROL FLOOR
3RD FL. ARTIST LOUNGE FLOOR
2ND FL. PUBLIC RECEPTION FLOOR
ENTRANCE LOBBY FLOOR
6TH AVE

153

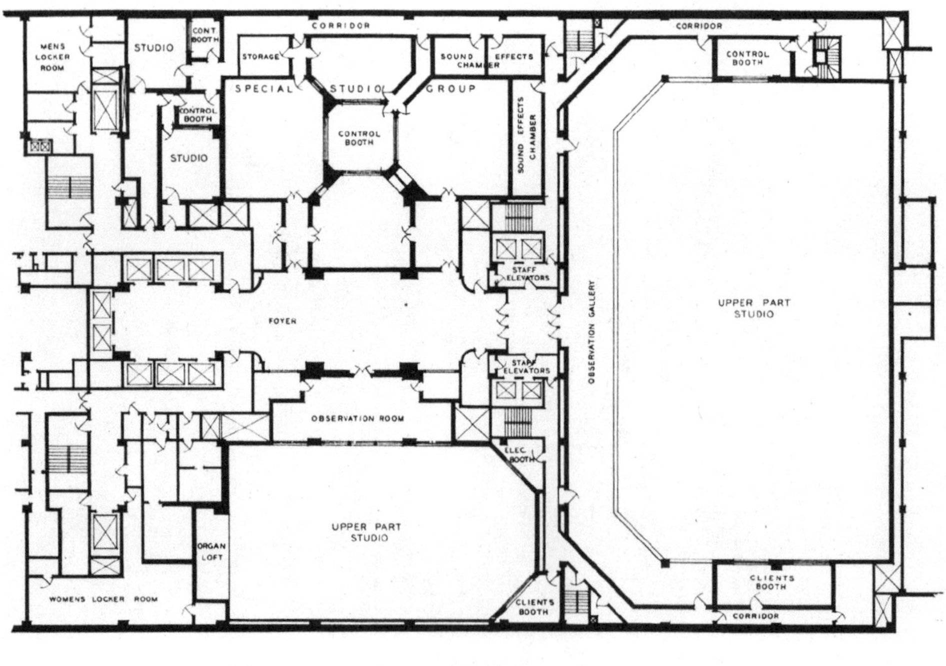
CORRIDOR
CORRIDOR
MENS LOCKER ROOM
STUDIO
CONT BOOTH
STORAGE
SOUND CHAMBER
EFFECTS
CONTROL BOOTH
SPECIAL STUDIO GROUP
CONTROL BOOTH
CONTROL BOOTH
SOUND EFFECTS CHAMBER
STUDIO
STAFF ELEVATORS
FOYER
OBSERVATION GALLERY
UPPER PART STUDIO
STAFF ELEVATORS
OBSERVATION ROOM
ELEC BOOTH
UPPER PART STUDIO
ORGAN LOFT
WOMENS LOCKER ROOM
CLIENTS BOOTH
CLIENTS BOOTH
CORRIDOR

154

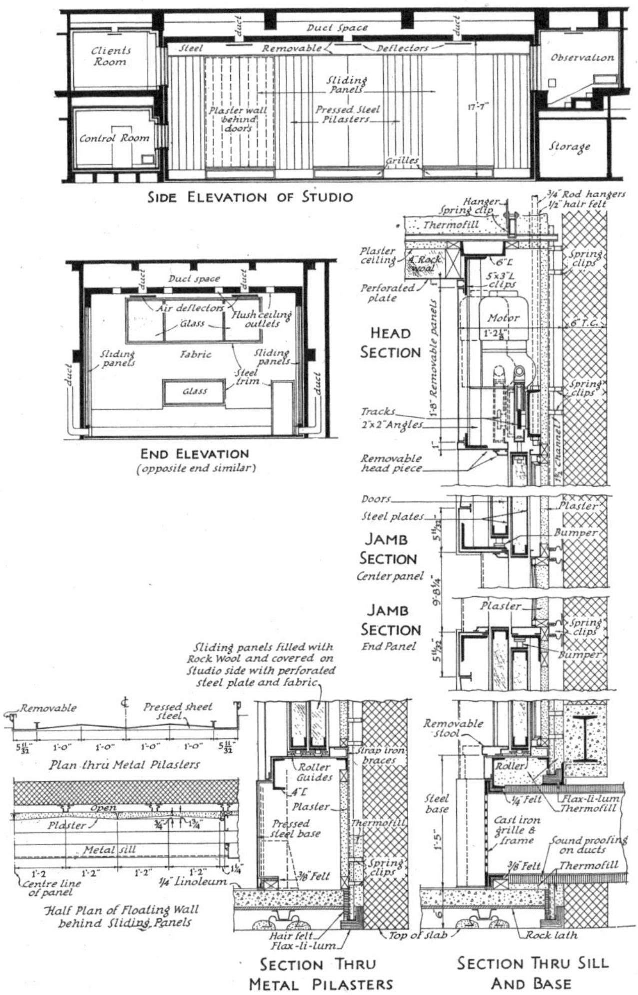

ELEVATIONS AND SECTIONS OF A STUDIO TYPE
NATIONAL BROADCASTING COMPANY STUDIOS

155

156

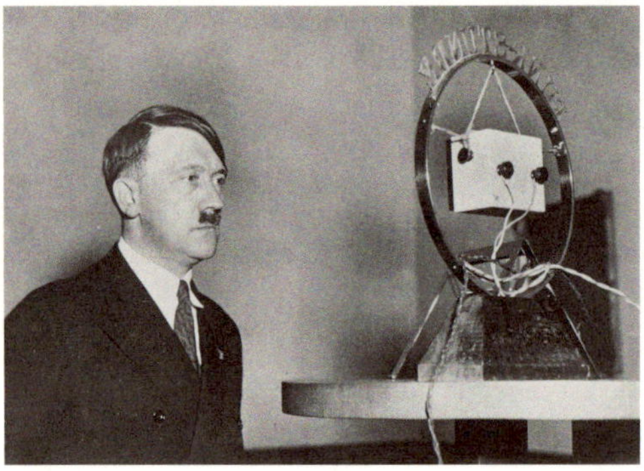

157

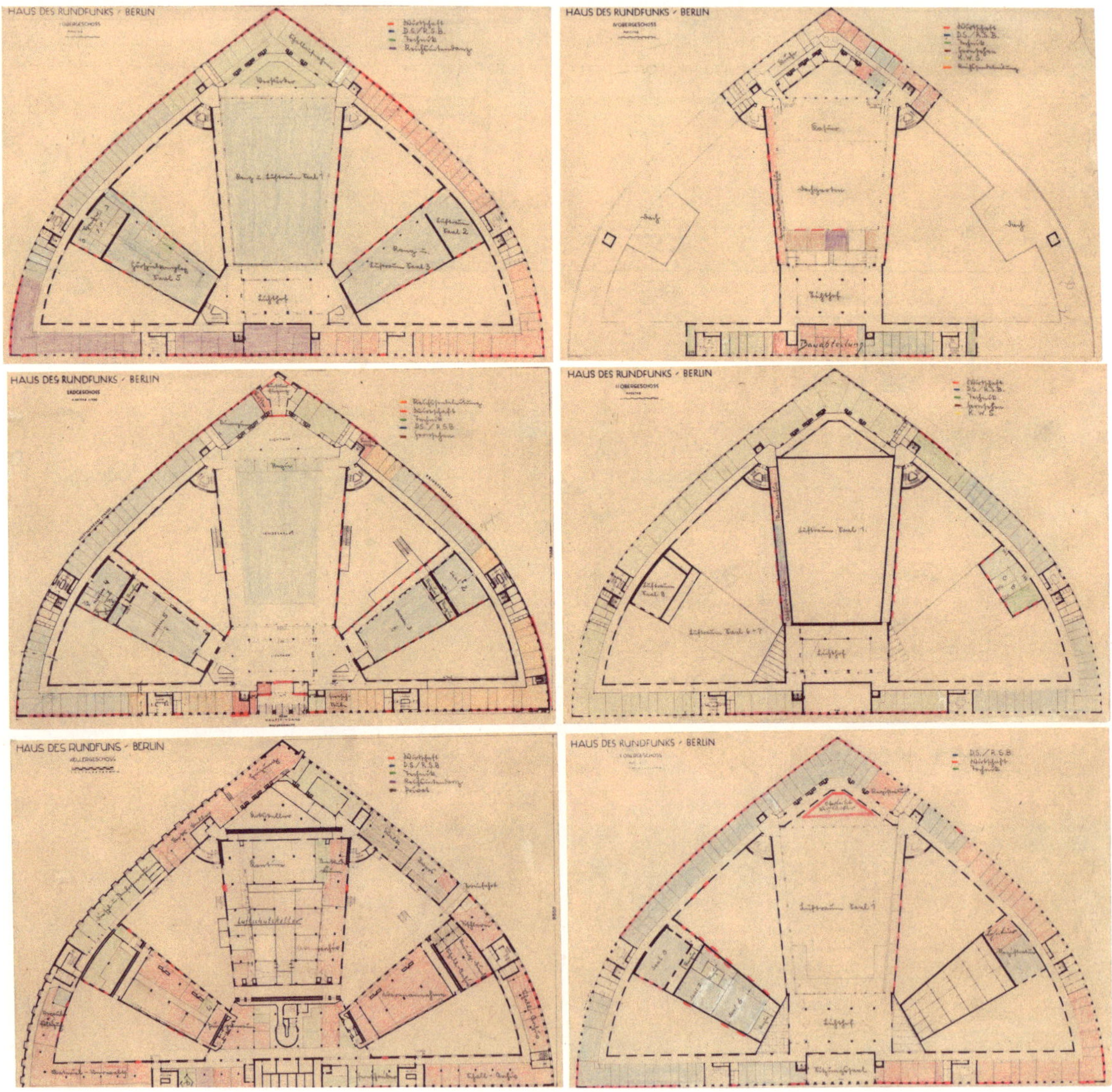
HAUS DES RUNDFUNKS / BERLIN
HAUS DES RUNDFUNKS / BERLIN
HAUS DES RUNDFUNKS / BERLIN
ERDGESCHOSS
HAUS DES RUNDFUNKS / BERLIN
HAUS DES RUNDFUNS / BERLIN
KELLERGESCHOSS
HAUS DES RUNDFUNKS / BERLIN

158

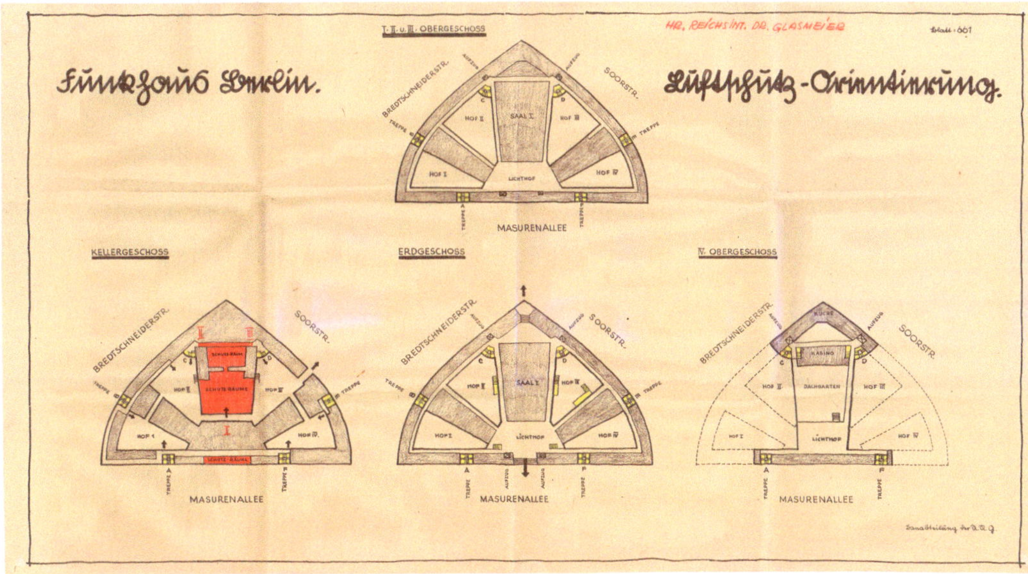
Funkhaus Berlin.
Luftschutz-Einrichtung.
T. II. u. III. OBERGESCHOSS
KELLERGESCHOSS
ERDGESCHOSS
IV. OBERGESCHOSS
MASURENALLEE
BREDTSCHNEIDERSTR.
SOORSTR.

159

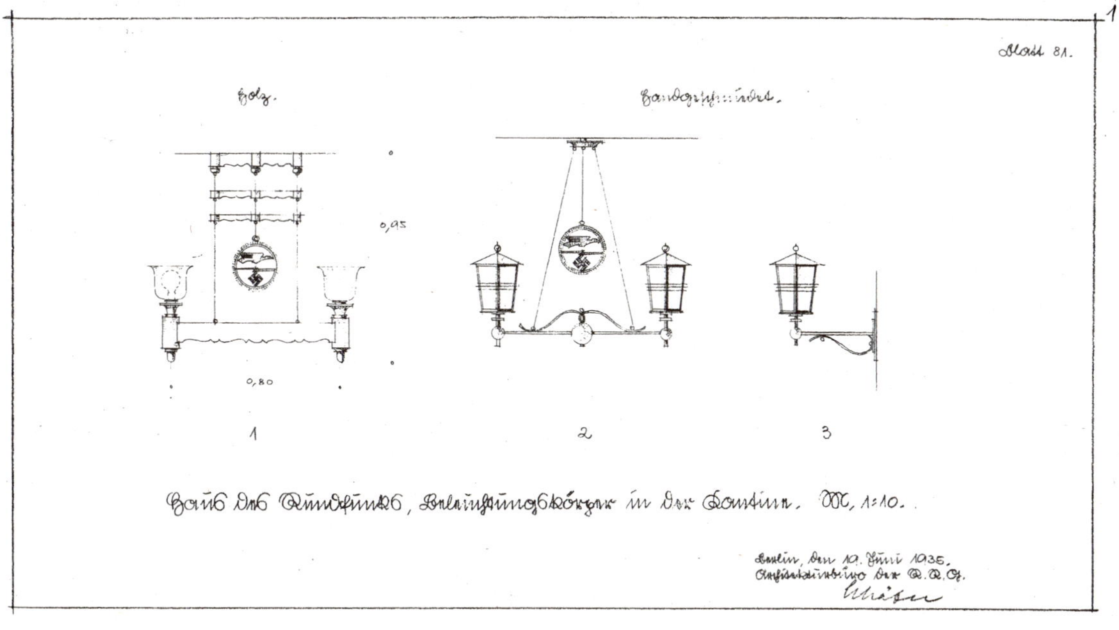

160

Fig. 144 BBC Broadcasting House, exterior view. Source: British Broadcasting Corporation, *The Broadcasting House* (Plaistow: Curwen Press, 1932), 3.

Fig. 145 BBC Broadcasting House, interior of chapel designed by Edward Maufe. Source: Robert Byron, "The Broadcasting House," *Architectural Review* 72, no. 429 (1932): 52.

Fig. 146 BBC Broadcasting House, level 0. Source: British Broadcasting Corporation, *A Technical Description of Broadcasting House* (Bungay: Chaucer Press, 1932), 3.

Fig. 147 BBC Broadcasting House, detailed plan of the special-effects studio designed by Wells Coates. Source: Canadian Centre for Architecture Archival Collection. Wells Coates Fonds. 30/02 B20-1-01.

Fig. 148 BBC Broadcasting House, interior view of the special-effects studio designed by Wells Coates. Source: Canadian Centre for Architecture Archival Collection. Wells Coates Fonds. 30/02 B20-1-01.

Fig. 149 NBC Studios in the RCA Tower at the Rockefeller Center in Manhattan, plans of levels 5 and 8. Source: O. B. Hanson, "Planning the NBC Studios for Radio City," *Proceedings of the Institute of Radio Engineers* 20, no. 8 (1932): 1305–1307.

Fig. 150 NBC Studios in the RCA Tower at the Rockefeller Center in Manhattan, plans of level 9. Source: O. B. Hanson, "Planning the NBC Studios for Radio City," *Proceedings of the Institute of Radio Engineers* 20, no. 8 (1932): 1305–1307.

Fig. 151 The RCA Tower of the Rockefeller Center, January 27, 1937. Photograph by Berenice Abbott, gelatin silver print, 33.6 x 26.1 cm. Source: Canadian Centre for Architecture. © Berenice Abbott /Commerce Graphics, NYC.

Fig. 152 View of Rockefeller Center from the Shelton Hotel in 1935. Photograph by Alfred Stieglitz, gelatin silver print mounted on cardboard, 24.2 x 19 cm. Source: Canadian Centre for Architecture.

Fig. 153 NBC Studios, longitudinal section. Source: O. B. Hanson, "Planning the NBC Studios for Radio City," *Proceedings of the Institute of Radio Engineers* 20, no. 8 (1932): 1302.

Fig. 154 NBC Studios in the RCA Tower at the Rockefeller Center in Manhattan, plans of level 10. Source: O. B. Hanson, "Planning the NBC Studios for Radio City," *Proceedings of the Institute of Radio Engineers* 20, no. 8 (1932): 1305–1307.

Fig. 155 NBC Studios, detailed wall sections. Source: O. B. Hanson, "The Plan and Construction of the National Broadcasting Company Studios," *Architectural Forum* 57 (1932): 158.

Fig. 156 Photo collage depicting the RCA Tower from different viewpoints. Source: Sigfried Giedion, *Space, Time and Architecture: The Growth of a New Tradition* (Cambridge, MA: Harvard University Press, 1962), 754.

Fig. 157 Adolf Hitler standing in front of the Berliner Funkstunde's microphone. Source: Bundesarchiv. Bild 183-1987-0703-506.

Fig. 158 Haus des Rundfunks, series of floor plans showing the reorganization of the interior to accommodate new programs, including television. Source: Deutsche Rundfunkarchiv Frankfurt am Main. A48 RRG Bauzeichnungen, 413.

Fig. 159 Haus des Rundfunks, circulation diagrams in the event of an air raid: the basement was converted into a bomb shelter. Source: Deutsche Rundfunkarchiv Frankfurt am Main. A48 RRG Bauzeichnungen, 412-661.

Fig. 160 Haus des Rundfunks, detail drawings of new light fixtures for the canteen featuring Nazi iconography. Source: Deutsche Rundfunkarchiv Frankfurt am Main. A48 RRG Bauzeichnungen, 394-81.

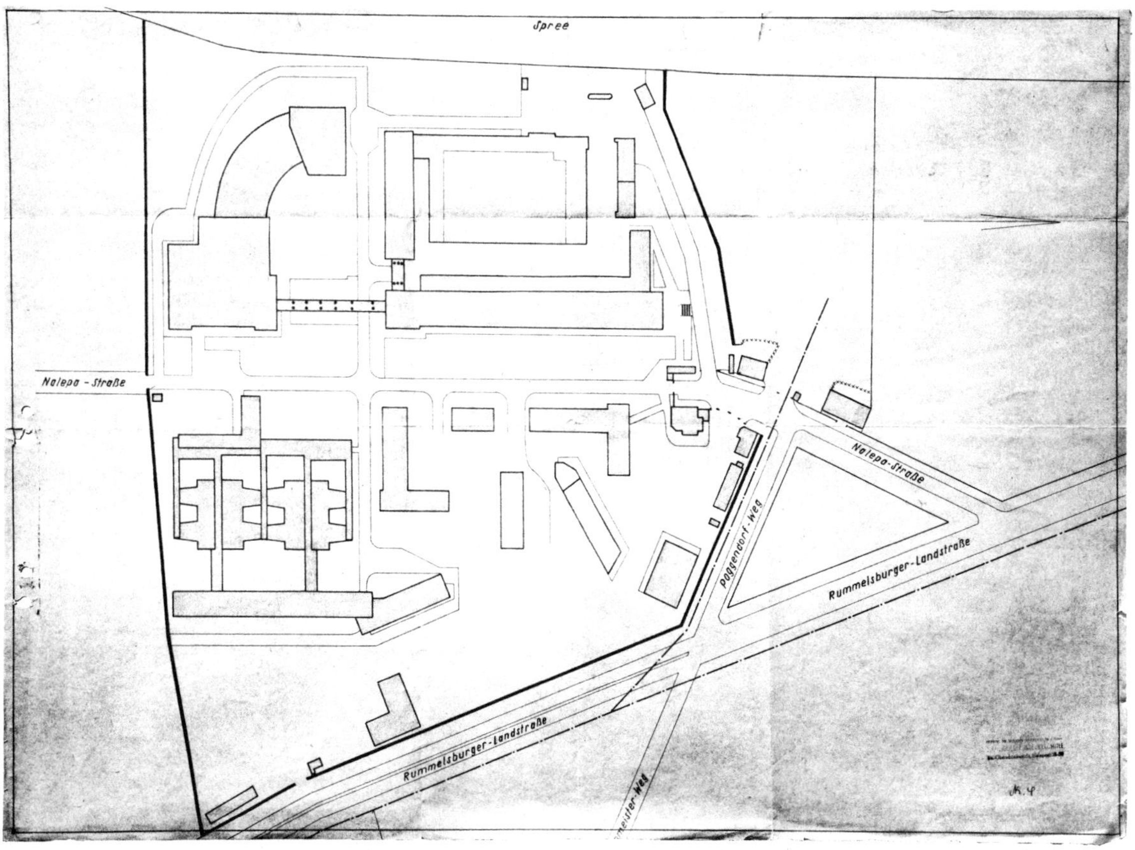
Spree
Nalepa - Straße
Nalepa-Straße
Poggendorf-Weg
Rummelsburger-Landstraße
Rummelsburger-Landstraße

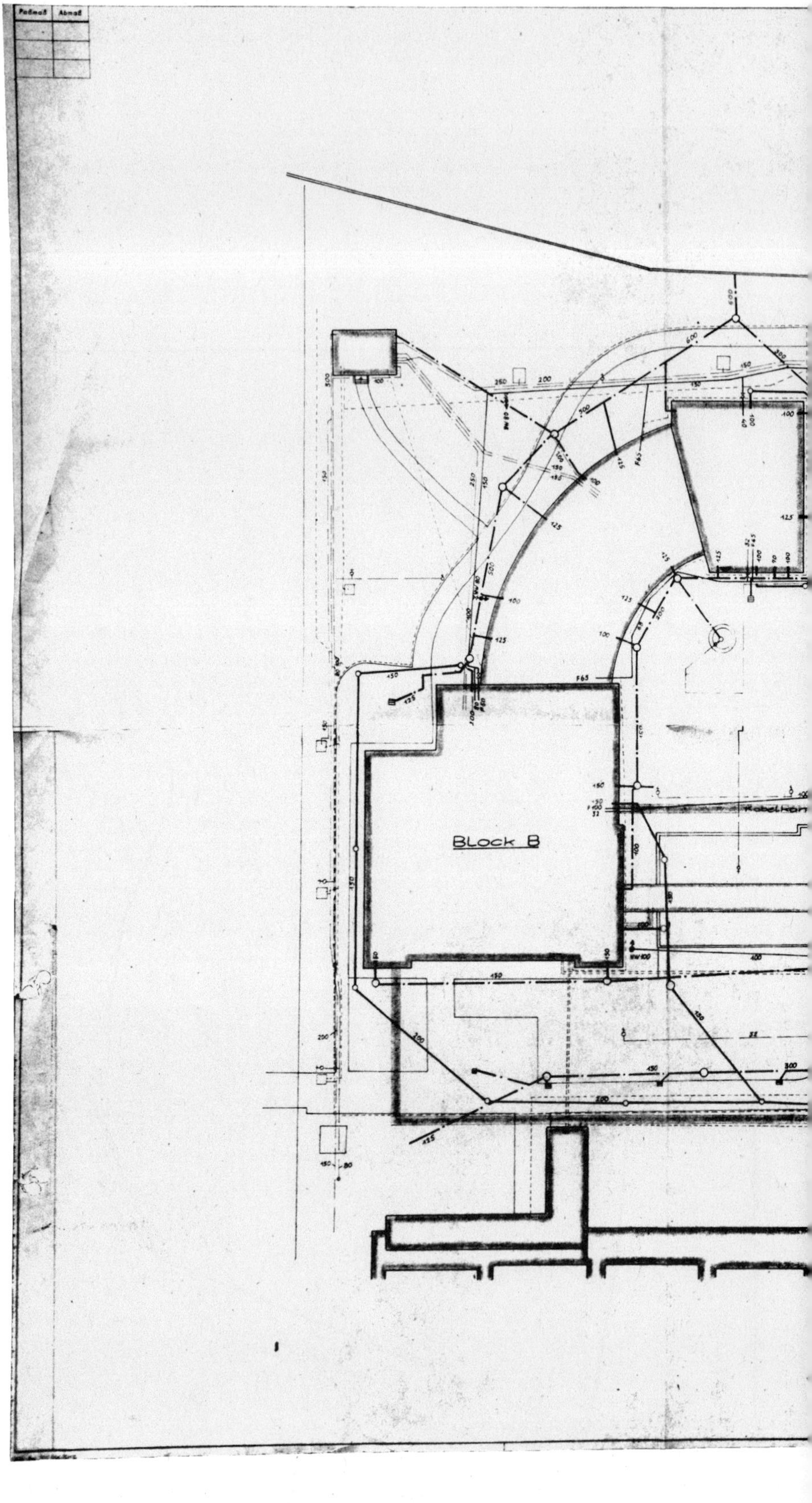
Paßmaß
Abmaß
Block B
Kabelkan

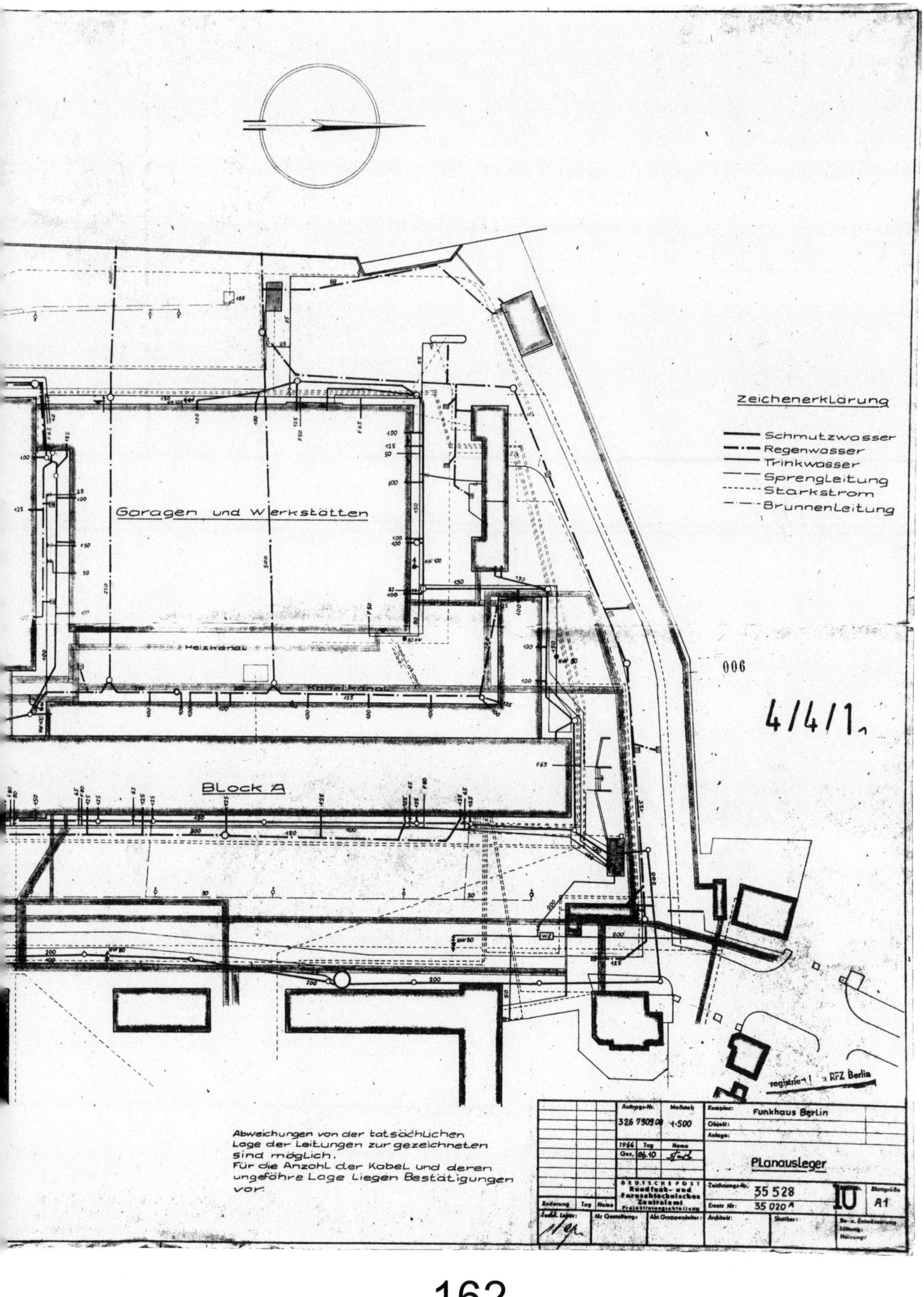
Zeichenerklärung
Schmutzwasser
Regenwasser
Trinkwasser
Sprengleitung
Starkstrom
Brunnenleitung
Garagen und Werkstätten
Heizkanal
Kabelkanal
Block A
006
4/4/1
registriert ... RFZ Berlin
Abweichungen von der tatsächlichen
Lage der Leitungen zur gezeichneten
sind möglich.
Für die Anzahl der Kabel und deren
ungefähre Lage liegen Bestätigungen
vor.
Auftrags-Nr. 326 7309 00
Maßstab 1:500
Komplex: Funkhaus Berlin
Objekt:
Anlage:
1966 Tag Name
Gez. 06.10
Planausleger
DEUTSCHE POST
Rundfunk- und
Fernsehtechnisches
Zentralamt
Projektierungsabteilung
Zeichnungs-Nr. 35 528
Ersatz für: 35 020 A
Blattgröße A1
Änderung Tag Name
Archiktekt:
Statiker:

HAUS DES RUNDFUNKS · BERLIN
ERDGESCHOSS
MASSTAB 1:100
BREDTSCHNEIDER STRASSE
PROBESAAL IIa
SENDESAAL II

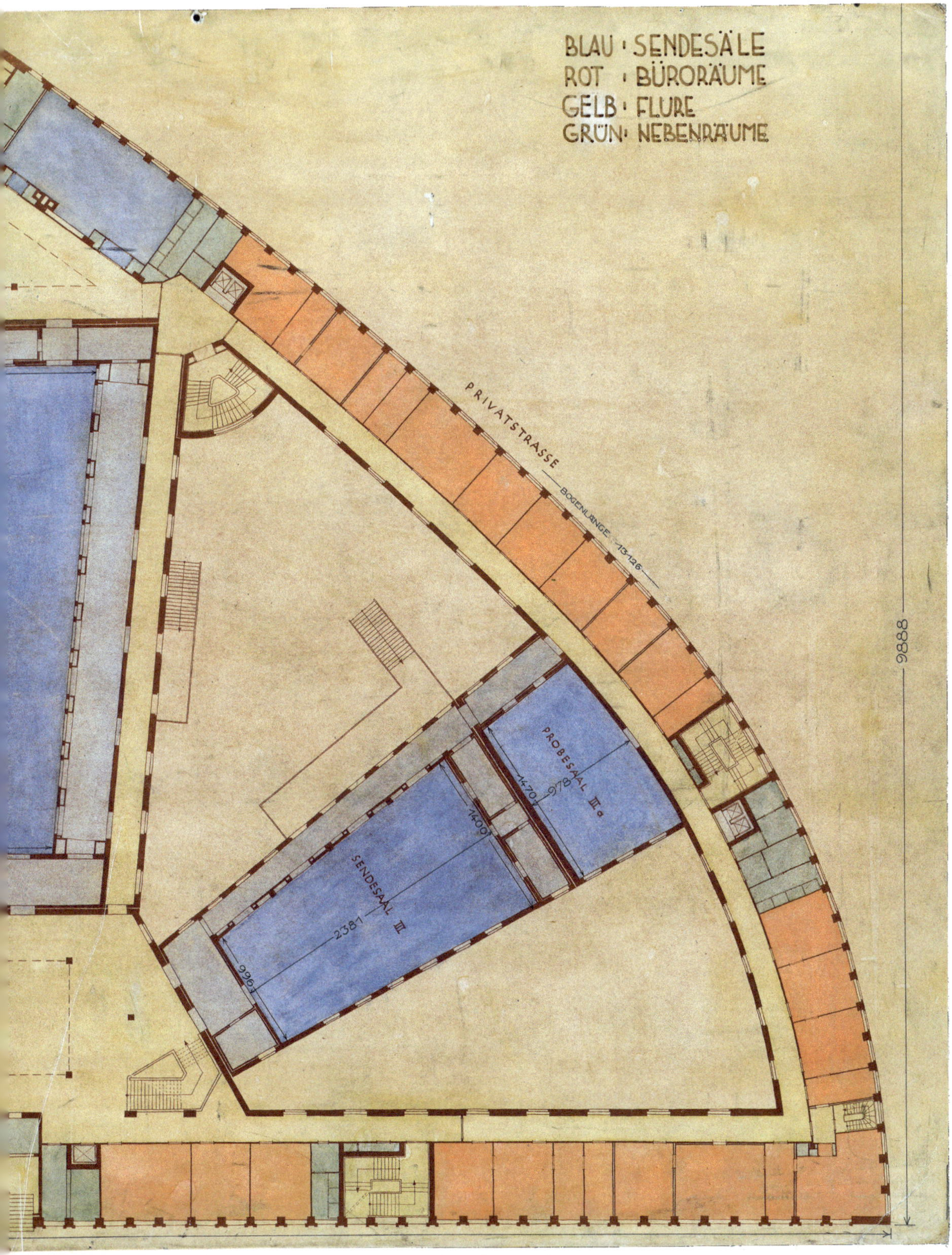
BLAU : SENDESÄLE
ROT : BÜRORÄUME
GELB : FLURE
GRÜN : NEBENRÄUME
PRIVATSTRASSE
BOGENLÄNGE 13.126
9888
PROBESAAL IIIa
1470
978
1400
SENDESAAL III
2381
996

163

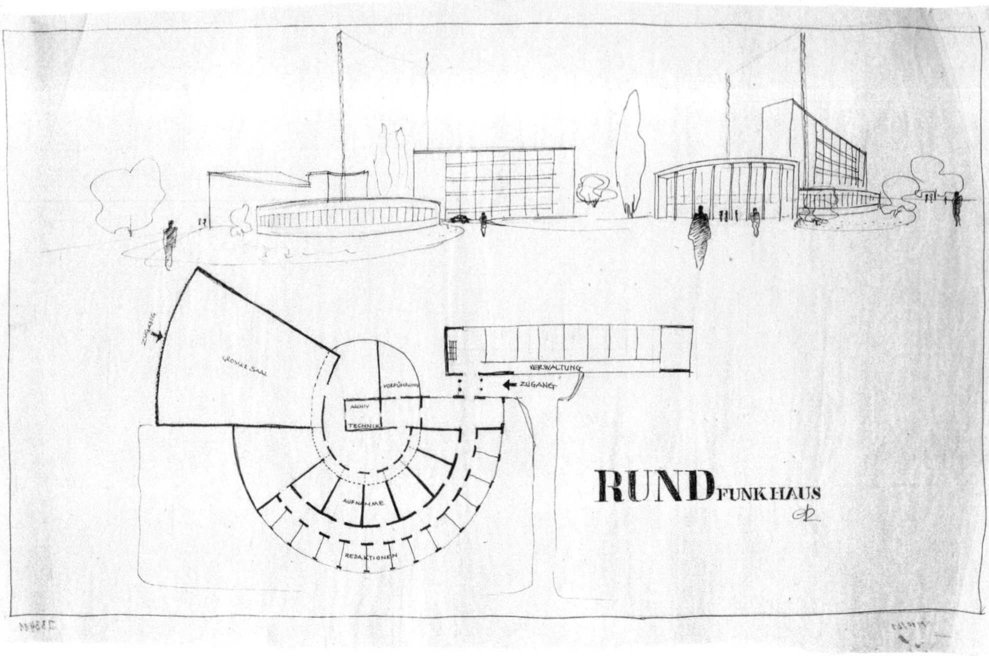
ZUGANG
GROSSER SAAL
ARCHIV
TECHNIK
AUFNAHME
REDAKTIONEN
VERWALTUNG
ZUGANG
RUNDFUNKHAUS

164

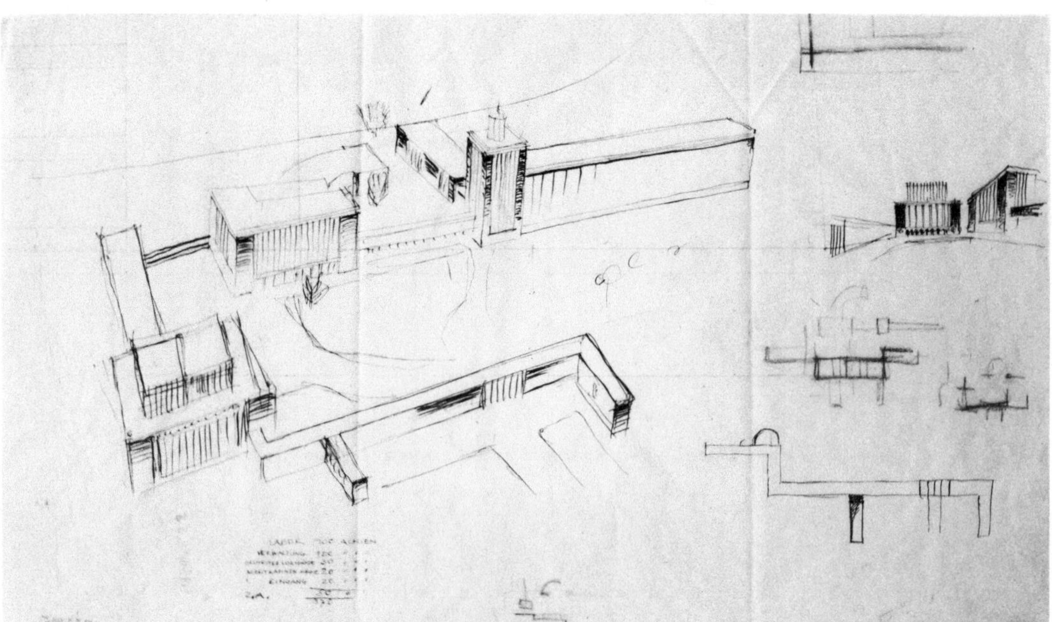

165

166

167

168

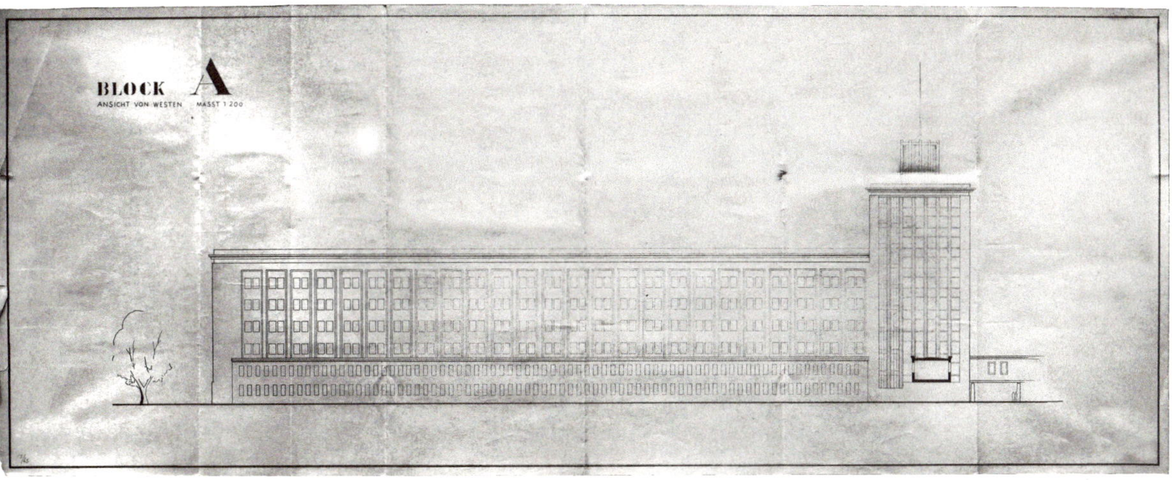

169

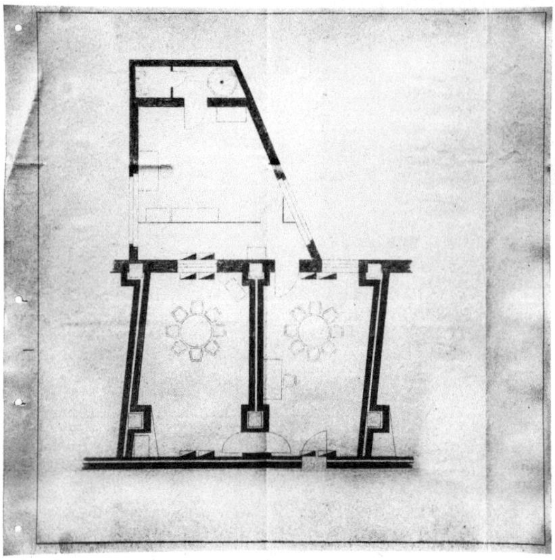

170

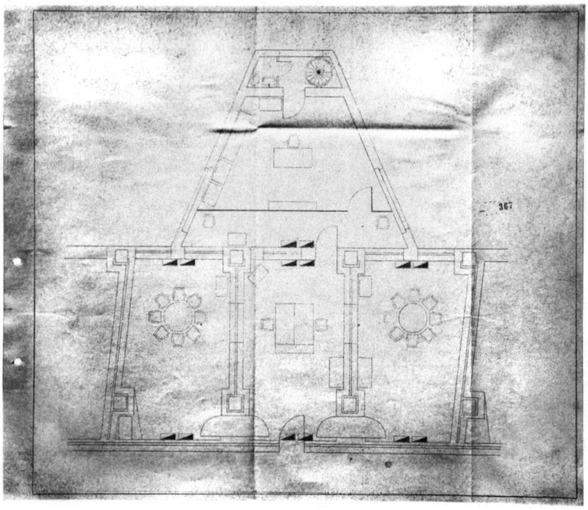

171

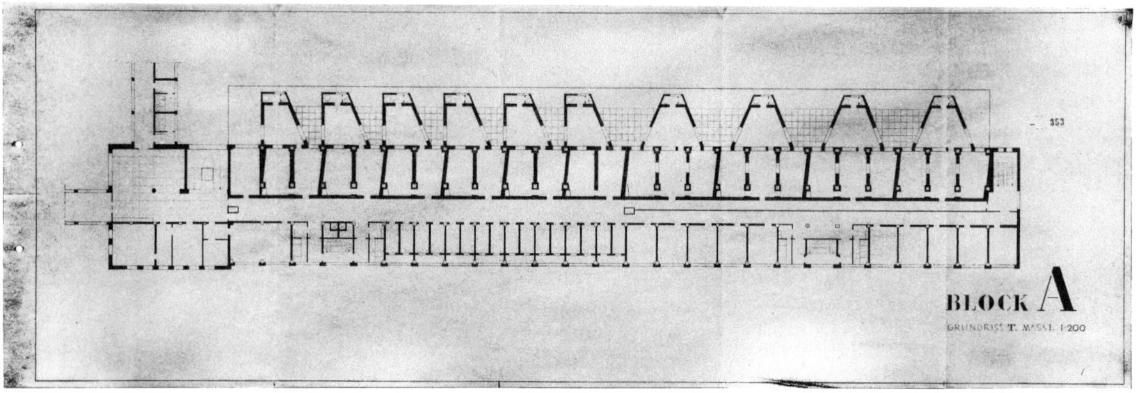

172

173

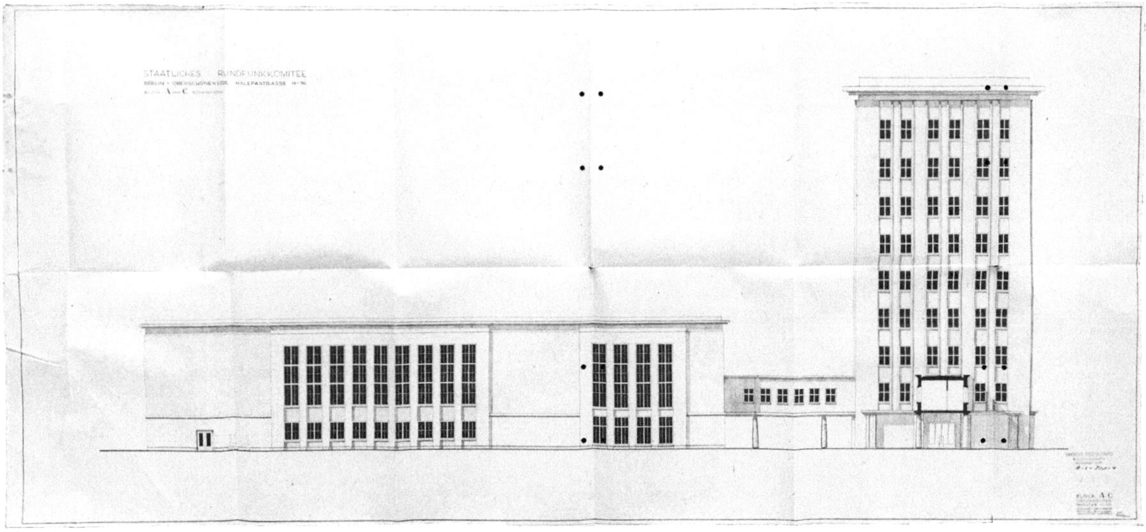
STAATLICHES RUNDFUNKKOMITEE

174

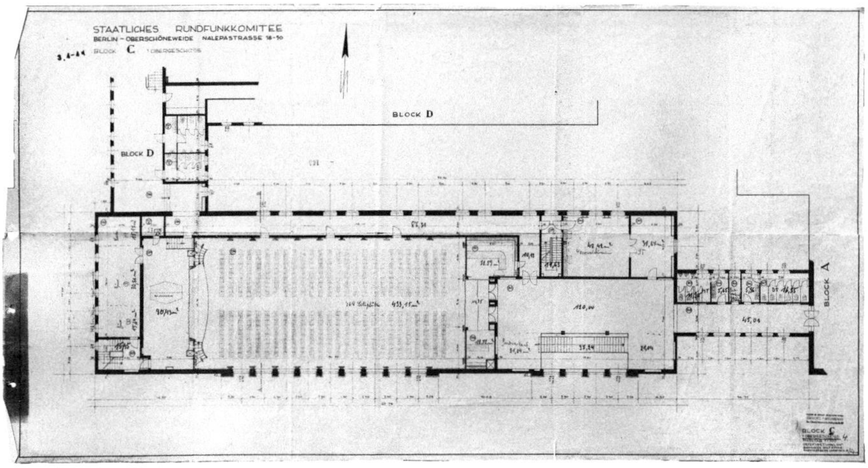
STAATLICHES RUNDFUNKKOMITEE
BERLIN - OBERSCHÖNEWEIDE NALEPASTRASSE 18-50
BLOCK C
BLOCK D
BLOCK A

175

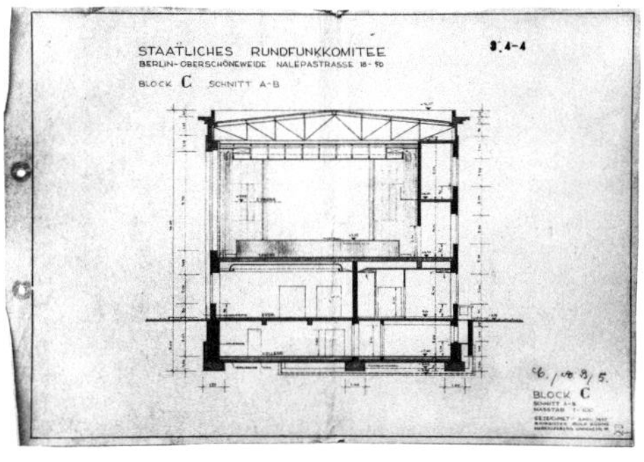
STAATLICHES RUNDFUNKKOMITEE
BERLIN - OBERSCHÖNEWEIDE NALEPASTRASSE 18-50
BLOCK C SCHNITT A-B
BLOCK C

176

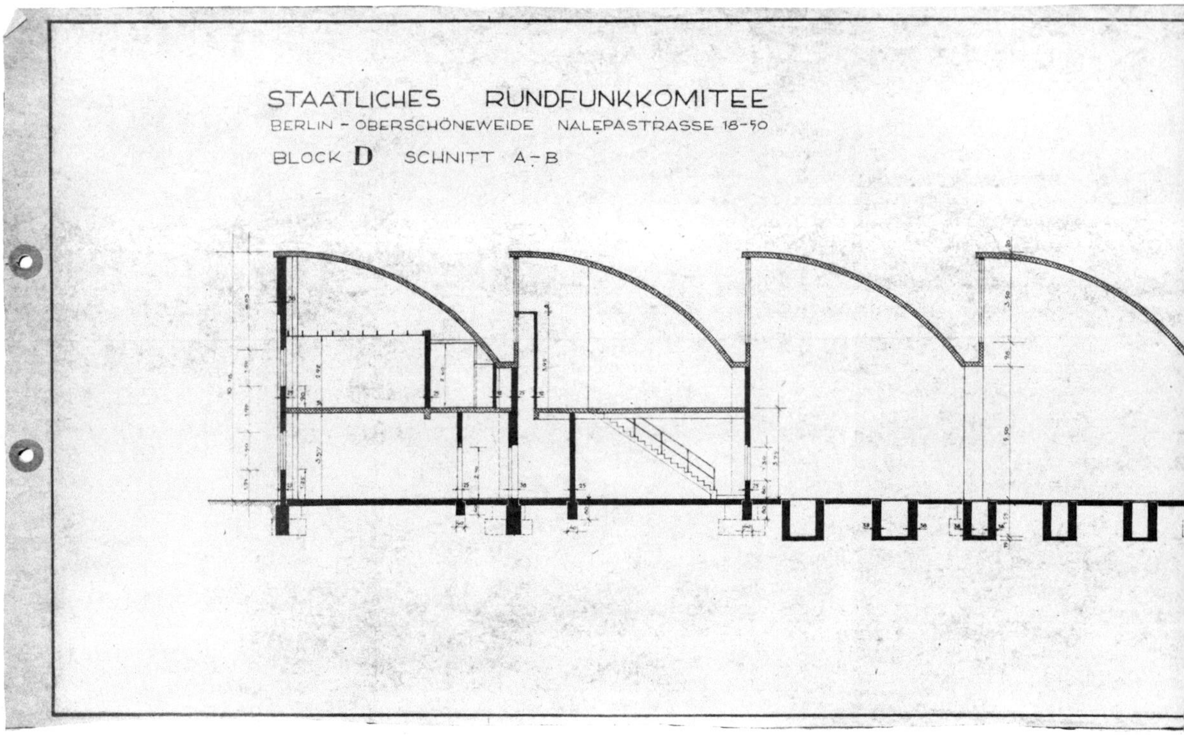
STAATLICHES RUNDFUNKKOMITEE
BERLIN - OBERSCHÖNEWEIDE NALEPASTRASSE 18-50
BLOCK D SCHNITT A-B

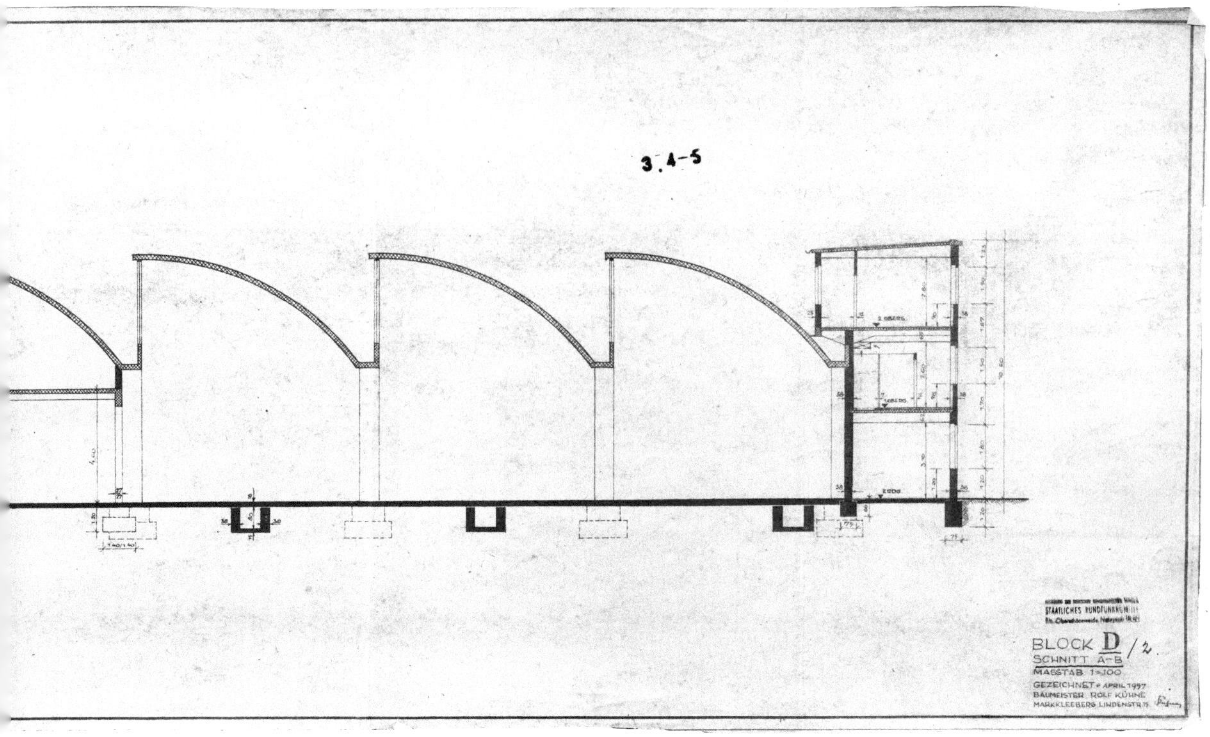
3.4-5
BLOCK D /2
SCHNITT A-B
MASSTAB 1:100
BAUMEISTER ROLF KÜHNE
MARKKLEEBERG LINDENSTR. 15

177

178

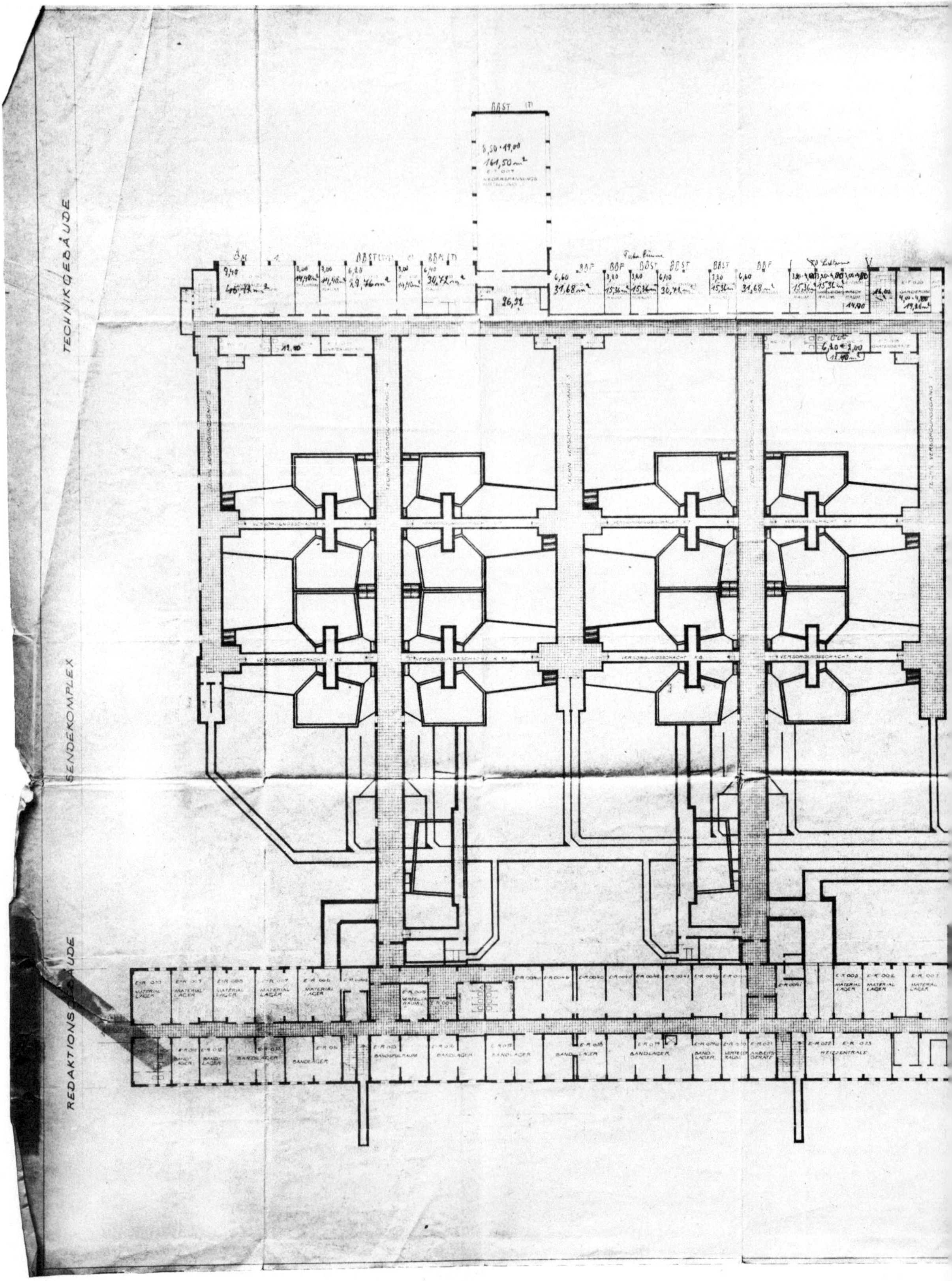

179

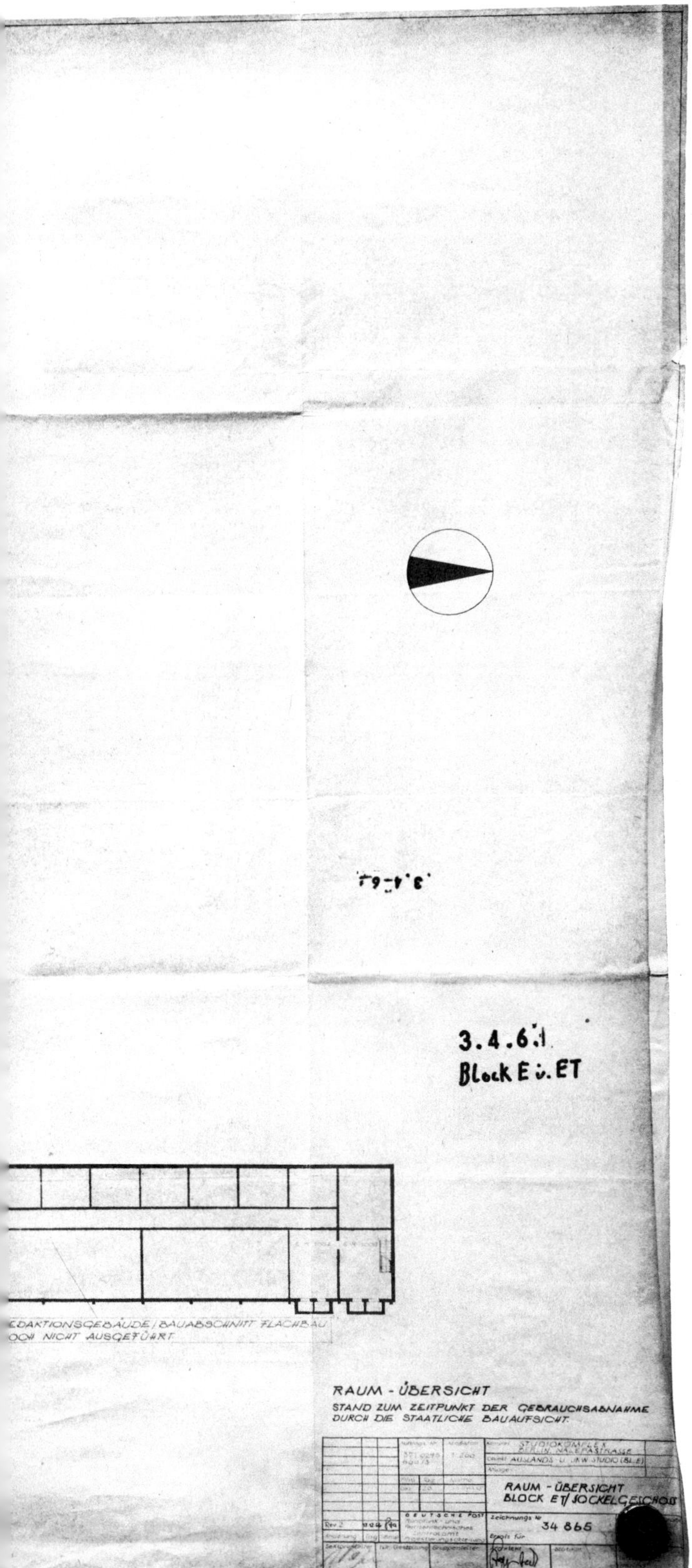
3.4.6.1
Block E u. ET
RAUM - ÜBERSICHT
STAND ZUM ZEITPUNKT DER GEBRAUCHSABNAHME
DURCH DIE STAATLICHE BAUAUFSICHT.
STUDIOKOMPLEX
BERLIN, NALEPASTRASSE
AUSLANDS U. JKW STUDIO (BL.E)
RAUM - ÜBERSICHT
BLOCK ET/ SOCKELGESCHOSS
DEUTSCHE POST
34 865

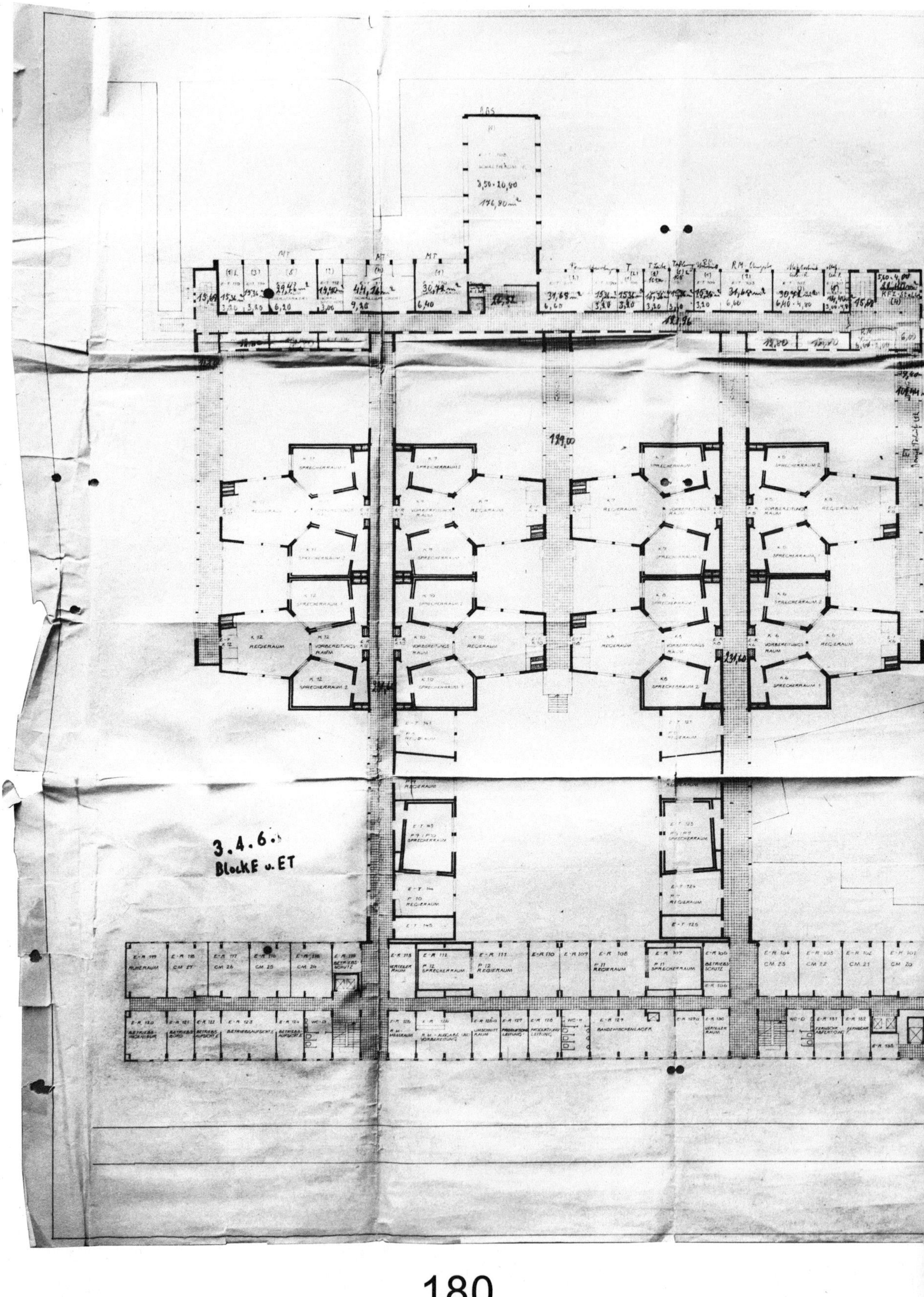
3.4.6.
BlocKE u. ET

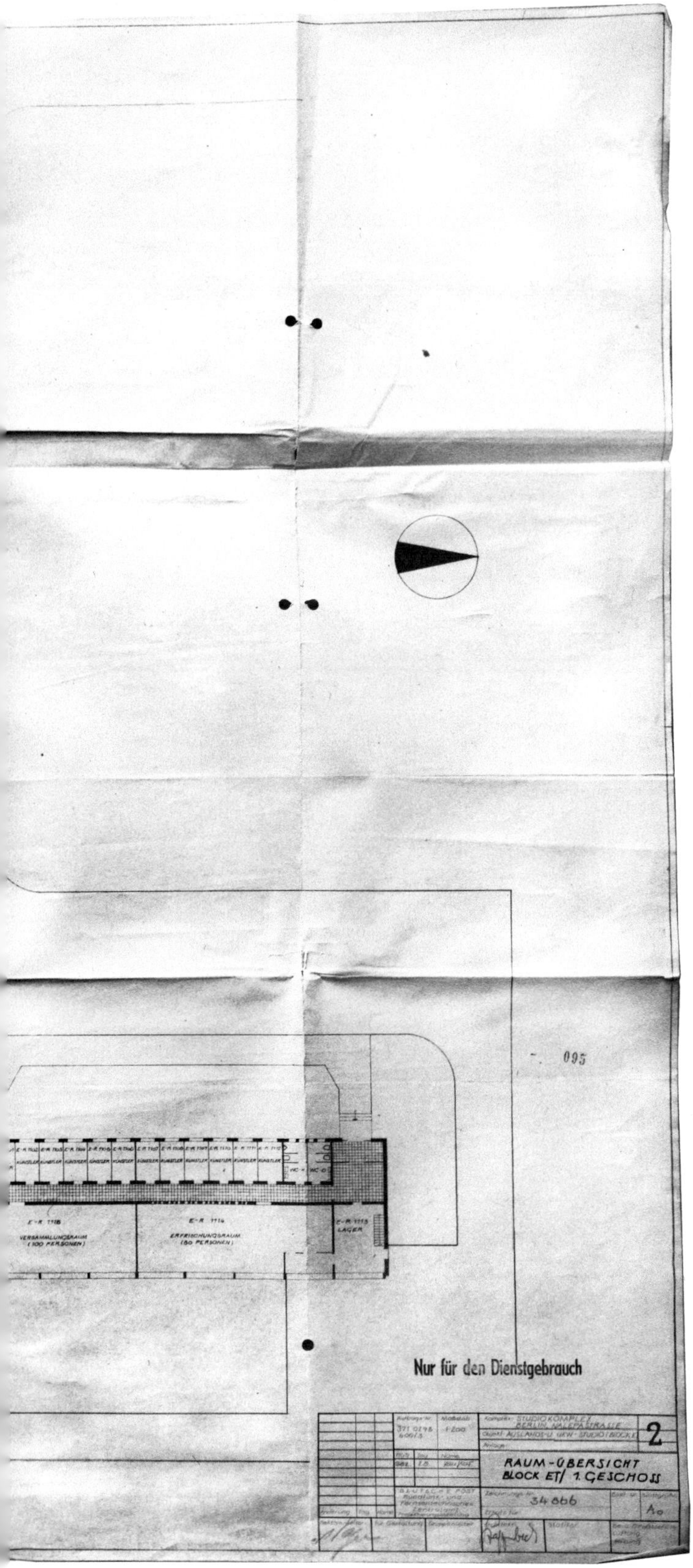
095
Nur für den Dienstgebrauch
STUDIOKOMPLEX BERLIN, NALEPASTRASSE
2
RAUM-ÜBERSICHT
BLOCK E1/ 1. GESCHOSS
34 866
A0

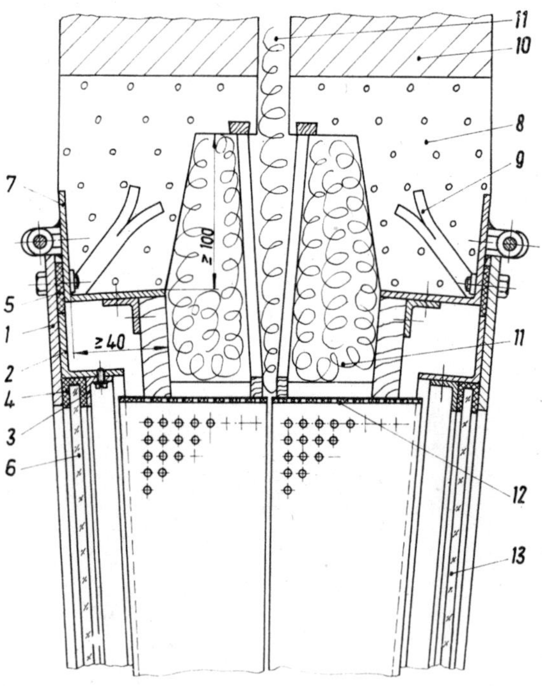

Abb. 4 Zweischaliges Regiefenster
1) Stahlrahmen
2) Stahlrahmen
3) Winkelschiene
4) Profilgummi
5) Dichtungsgummi
6) 6 mm dicke Scheibe
7) Stahlzarge
8) Beton
9) Maueranker
10) Mauerwerk
11) Glaswolle
12) gelochte Alu-Platte (Perforationsgrad 27 %)
13) 8 mm dicke Scheibe

181

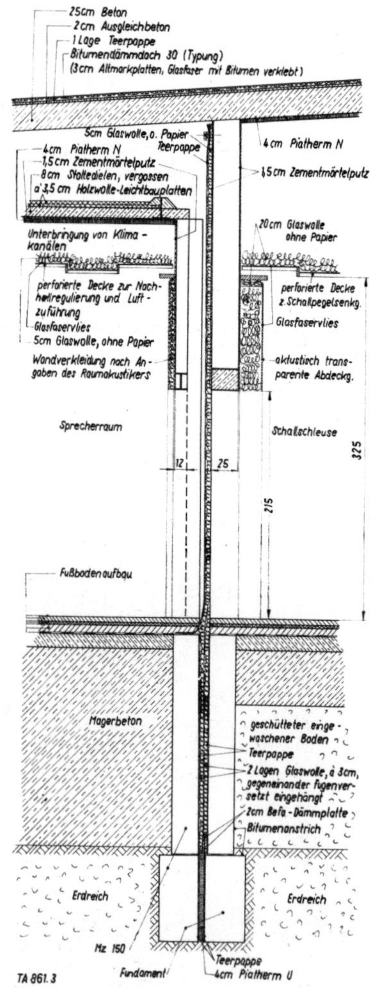

Abb. 3 Zweischalige Wand- und Deckenkonstruktion (Haus-in-Haus-Bauweise für die Sprecherräume)

182

183

184

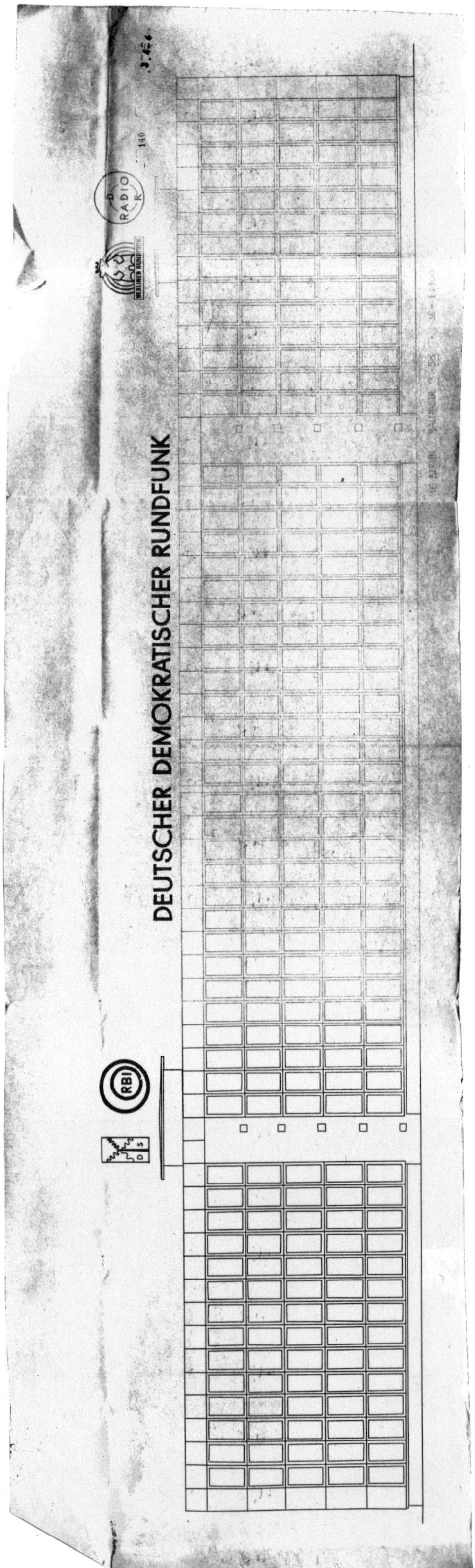
DEUTSCHER DEMOKRATISCHER RUNDFUNK
RADIO
RBI

185

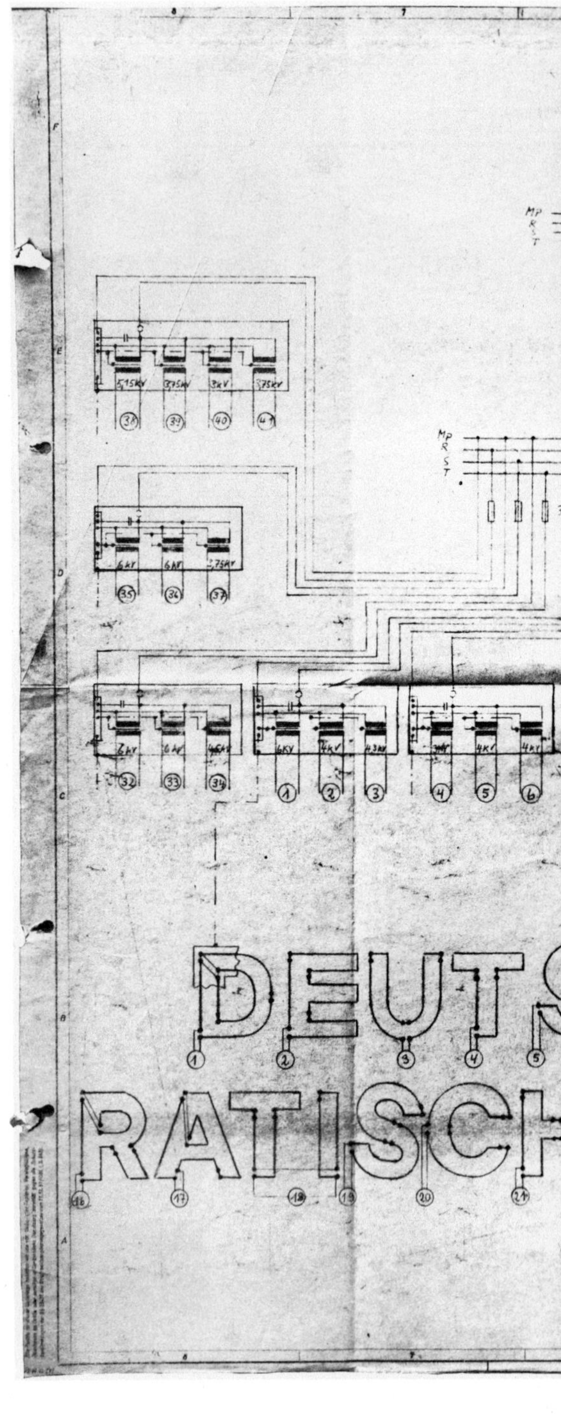
DEUTS
RATISCH

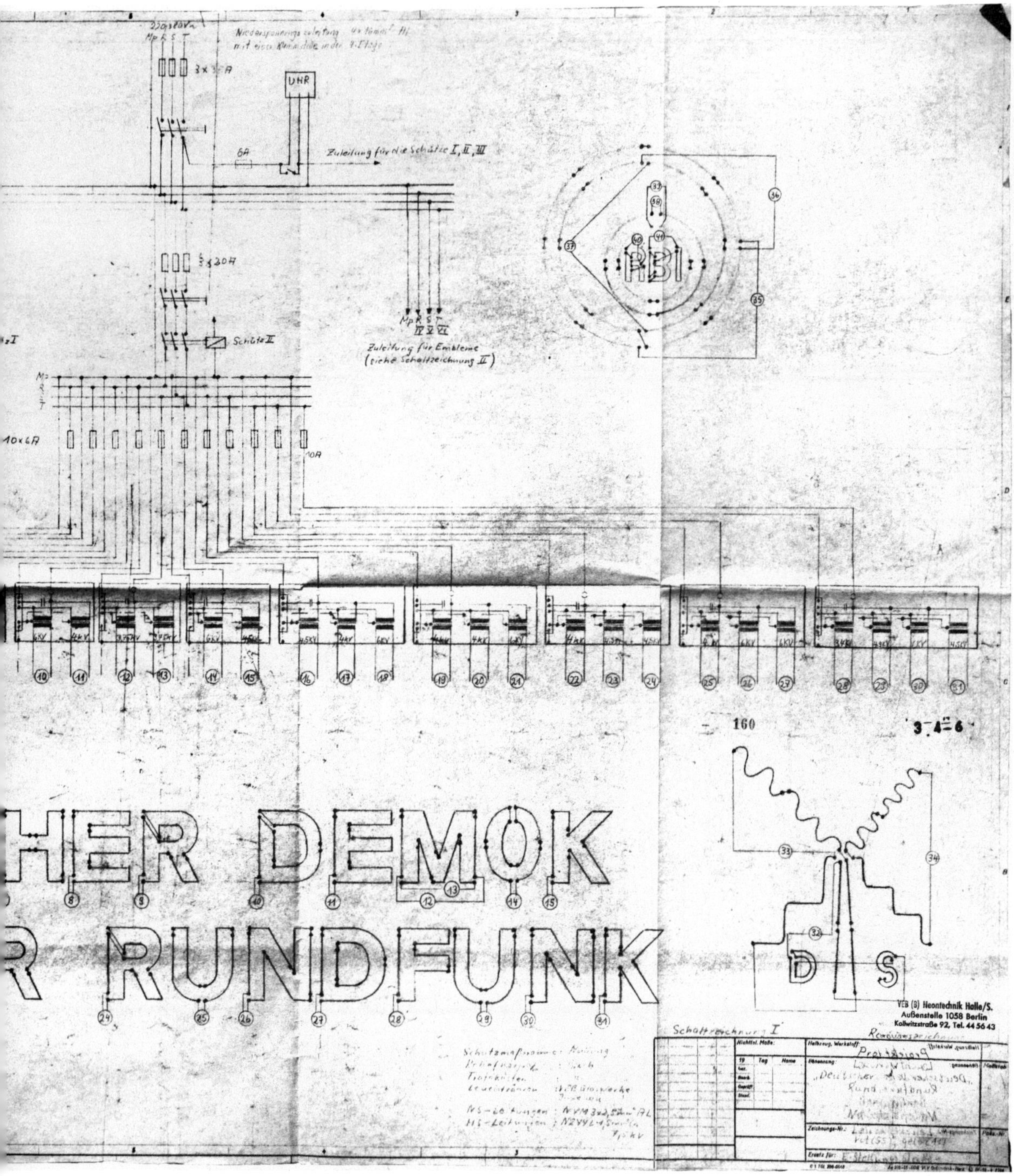
Zuleitung für die Schütze I, II, III
UHR
Schütz II
Zuleitung für Embleme (siehe Schaltzeichnung II)
HER DEMOK
R RUNDFUNK
D S
160
VEB (B) Neontechnik Halle/S.
Außenstelle 1058 Berlin
Kollwitzstraße 92, Tel. 44 56 43
Schaltzeichnung I

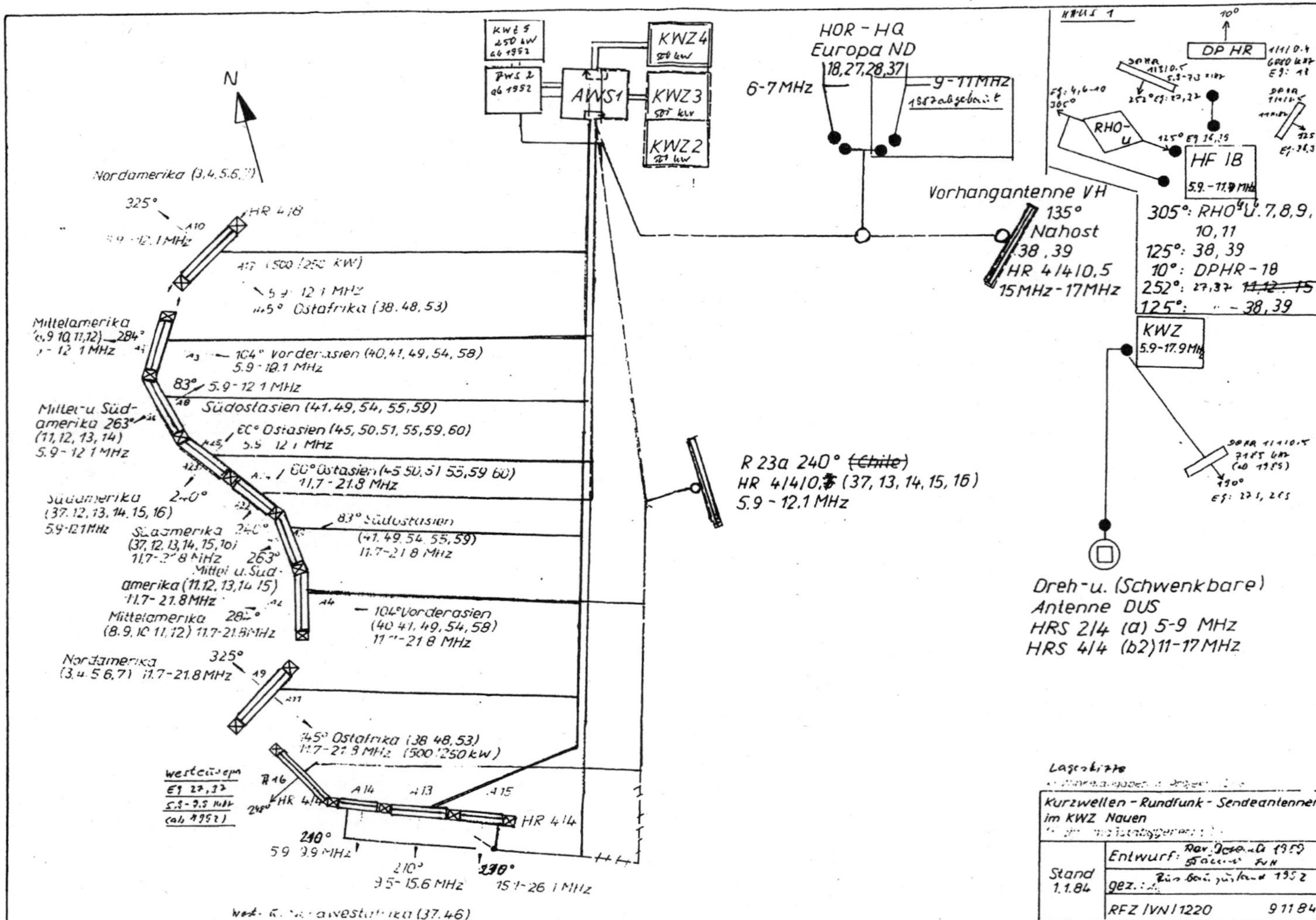
N
KWZ4
AWS1
KWZ3
KWZ2
HOR - HQ
Europa ND
18,27,28,37
6-7 MHz
9-11 MHz
Vorhangantenne VH
135°
Nahost
38, 39
HR 4/4 10,5
15 MHz - 17 MHz
DP HR
RHO-u
HF 1B
5.9 - 11.9 MHz
305°: RHO u. 7,8,9,
10,11
125°: 38, 39
10°: DPHR - 18
252°: 27,37
125°: - 38,39
KWZ
5.9-17.9 MHz
R 23a 240° (Chile)
HR 4/4 10,5 (37, 13, 14, 15, 16)
5.9 - 12.1 MHz
Dreh-u. (Schwenkbare)
Antenne DUS
HRS 2/4 (a) 5-9 MHz
HRS 4/4 (b2) 11-17 MHz
Nordamerika (3,4,5,6,7)
325°
HR 4/8
45° Ostafrika (38, 48, 53)
Mittelamerika
104° Vorderasien (40, 41, 49, 54, 58)
83° 5.9 - 12.1 MHz
Südostasien (41, 49, 54, 55, 59)
Mittel-u. Süd-
amerika 263°
(11, 12, 13, 14)
60° Ostasien (45, 50, 51, 55, 59, 60)
60° Ostasien (45, 50, 51, 55, 59, 60)
11.7 - 21.8 MHz
Südamerika 240°
(37, 12, 13, 14, 15, 16)
83° Südostasien
(41, 49, 54, 55, 59)
Mittel u. Süd-
amerika (11, 12, 13, 14, 15)
11.7 - 21.8 MHz
Mittelamerika 284°
(8, 9, 10, 11, 12) 11.7 - 21.8 MHz
104° Vorderasien
(40, 41, 49, 54, 58)
Nordamerika 325°
(3, 4, 5, 6, 7) 11.7 - 21.8 MHz
45° Ostafrika (38, 48, 53)
11.7 - 21.9 MHz (500/250 kW)
HR 4/4
240°
210°
9.5 - 15.6 MHz
15.1 - 26.1 MHz
Kurzwellen - Rundfunk - Sendeantennen
im KWZ Nauen
Entwurf:
Stand
1.1.84
gez.:
RFZ IVN I 1220
9 11 84

Fig. 161 Funkhaus Berlin-Nalepastrasse, site plan. Source: Deutsche Rundfunkarchiv Babelsberg. F 088-01-02/0003.

Fig. 162 Funkhaus Berlin-Nalepastrasse, site plan showing location of utilities. Source: Deutsche Rundfunkarchiv Babelsberg. F 088-01-02/0005.

Fig. 163 Haus des Rundfunks, level 0. Source: TU Berlin Architekturmuseum, Inv. Nr. 4627.

Fig. 164 Funkhaus Berlin-Nalepastrasse, early sketch of the general scheme by Franz Ehrlich. Source: Archives of the Bauhaus Dessau Foundation. Franz Ehrlich, 233411.

Fig. 165 Funkhaus Berlin-Nalepastrasse, early sketch of the general scheme by Franz Ehrlich. Source: Archives of the Bauhaus Dessau Foundation. Franz Ehrlich, 233405.

Fig. 166 Funkhaus Berlin-Nalepastrasse, exterior view of Block A. Photograph by author.

Fig. 167 Funkhaus Berlin-Nalepastrasse, exterior view. Source: Bundesarchiv. Bild 183-23703-0001.

Fig. 168 Funkhaus Berlin-Nalepastrasse, south elevation of Block A. Source: Deutsche Rundfunk Archiv Babelsberg. F 088-01-02/0005.

Fig. 169 Funkhaus Berlin-Nalepastrasse, east elevation of Block A. Source: Deutsche Rundfunk Archiv Babelsberg. F 088-01-02/0005.

Fig. 170 Funkhaus Berlin-Nalepastrasse, plan fragment of production studios in Block A. Source: Deutsche Rundfunk Archiv Babelsberg. F 088-01-02/0005.

Fig. 171 Funkhaus Berlin-Nalepastrasse, plan fragment of production studios in Block A. Source: Deutsche Rundfunk Archiv Babelsberg. F 088-01-02/0005.

Fig. 172 Funkhaus Berlin-Nalepastrasse, general plan of Block A depicting the series of production studios after the transformation envisaged by Franz Ehrlich. Source: Deutsche Rundfunk Archiv Babelsberg. F 088-01-02/0005.

Fig. 173 Funkhaus Berlin-Nalepastrasse, interior view through the windows of production studios in Block A. Source: Bundesarchiv. Bild 183-B1028-0015-005.

Fig. 174 Funkhaus Berlin-Nalepastrasse, south elevation showing Blocks A and C. Source: Deutsche Rundfunk Archiv Babelsberg. F 088-01-02/0005.

Fig. 175 Funkhaus Berlin-Nalepastrasse, plan of auditorium in Block C. Source: Deutsche Rundfunk Archiv Babelsberg. F 088-01-02/0003.

Fig. 176 Funkhaus Berlin-Nalepastrasse, cross section through Block C. Source: Deutsche Rundfunk Archiv Babelsberg. F 088-01-02/0003.

Fig. 177 Funkhaus Berlin-Nalepastrasse, longitudinal section through Block D. Source: Deutsche Rundfunk Archiv Babelsberg. F 088-01-02/0003.

Fig. 178 Funkhaus Berlin-Nalepastrasse, interior view of the concrete vaults defining Block D. Photograph by the author.

Fig. 179 Funkhaus Berlin-Nalepastrasse, typical floor plan of Block E. Source: Deutsche Rundfunk Archiv Babelsberg. F 088-01-02/0003.

Fig. 180 Funkhaus Berlin-Nalepastrasse, detail of plan of recording studios in Block E. Source: Deutsche Rundfunk Archiv Babelsberg. F 088-01-02/0003.

Fig. 181 Funkhaus Berlin-Nalepastrasse, wall section detail defining recording studios in Block E after the design of Gisela Herzog. Source: Gisela Herzog, "Der Studioneubau des Funkhauses Berlin," *Technische Mitteilungen des RFZ* 10, no. 4 (1966): 144. Scan by PTT-Archiv, Köniz.

Fig. 182 Funkhaus Berlin-Nalepastrasse, window detail for recording studios in Block E after the design of Gisela Herzog. Source: Gisela Herzog, "Der Studioneubau des Funkhauses Berlin," *Technische Mitteilungen des RFZ* 10, no. 4 (1966): 143. Scan by PTT Archiv, Köniz.

Fig. 183 Funkhaus Berlin-Nalepastrasse, interior view of recording studios in Block E. Source: Gisela Herzog, "Der Studioneubau des Funkhauses Berlin," *Technische Mitteilungen des RFZ* 10, no. 4 (1966): 148. Scan by PTT Archiv, Köniz.

Fig. 184 Funkhaus Berlin-Nalepastrasse, interior view of recording studios in Block E. Source: Gisela Herzog, "Der Studioneubau des Funkhauses Berlin," *Technische Mitteilungen des RFZ* 10, no. 4 (1966): 145.

Fig. 185 Funkhaus Berlin-Nalepastrasse, street elevation of Block E. Source: Deutsche Rundfunk Archiv Babelsberg. F 088-01-02/0003.

Fig. 186 Funkhaus Berlin-Nalepastrasse, detail of neon sign built over Block E. Source: Deutsche Rundfunk Archiv Babelsberg. F 088-01-02/0003.

Fig. 187 Plan depicting the transmission antennas of Radio Berlin International, the international radio of the GDR. The plan shows the antenna built specifically to transmit to Chile after the coup d'état in 1973. Source: Bundesarchiv DM303, 1309.

188

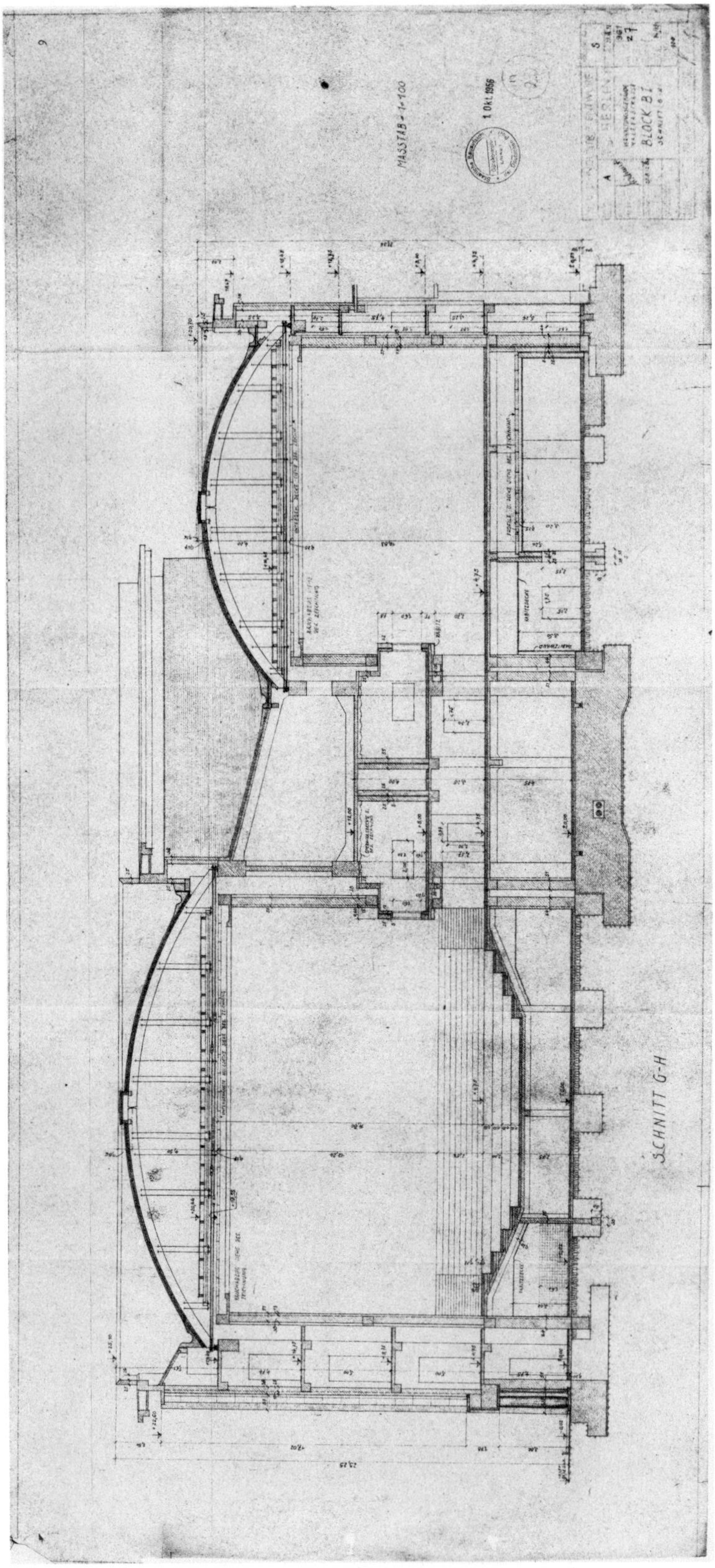

189

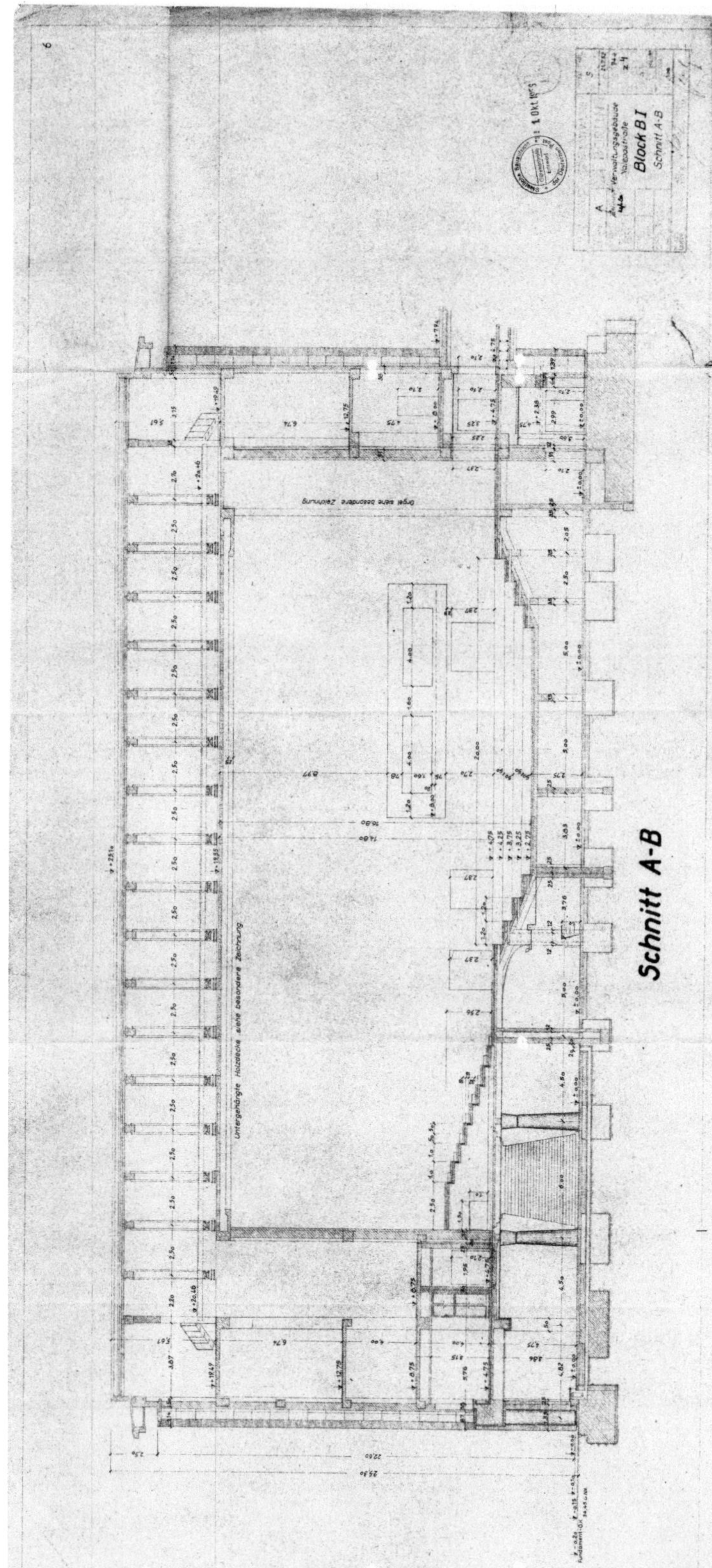
Block B I
Schnitt A-B
Schnitt A-B

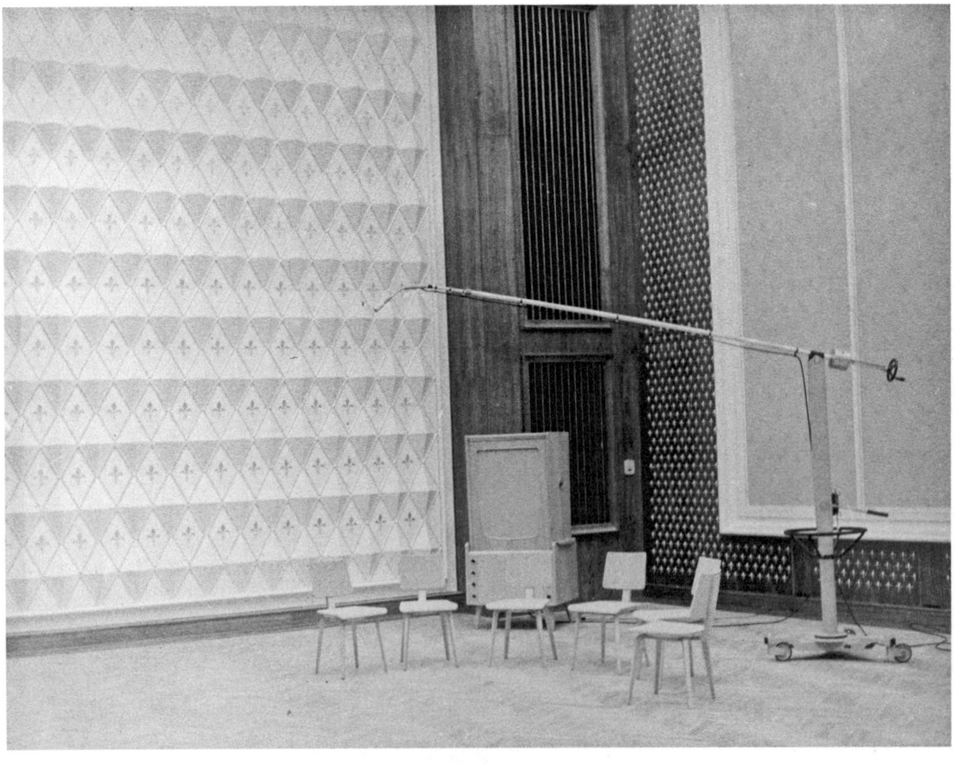

191

196

197

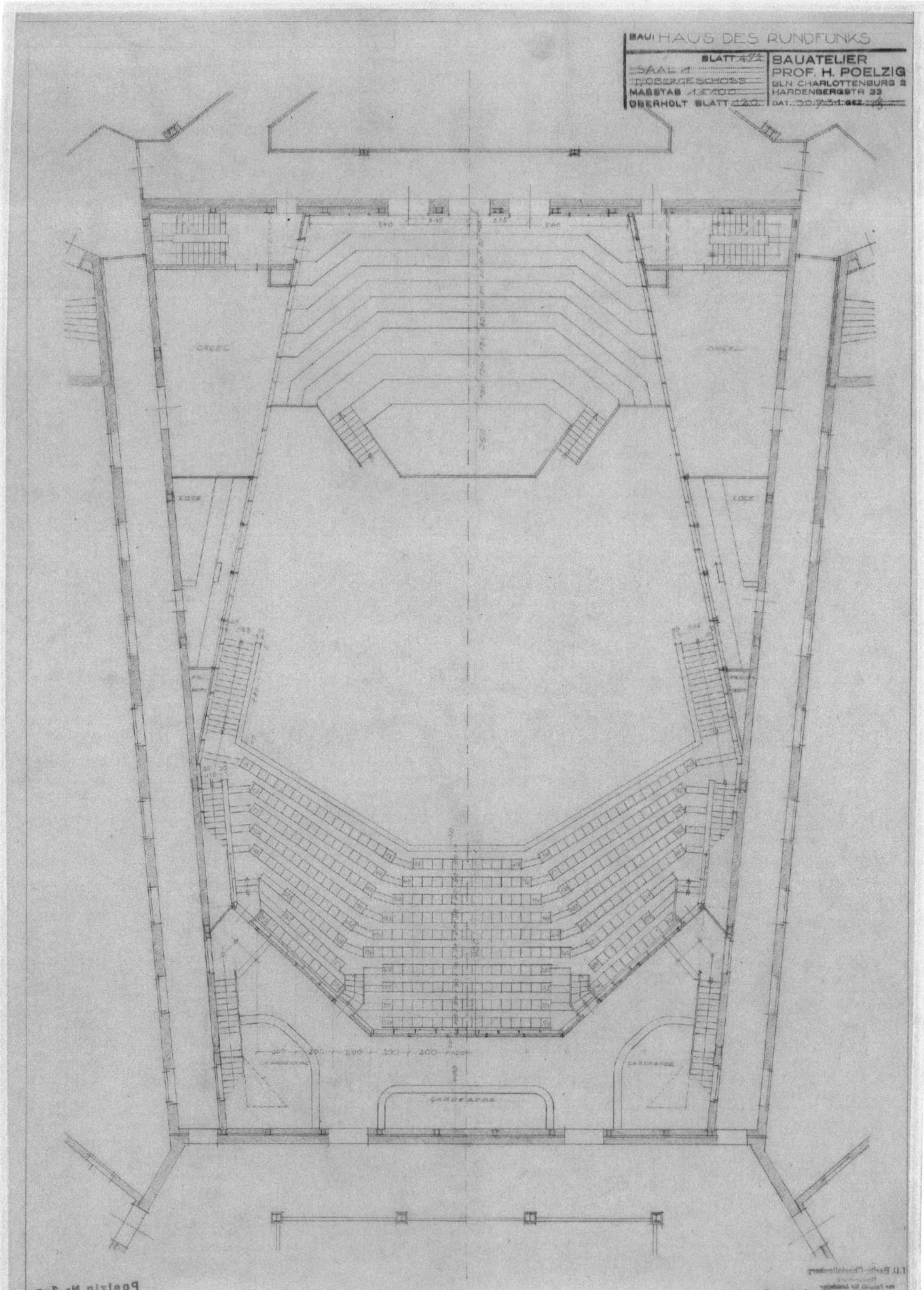
BAU: HAUS DES RUNDFUNKS
BLATT 472
SAAL 1
I. OBERGESCHOSS
MASSTAB 1:200
ÜBERHOLT BLATT 120
BAUATELIER
PROF. H. POELZIG
BLN CHARLOTTENBURG 2
HARDENBERGSTR 33
DAT. 30.7.31. GEZ.
ORGEL
ORGEL
LOGE
LOGE
GARDEROBE
Poelzig Nr. 2043

198

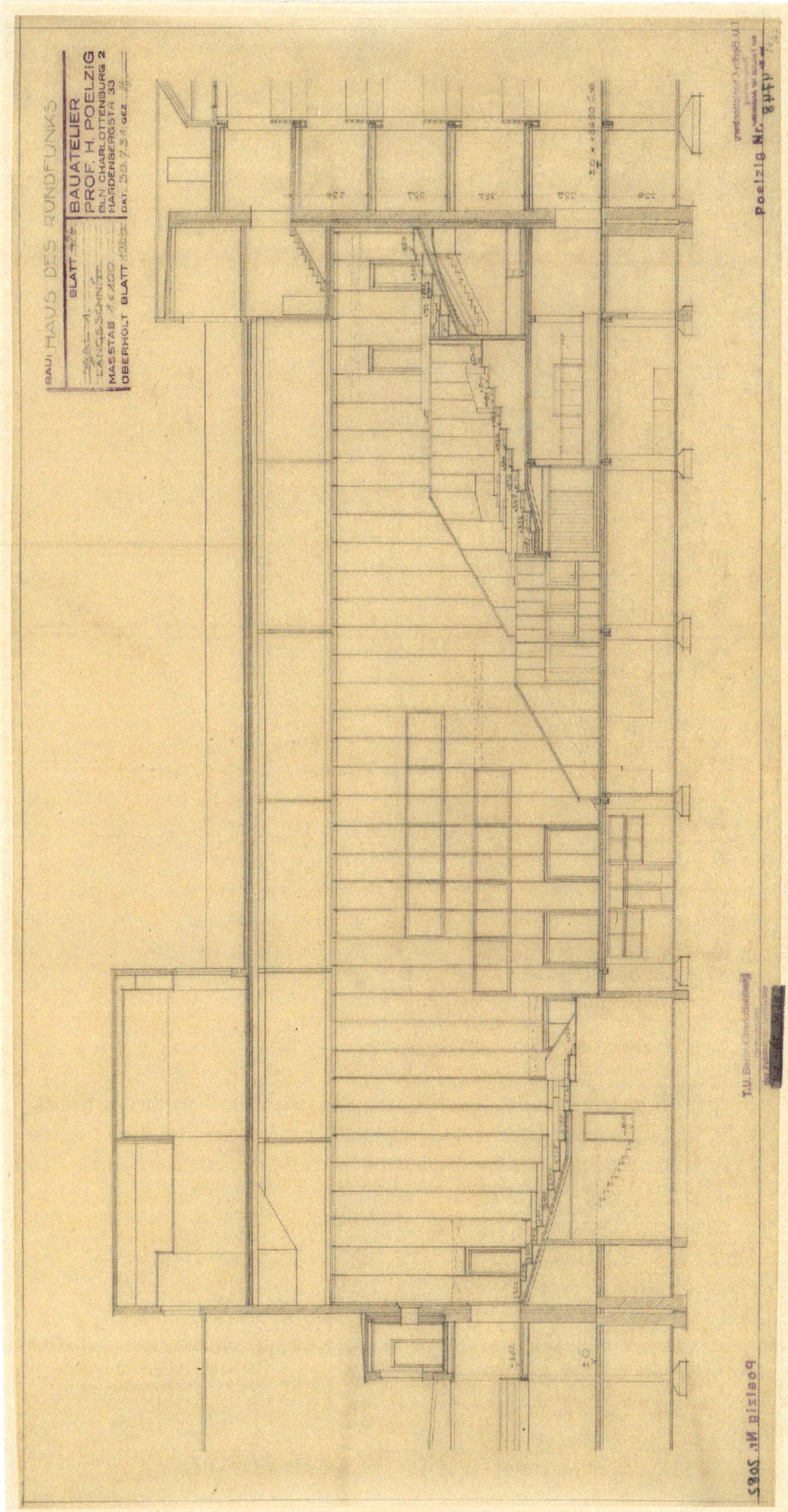
BAU: HAUS DES RUNDFUNKS
LÄNGSSCHNITT
BAUATELIER
PROF. H. POELZIG
BLN CHARLOTTENBURG 2
HARDENBERGSTR 33
Poelzig Nr. 2082

199

Fig. 188 Funkhaus Berlin-Nalepastrasse, interior view of radio play complex in Block B. Source: Deutsche Rundfunk Archiv Babelsberg. F 088-01-02/0003.

Fig. 189 Funkhaus Berlin-Nalepastrasse, cross section through Saal I and Saal II in Block B. Source: Bundesarchiv. DM 311 PLAN/99.

Fig. 190 Funkhaus Berlin-Nalepastrasse, cross section through Saal I. Source: Bundesarchiv. DM 311 PLAN/99.

Fig. 191 Funkhaus Berlin-Nalepastrasse, interior view of Saal IV in Block B. Source: Bundesarchiv. Bild 183-36157-0002.

Fig. 192 Funkhaus Berlin-Nalepastrasse, level 0 of Block B. Source: Photographic reconstruction done by the author after original construction drawings. For the original drawings, see Bundesarchiv. DM 311 PLAN/99.

Fig. 193 Funkhaus Berlin-Nalepastrasse, detail of the wall surface of Saal IV in Block B. Source: Bundesarchiv. Bild 183-36132-0004.

Fig. 194 Funkhaus Berlin-Nalepastrasse, interior view from the control room into Saal I in Block B. Source: Bundesarchiv. Bild 183-36157-0009.

Fig. 195 Funkhaus Berlin-Nalepastrasse, interior view of Saal III in Block B. Source: Bundesarchiv. DM 3/27362.

Fig. 196 Funkhaus Berlin-Nalepastrasse, interior view of the staircase leading to the foyer of Saal I in Block B. Source: Bundesarchiv. Bild 183-36157-0006.

Fig. 197 Funkhaus Berlin-Nalepastrasse, interior view from the foyer of Saal I in Block B. Source: Bundesarchiv. Bild 183-36157-0007.

Fig. 198 Haus des Rundfunks, West Berlin, plan detail of main recording studio. Source: TU Berlin Architekturmuseum. Inv. Nr. 4743.

Fig. 199 Haus des Rundfunks, West Berlin, section detail of main recording studio. Source: TU Berlin Architekturmuseum. Inv. Nr. 4743.

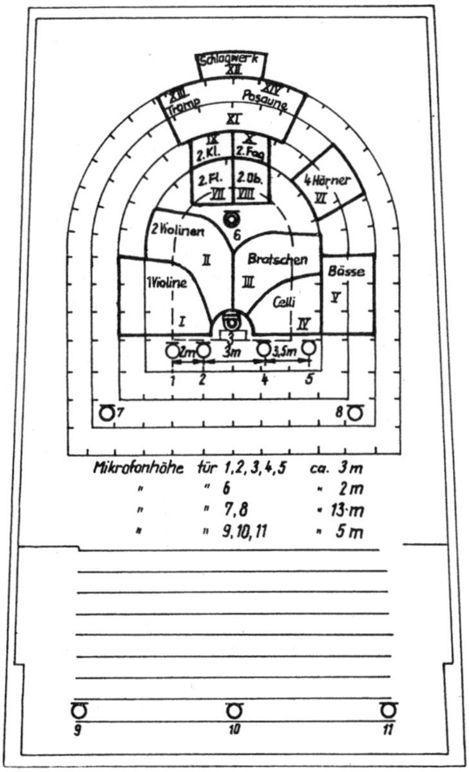

Schlagwerk XII
XIII Tromp.
XIV Posaune
XI
IX 2.Kl.
X 2.Fag
2.Fl. VII
2.Ob. VIII
4 Hörner VI
2 Violinen
II
6
Bratschen
III
1 Violine
I
Celli
IV
Bässe
V
3
2m
3m
3,5m
1 2 4 5
7
8
Mikrofonhöhe für 1,2,3,4,5 ca. 3m
" " 6 " 2m
" " 7,8 " 13 m
" " 9,10,11 " 5m
9 10 11

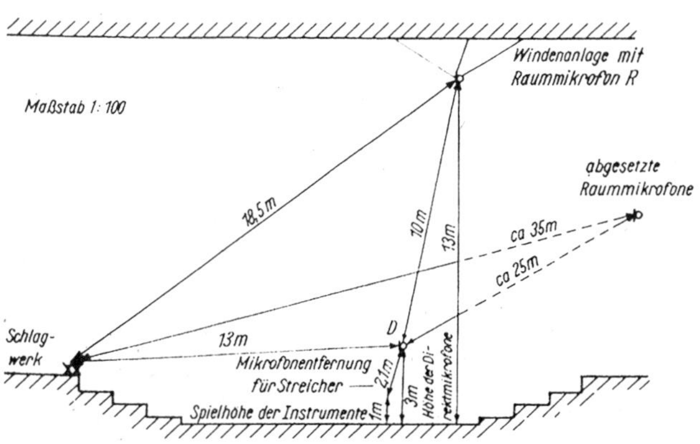

Windenanlage mit Raummikrofon R
Maßstab 1:100
abgesetzte Raummikrofone
18,5m
10m
13m
ca 35m
ca 25m
Schlagwerk
13m
D
Mikrofonentfernung für Streicher
2,1m
3m
Höhe der Direktmikrofone
1m
Spielhöhe der Instrumente

200

201

202

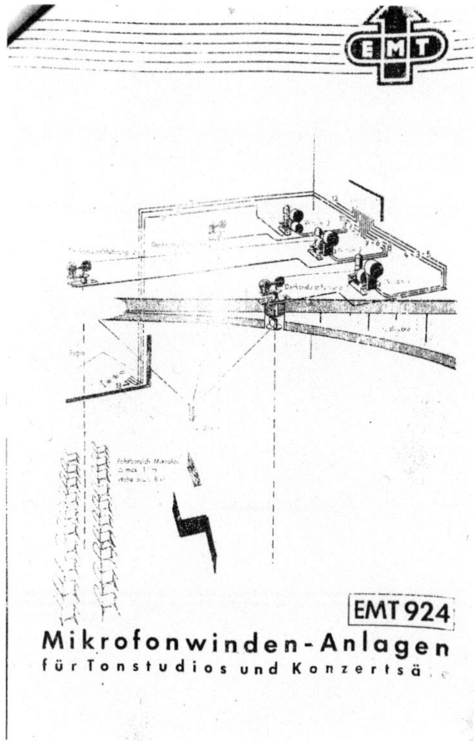
EMT
EMT 924
Mikrofonwinden-Anlagen
für Tonstudios und Konzertsä

203

204

205

206

207

208

209

210

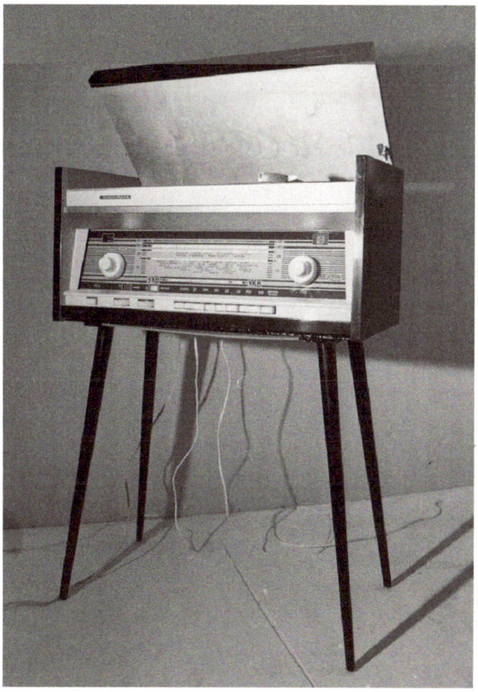

211

212

Fig. 200 Funkhaus Berlin-Nalepastrasse, plan and section scheme depicting the organization of microphones in Saal I, Block B. Source: Klaus Wagner and Wolfgang Hoeg, *Stereofonie-Aufnahmetechnik* (Berlin: VEB Technik, 1970), 82–84.

Fig. 201 Funkhaus Berlin-Nalepastrasse, interior view of Saal I in Block B showing the organ and the microphones replacing the live audience. Source: Bundesarchiv. DM 3/27362.

Fig. 202 Funkhaus Berlin-Nalepastrasse, exterior perspective by Franz Ehrlich depicting Block B and the elevated circulation leading to Block A. Source: Archives of the Bauhaus Dessau Foundation. Franz Ehrlich.

Fig. 203 Brochure for the microphone's automatic mechanical system for Saal I in Block B of the Funkhaus Berlin-Nalepastrasse. Source: Bundesarchiv. DM 301/11250.

Figs. 204–205 Funkhaus Berlin-Nalepastrasse: a staircase in the radio play complex in Block B served as a resonating device for recording footsteps. Source: Bundesarchiv. Bild 183-36965-0006 (Fig. 204). Bild 183-36157-0003 (Fig. 205).

Fig. 206 Funkhaus Berlin-Nalepastrasse, exterior view depicting the elevated circulation connecting Blocks B and A. Source: Franz Ehrlich, "Aufnahme- und Studiogebäude des Staatlichen Rundfunkkomitees," *Deutsche Architektur* 9 (1956): 403.

Fig. 207 Funkhaus Berlin-Nalepastrasse, Saal I could accommodate a symphonic orchestra. Source: Bundesarchiv. Bild 183-36965-0001.

Fig. 208 Funkhaus Berlin-Nalepastrasse, view of a control room in Block B. Source: Bundesarchiv. Bild 183-36965-0003.

Fig. 209 Funkhaus Berlin-Nalepastrasse, view of switch room (*Schaltraum*). Source: Bundesarchiv. Bild 183-J070-0301-007.

Fig. 210 Shortwave antenna built by VEB Funkwerk Köpenick in Nauen. Source: Bundesarchiv. Bild 183-J0701-0301-006.

Fig. 211 Soviet radio receiver inside a domestic space in 1963. As the last link in the chain of the *architecture of radio*, the device stands in its own structure and arranges space inside and around it. Source: DDR Bildarchiv ID 57777.

Fig. 212 Interior view of the SM8/H1 transmission station in Köpenick, East Berlin. Source: Museum Funkeberg.

PART 2
THE BROADCASTING HOUSE

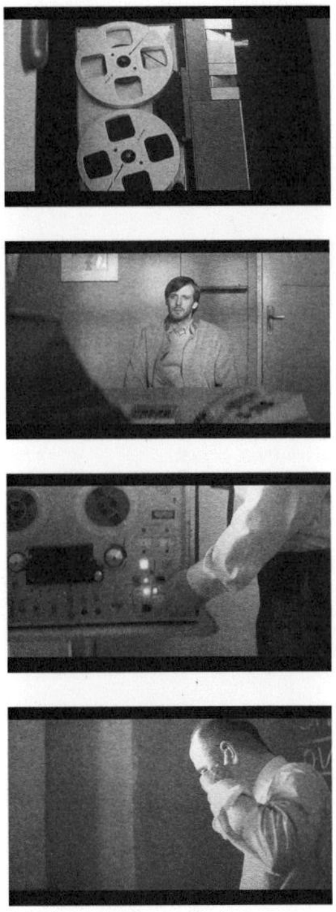

99 → P. 172

Florian Henckel von Donnersmarck's film *Das Leben der Anderen* (*The Lives of Others*) opens with a scene of a Stasi officer interrogating a political prisoner.[1] The action ostensibly takes place in the Stasi's main prison, Hohenschönhausen, but the scene was actually staged in the former GDR broadcasting house, Funkhaus Berlin-Nalepastrasse. With a particular eye—and ear—the film's director collapses into a single space both the fictional setting for the interrogation and the real location for sound broadcasting. (Fig. 99) The foregrounding of these sound-based spaces used for torture, or for radio production, evokes the nature of the built environment in a city divided by the Cold War. The reverberation modulated by the building surfaces faithfully re-enacts a specific sensorial atmosphere and a time, especially in Berlin, when everyday life was saturated in competing reverberations, beats, and frequencies, and architecture played a significant role in their creation and performance.

Construction of the building that partially served as the visually potent and acoustically accurate setting for von Donnersmarck's film began in 1952, after the GDR's radio stations were forced to vacate their former headquarters in the British sector of Berlin. As discussed in the previous chapter, Soviet troops took over the Haus des Rundfunks designed by Hans Poelzig almost immediately after the Red Army entered Berlin in May 1945. That takeover was just the beginning of a larger dispute over material infrastructures that were seen by both sides as key to conquering the apparently immaterial realm of information and communications. In the previous chapter, I analyzed how postwar Berlin was divided not only by the Wall but also by the overlapping effects of radio signals, transmitted and received on both sides of the divide. At the territorial scale, radio delineated another set of thresholds defining the space of the city. Its apparently immaterial constitution and agency were determined by material artifacts on the ground. If the previous chapter focused on radio activities at the scale of Berlin's overall urban territory,

1 *Das Leben der Anderen/The Lives of Others*, dir. Florian Henckel von Donnersmarck (Germany: Wiedemann & Berg Film GmbH, 2006).

this chapter will examine the architectural scale, looking at how ideas about technology and typology became entangled in—and through—the broadcasting house.

I

In the extensive war of information that defined to a large degree the structure of Berlin, the GDR's State Radio Committee, established in 1952, had a significant impact at different scales of operation.[2] On one level, it orchestrated the creation and dissolution of radio stations and institutions: all radio production was centralized in Berlin, with the entire workforce and all technological resources being moved to the politically divided capital. But at the same time, it oversaw the building of a large number of transmission stations across the territory of the GDR, under the aegis of the first Five-Year Plan.[3] This double move, involving the centralization of production and the expansion of transmission, aimed to construct a specific kind of space—one characterized not only by the sonic signature with which radio signals were encoded, but by the intensity at which those signals were transmitted. The greater the intensity, the more contested the space.

The legal framing of the reform of the GDR's media institutions clearly states that the centralization of radio production in Berlin was intended to give "the population in the German Democratic Republic and in West Germany the opportunity to receive three different, carefully coordinated, high-quality and all day programs of German democratic broadcasting."[4] To this end, the State Radio Committee repurposed and renamed three stations in 1952. Berliner Rundfunk, originally founded in 1945, just a few days after the end of the war, was refocused to provide political commentary. The content of Radio DDR was reoriented towards culture, science, and educational programs. Finally, the broadcasts of Deutschlandsender were redirected towards audiences in the West.[5] Within the new program proposed by the

2 For more on the formation of the State Radio Committee in 1952, see "Verordnung über die Bildung des Staatlichen Rundfunkkomitees," *Gesetzblatt der Deutschen Demokratischen Republik*, no. 112 (August 21, 1952): 733–734. See also chapter 1.

3 The Five-Year Plans in the GDR were a set of economic goals aiming to reduce the development lag vis-à-vis the Federal Republic. They were implemented after the State Planning Commission was formed in 1950 and they covered the periods 1951–1955 and 1956–1960. See André Steiner, "From Soviet Occupation Zone to 'New Eastern States,' in *The East German Economy, 1945–2010: Falling Behind or Catching Up,* ed. Hartmut Berghoff and Uta Andrea Balbier (Cambridge: Cambridge University Press, 2013), 17–52.

4 "Verordnung über die Bildung des Staatlichen Rundfunkkomitees," 733. Author's translation.

5 Initially the three stations were renamed Berlin I, II, and III, but those names lasted only for one year and, in 1953, the names of Berliner Rundfunk and Deutschlandsender were reinstated while Berlin II was renamed Radio DDR. For more on the institutional reorganization of radio institutions in Berlin between 1945 and 1952, see Maral Herbst, *Demokratie und Maulkorb: Der Deutsche Rundfunk in Berlin zwischen Staatsgründung und Mauerbau* (Berlin: Vistas, 2001), 37–55. In 1959, another radio station was created in the GDR, Radio Berlin International. Also centrally directed by the State Radio Committee, it was intended to serve as the international voice of the GDR. See Heinz Odermann, *Wellen mit tausend Klängen: Geschichten rund um den Erdball in Sendungen des Auslandsrundfunks der DDR Radio Berlin International* (Berlin: Vistas, 2003). For more on the transmission of Radio Berlin International from the station in Nauen, see James Wood, *History of International Broadcasting*, 2 vols. (London: Institution of Electrical Engineers, 2000), vol. 2, 56–59.

100 → P. 172

6 For more on the role of music in the GDR, see Christoph Classen and Klaus Arnold, eds., *Zwischen Pop und Propaganda: Radio in der DDR* (Berlin: Christoph Links, 2004). Particularly relevant for a young audience was the launching in 1964 of the program DT64 broadcast through Radio DDR. See also Christopher Classen, "Captive audience? GDR Radio in the Mirror of Listener's Mail," *Cold War History* 13, no. 2 (2013): 239–254; Edward Larkey, "GDR Rock Goes West: Finding a Voice in the West German Market," *German Politics & Society* 23, no. 4 (Winter 2005): 45–68. For more on the relationship between music and national identity in East Germany, see Laura Silverberg, "East German Music and the Problem of National Identity," in *The Communist Quest for National Legitimacy in Europe, 1918–1989*, ed. Martin Mevius (New York: Routledge, 2011), 123–145.

committee, music was to play a particularly significant role in constructing and disseminating political ideology: to a certain extent, it would become *the* vehicle for the symbolic transmission of information. On all three GDR stations, speech-based programs were limited to thirty-six percent of the total airtime: the rest had to be devoted to playing music. This apparently superficial change in programming in fact signaled a deep transformation in the development of radio during the Cold War period. As political speech was particularly unpopular with audiences (who could switch freely between the rival radiophonic spaces of the East and West), music—and a particular approach to transmitting music at that—became the state's preferred messaging medium on both sides of the Wall. The idea of the media reform in the GDR was to educate the population and, at the same time, to create content that was culturally distinct from the "capitalist," consumer-oriented broadcasts from the West. This ambition called for a new type of building that, in creating a specific kind of music, would also engender a specific sonic signature of socialist broadcasting space.[6]

One building in particular would give material form to the complex process of institutional reform of the media, which reflected equally complex transformations in both politics and technology. Construction of the Funkhaus Berlin-Nalepastrasse, the new headquarters for radio in the GDR, would begin in 1952, the same year the GDR's State Radio Committee was formed. (Fig. 100) The process, however, was far from linear. Recognizing the precarity of the GDR's broadcasting base in the Western sector, plans for an alternative facility to the Poelzig building in Masurenallee were made as early as 1949. The project was lent added urgency in 1950, when the power supply to the building was cut off, forcing the GDR stations to rely on an independent electricity generator. The following year, the architect Franz Ehrlich was commissioned to design a new media complex in the Köpenick area of East Berlin. By the summer of 1952, when the GDR was forced to abandon its established base, the new building was already taking shape.[7] In a letter sent to the Ministry of Construction, Ehrlich proudly stated, "I came to this commission with all my previous experience, and that allowed me to devise a design scheme in only eight days."[8] The "previous experience" that had facilitated this quick response spoke of an unorthodox training, with the biography of the architect being just as complex and multilayered as the Funkhaus on Nalepastrasse itself.

Born in Leipzig in 1907, Franz Ehrlich was the first of his family to attend university. He studied at the Bauhaus in Dessau between 1927 and 1930 and was a student of Moholy-Nagy in the period when he took his celebrated photographs of the recently completed Berlin radio tower.[9] (Fig. 34) Another Bauhaus master, Oskar Schlemmer, was eagerly working to "dissolve the subject" into circular radiating patterns, as expressed in the design of the costumes for his performance pieces.[10] For Ehrlich, "Schlemmer's thoughts about man and space were a natural prerequisite for all work done at the Bauhaus."[11] At Dessau at the same time, Siegfried Ebeling was reconsidering the notion of the wall in terms of a permeable and responsive membrane, thinking of space in its relation to electromagnetic radiation.[12] Ehrlich took classes on color with Wassily Kandinsky and on form with Paul Klee.[13] He had to work to finance his studies, since there were no fellowships at the Bauhaus.[14] In 1928, he worked for Walter Gropius on the Total Theater for Erwin Piscator,[15] and in 1930 he interned in the office of Hans Poelzig: while he mostly worked on the project for the Ukrainian State Theater in Kharkiv, he would undoubtedly have been familiar with the Haus des Rundfunks, which was being developed at the time.[16] His last documented project at the Bauhaus, a collaboration with his typography tutor Joost Schmidt, was for the design of the Junkers installation for the 1929 "Gas and Water" exhibition in Berlin.[17]

7 For more on the early alternative plans to replace the building in Masurenallee, see the confidential letter addressed to "Genosse Eisler," December 10, 1949, in Bundesarchiv, DR 6, 323.

8 Correspondence between Franz Ehrlich and the Staatssekretariat für Bauwirtschaft, Mr. Mayer, November 12, 1952, in Bundesarchiv, DH 1/43892. Author's translation. The selection of the site for the new broadcasting house and the initial conversations with Franz Ehrlich took place in March and April 1951. For more on the legal protocols and early design studies for the site, see "Vorprojektierungen 1951–1952 Verwaltungsgebäaude HV Bau-Industrie Nalepastrasse VVB Industrie-Entwurf, June 7, 1951," in Deutsches Rundfunkarchiv Babelsberg, F 088-01-02/0005.

9 Two of Franz Ehrlich's works are featured in László Moholy-Nagy's *Von Material zu Architektur* (Mainz; Berlin: Florian Kupferberg, 1929), 116.

10 Noam M. Elcott, *Artificial Darkness: An Obscure History of Modern Art and Media* (Chicago: University of Chicago Press, 2016), 175–182.

11 Stiftung Bauhaus Dessau, I 11632/67D-69D. Author's translation.

12 Siegfried Ebeling, *Der Raum als Membran* (Dessau: C. Dünnhaupt, 1926). For an English translation, read Siegfried Ebeling, Spyros Papapetros, Walter Scheiffele, and Pamela Johnston, *Space as Membrane* (London: AA Publications, 2010).

13 Lutz Schöbe, "Von der Fläche zum Raum—Franz Ehrlich," *Form+Zweck* 3 (1986): 27–36.

14 Correspondence between Franz Ehrlich and the Staatssekretariat für Bauwirtschaft, Mr. Mayer, November 12, 1952, in Bundesarchiv, DH 1/43892.

15 Franz Ehrlich, "Bauhaus und Totaltheater," *Wissenschaftliche Zeitschrift der Hochschule für Architektur und Bauwesen* 29, nos. 5/6 (1983): 424. For models and original drawings related to the project Walter Gropius developed for Erwin Piscator, see Total Theater for Erwin Piscator, Berlin, 1927, Harvard Art Museums, HUAM169343.

16 The curriculum vitae elaborated by Franz Ehrlich can be consulted at Stiftung Bauhaus Dessau, I 11632/1D. Bernhard Kohlenbach, "Franz Ehrlich—ein Architekt zwischen Bauhaustradition und DDR-Baudoktrin," *ICOMOS—Hefte des Deutschen Nationalkomitees* 20, (2015): 44–47, here 45.

17 Whether Ehrlich officially graduated from the Bauhaus remains unclear. The

In 1930, Ehrlich left the Bauhaus in protest because he no longer agreed with the design principles and working methods taught there.[18] His departure coincided with a spate of protests at the school, all sparked by the ousting of its second director, Hannes Meyer, due to his political alignment with the left.[19] Ehrlich moved to Berlin and set up an independent practice, Studio Z, together with the photographer Heinz Loew and the painter Fritz Winter. He worked as an artistic director for the publishing company of Otto Bayer and as a consultant to the trade fair in Leipzig. Like many others from the Bauhaus, he was politically aligned with the left. Though not a member of either the German Communist Party (KPD) or the militant communist student faction, Kostufra, he did work in Leipzig for the illegal communist journal *Junge Garde*, which for the Gestapo was sufficient reason to arrest him in August 1934 and to charge him with high treason—a charge to which he pleaded guilty.[20] Ehrlich would spend the next eleven years, from 1935 till the end of World War II, in various forms of imprisonment. And it was those years that formed his later career.[21] If the Bauhaus sparked early ideas and defined the character of a young idealist, then the war made him a professional architect.

From 1935 to 1937, Ehrlich was imprisoned in Schloss Osterstein in Zwickau, Saxony. In those early years, he continued his graphic and furniture design, at least on paper, as a form of therapy to avoid a mental breakdown. In 1937, he was transferred to Buchenwald concentration camp near Weimar, which was then under construction.[22] Exempted from heavy labor at the stone quarry, he worked as a carpenter and, in 1938, was transferred to the building unit, along with other inmates from a professional background, who were forced by the SS to help with the construction, maintenance, and operation of the camp. His unit was led by the architect, and committed National Socialist, Robert Riedl. Ehrlich initially designed the office interiors and furniture, and also assisted in the general planning of the infrastructure and water, and energy supply

most accurate and recently published biography, however, states that that he did not get a diploma. See Friedrich von Borries and Jens-Uwe Fischer, *Gefangen in der Titotalitätsmaschine: Der Bauhäusler Franz Ehrlich* (Berlin: Suhrkamp, 2022), 83. See also Schöbe, "Von der Fläche zum Raum," 27–36.

18 Correspondence between Franz Ehrlich and the Staatssekretariat für Bauwirtschaft, Mr. Mayer, November 12, 1952, in Bundesarchiv, DH 1/43892.

19 Philipp Oswalt, ed., *Hannes Meyer neue Bauhauslehre: Von Dessau bis Mexiko* (Basel: Birkhäuser, 2019); and Zeynep Çelik Alexander, *Kinaesthetic Knowing: Aesthetics, Epistemology, Modern Design* (Chicago: University of Chicago Press, 2017), 127–201. For a condensed version of Meyer's vision of architecture for the Bauhaus see Hannes Meyer, "Bauen," in *Bauhaus Zeitschrift für Gestaltung* 2, no. 4 (1928): 12–13. For the English translation, see Hannes Meyer, "Building," in *Programs and Manifestoes on Twentieth-Century Architecture*, ed. Ulrich Conrads (Cambridge, MA: MIT Press, 1970), 117–120.

20 For an early and detailed biographical account on Franz Ehrlich, see Lutz Schöbe, "Franz Ehrlich: Beitrag zu einer Monographie" (Master's thesis, Humboldt University, 1983).

21 For a historical account on how war has shaped the development of architects' careers and the way in which architecture is practiced, see Jean-Louis Cohen, *Architecture in Uniform: Designing and Building for the Second World War* (Paris: Hazan, 2011).

22 See Volkhard Knigge et al., *Franz Ehrlich: Ein Bauhäusler in Widerstand und Konzentrationslager* (Weimar: Stiftung Gedenkstätten Buchenwald und Mittelbau-Dora, 2009). See also Michael H. Kater, *Weimar: From Enlightenment to the Present* (New Haven: Yale University Press, 2014), 247.

for the camp. His most visible work from this period is the infamous sign, JEDEM DAS SEINE (To Each His Due), placed over the entrance gate. The words are spelled out using the typographical language he had developed at the Bauhaus under the guidance of Joost Schmidt. The exact role played by Ehrlich in the overall design of the concentration camp has yet to be established. From his own description of his years in Buchenwald, however, it is clear that he gained his first professional experience as an architect there.[23]

"Terror wave 4. Continuation of the stake-hanging method, going over the 'trestle,' gate-standing, first attempt at gassing."[24] This is from Ehrlich's lapidary, but thorough, account of the torture and killing mechanisms he observed at Buchenwald. Going over the "trestle" meant the prisoner's buttocks were beaten with a bullwhip until the skin burst open. "Gate-standing" required the victim to stand in the open with their arms raised. If they lowered their arms or fell over, they were beaten until they stood up again or died. He describes the escalation of horror: "Terror wave 5. Jew Action. Goebbels admits at the party congress that sixty Jews die a day at a concentration camp near Weimar because they're working from morning to night. Prisoners are nailed into boxes covered with barbed wire, which are lined up at the gate until they die of starvation. Shooting prisoners. Guards order them to be carried over the guard chain so they can shoot them in greater comfort. Number of dead in this wave not possible to determine."[25] After noting the "first attempts by camp physicians to kill by air injection into the bloodstream," he states that in "October 1939, the first Poles were brought to the camp and completely exterminated by just abandoning them on the ground."[26] Bearing witness to the scale of this extermination, Ehrlich describes how "it happened that corpses fell from a truck loaded with more than one hundred of them, and they remained on the streets of Weimar for a long time. By the end of 1939, all Weimar residents knew what was going on in Buchenwald."[27] Ehrlich also gives an account of the technological advances in the

23 For more on the role of Ehrlich in the concentration camp, see Knigge et al., *Franz Ehrlich*, and von Borries, *Gefangen in der Titotalitätsmaschine*, 86–149. For a list of his works at Buchenwald concentration camp, see "Handwerk im Buchenwald, undated," in Stiftung Bauhaus Dessau, I 11632/24D-25D.

24 Franz Ehrlich's description of his experience in Buchenwald, undated, in Stiftung Bauhaus Dessau, I 11634/31D-36D, here 31D. This and subsequent quotes are translated by the author.

25 Stiftung Bauhaus Dessau, I 11634/31D.

26 Stiftung Bauhaus Dessau, I 11634/32D.

27 Stiftung Bauhaus Dessau, I 11634/32D.

machinery of death, writing about the "prototype of a portable crematorium developed by Topf & Sons in Erfurt. Involved all the leading engineers and technicians of the company."[28] Though he positions himself throughout as an outsider to these actions, it is clear from his descriptions that he learned a great deal about architecture and its capacity to do great harm, as well as its potentially subversive power. He describes the "countermeasures" he engaged in, relating that "political prisoners employed in the photography department [Foto Abteilung] took over the running of the construction program and set up a radio transmitter powered by the electricity generator for the commandant's office. By both means, with the involvement of SS men, messages about the situation here reached the public."[29] As crude as it might sound, Ehrlich's experience at the concentration camp taught him something he had not learned at the Bauhaus: namely, how to professionally conceive architecture in terms of processes and functions. The architect who would go on to redesign the entire media infrastructure of East Germany after the war was, for better or worse, educated under these conditions.

28 Stiftung Bauhaus Dessau, I 11634/32D.

29 Stiftung Bauhaus Dessau, I 11634/31D.

In 1943, Ehrlich was transferred from Buchenwald to Strafbataillon 999, a punishment battalion formed the previous year to put political prisoners to work in the service of the Wehrmacht. He was sent to the Peloponnese, occupied since 1941, at a particularly difficult time, when the Nazis were losing territory to the Allies. Based in Amaliada, in northwestern Elis, he was "assigned to operate as a shooter and battalion's detector with the task of processing tactical operations of the IA [General Staff Officer of the battalion]. "In this function," Ehrlich writes:

> I led groups of dozens of soldiers throughout the defense area of Elis-South. In the uprising in May 1944, twelve of our comrades were shot and most of the comrades were transported back to Germany to the military training area in Baumholder. Then, my job was to develop

> the overall defense plan for the Elis-South area. Performing this role, I managed to master the tactical use of all the weapons of a division. My acquired knowledge was so extensive that the commander systematically asked me to give classes to the officers. The tactical and strategic experiences that I gained became the organizational basis for the Wehrmacht in the Peloponnese.[30]

Miraculously, Ehrlich survived his experience in the punishment battalion, and his words again reveal how formative the experience was for him. After the end of the war in 1945, he was sent as a POW to work on the reconstruction of Yugoslavia: "In this operation, we rebuilt all the railway facilities that were destroyed between Belgrade and Agram, we restored many factories and converted the synagogue in Panjewa into a theater. My responsibility here was to create all the necessary planning and construction documents for these projects."[31] Viewed as a loyal communist, Ehrlich would be sent back to East Germany in 1947 to oversee the reconstruction of Dresden. This was a role he considered himself well prepared to take on, on account of his "accumulated experience in the architectural, structural engineering, economic, and political fields" during the war.[32]

In 1950, Ehrlich became technical director of VVB Industrie-Entwurf, a satellite agency of the Ministry of Construction that focused on industrial buildings. Coinciding with the launch of the first of the GDR's Five-Year Plans, this was his most prestigious public appointment, gained after years of struggling for recognition in the GDR's architectural and cultural hierarchy.[33] A few years later, he would be absorbed into another sphere of the totalitarian regime. From February 1954, he worked as an informant for the Stasi, under the pseudonym of GI Neumann. Reporting to the Stasi's Third Division, devoted to "Radio Reconnaissance and Radio Defense," he was seen as useful for "processing and informing about hostile elements acting in the GDR construction

30 Correspondence between Franz Ehrlich and the Staatssekretariat für Bauwirtschaft, Mr. Mayer, November 12, 1952, in Bundesarchiv, DH 1/43892. This and subsequent quotes are the author's translation.

31 Bundesarchiv, DH 1/43892.

32 Bundesarchiv, DH 1/43892. Ehrlich also describes his experience in Dresden in his brief postwar correspondence with Walter Gropius, who in 1947 was already installed as chair at Harvard GSD. See Walter Gropius papers, MS Ger 208, (677). Houghton Library, Harvard College Library.

33 For more on the economic plans in East Germany after World War II, see Hartmut Berghoff and Uta Andrea Balbier, eds., *The East German Economy, 1945–2010* (Cambridge: Cambridge University Press, 2013). For a list of all the buildings Franz Ehrlich did while working for the VVB Industrie-Entwurf Berlin, see Stiftung Bauhaus Dessau, I 11632/126D-145D.

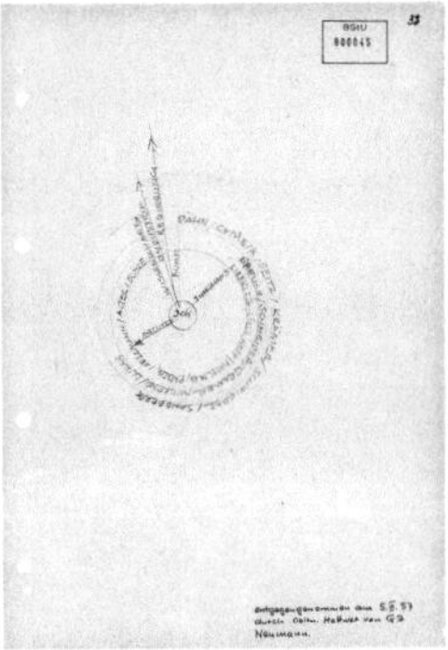

101 → P. 173

industry, in particular by providing information about leading figures in the Ministry and the Bauakademie."[34] (Fig. 101) In a diagram drawn by Ehrlich himself, we see the complex role he was playing, as an architect, within the entangled political and cultural circles of the GDR. He is positioned, of course, at the center of the action, using his knowledge both to craft the institutional reform of the media industry and to orchestrate the movements of the different players, artists, and scientists alike.[35]

34 Profile of Franz Ehrlich as GI Neumann signed by Mr. Kalipke, September 18, 1958, in Bundesarchiv, MfS AIM 6503/75, Band 1. Author's translation. For more on Ehrlich's role as an informant, see also Bundesarchiv, MfS AIM 6503/75, Band 2.

35 There is an uncanny parallel between the diagram drawn by Ehrlich and the diagrams of the expertise of "primitive man" drawn and published by László Moholy-Nagy on page 10 of his book, *Von Material zu Architektur*.

Spanning from the Weimar Republic to Cold War Germany, Ehrlich's career traces out transformations of the broadcasting house typology and the growing impact of radio on the way space was conceived, created, and perceived. Not by chance, the new Funkhaus on Nalepastrasse shared certain morphological and stylistic characteristics with Poelzig's building. The long frontal facade, with its echoes of brick expressionism, recalled its Western predecessor. Ehrlich's design even retained the studios attached on one side, with their characteristic curvilinear corridor, although they were no longer arranged symmetrically. As a whole, however, the building was shaped by the internal forces of the GDR's broadcasting program, by the external political forces prevailing at the time, and by new developments in the technology of broadcasting. Completed in 1956, the Funkhaus was the most complex iteration of the broadcasting house typology.

II

The Funkhaus Berlin-Nalepastrasee was the product of a turbulent period of political and institutional reform in Berlin following its political division in 1949. However, it should also be understood as the concretization of a much longer trajectory linking construction with the production and transmission of information through sound and electronic technical media. That process began in Berlin in 1928, when the Reich Broadcasting Corporation (Reichs-Rundfunk-Gesellschaft, RRG) launched an architectural competition to design what was touted as "the broadcasting house of the future."[36] Created in May 1925, RRG was the centralized body that not only administered local radio stations but also incorporated the national station, Deutsche Welle, which transmitted on longwave. From the early days of radio's transition from *technical medium* to *mass media*, broadcasting in Germany was established as a state monopoly.[37] Keeping radio under state control was seen as a way to prevent the misuse of the airwaves by radicalized political groups. It offered an alternative to the highly commercialized model of the United States, where from the very beginning radio had been operated by private stations in alliance with technological entrepreneurs—a set-up that, again, would have been likely to undermine the fragile political stability of the Weimar Republic. When the RRG was created, all local radio stations in Berlin, Leipzig, Frankfurt, Munich, Hamburg, Stuttgart, Breslau, Königsberg (now Kaliningrad), and Münster were required to transfer fifty-one percent of their stock to the new institution. In turn, fifty-one percent of the stocks of RRG were owned by the Reich Postal Ministry. The RRG's first chairman was Hans Bredow, a radio pioneer and early supporter of radio broadcasting with important ties to the industry. Bredow also had a seat on the board of each of the local stations in his capacity as radio commissioner (*Rundfunkkommissar*) of the Reich Postal Ministry. By these means, the Weimar Republic secured centralized control of the airwaves throughout its territory.[38]

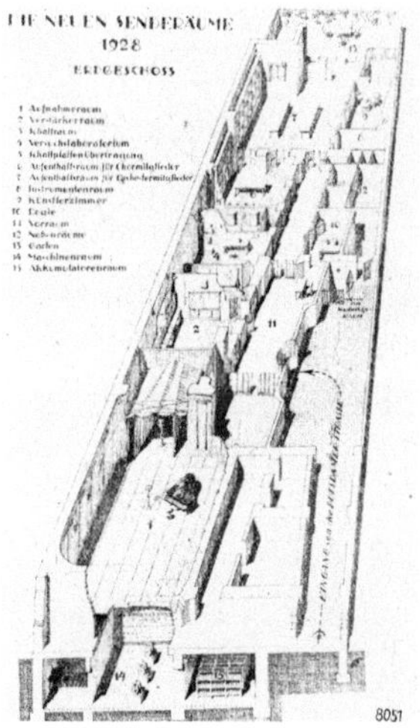

104 → P. 174

36 Werner Menzel, "Blick in Deutschlands Rundfunk-Zukunft," *Funk* 18 (1929): 75–76, here 75. Author's translation.

37 Hans Jürgen Koch and Hermann Glaser, *Ganz Ohr: Eine Kulturgeschichte des Radios in Deutschland* (Cologne: Böhlau, 2005), 5–66.

38 Peter Jelavich, *Berlin Alexanderplatz: Radio, Film, and the Death of Weimar Culture* (Berkeley: University of California Press, 2006).

39 Jelavich, *Berlin Alexanderplatz*, 41.

(Fig. 104) Before the creation of the RRG—and from the time of the very first radio transmission in 1923—radio production in Berlin was concentrated in Vox-Haus on Potsdamer Strasse. Vox, a phonograph company, financed the city's first local station, Funkstunde Berlin, with the aim of promoting sales of its records.[39] The station occupied a prototypical studio converted from the top two floors of a residential building. In 1928, the facility was renovated by Swiss architect Otto Rudolf Salvisberg

102 → P. 174

103 → P. 174

105 → P. 174

108 → P. 177

and the RRG's chief sound engineer Walter Schäffer.[40] (Fig. 102, 103, 105) The interiors Salvisberg crafted display Jugendstil features likely to appeal to the cultivated bourgeoisie. Photos of the renovation celebrate the geometric figures and patterns decorating the ceilings and the wood paneling of the walls. Strangely absent from the photographs are the tent-like forms inserted into the plan, which appear as structures-within-the-structure—a form of camouflage concealing the unruly presence of wires and waves. Salvisberg's project gives a representative image to a new institution—one that reflects the effort to control, through design, the still untamed presence of electronic technical media. The evident mismatch between this representation and the reality—the inability of the repurposed structure of Vox-Haus to keep pace with the ambitions of the growing radiophonic enterprise—came to the attention of Walter Benjamin, who was himself involved in the making of radio shows from the mid-1920s. Benjamin blamed the "technological and formal aspect" of the process of recording and transmitting sound—the fundamental separation between performer and audience—for making "the most interesting shows unbearable."[41]

There was a lot at stake, then, when the competition for the new purpose-built broadcasting house was launched by the RRG in 1928. The overall goal was to give identity, formal expression, and logic to an institution responsible for crafting a new form of communication with a reach that covered the entire nation—and beyond. The new building had to accommodate not only the production facilities of Funkstunde Berlin but also the administration and production spaces of the national station, Deutsche Welle, as well as the headquarters of the RRG. There were few, if any, precedents for such a program mix, which was neither purely industrial nor exclusively cultural.

(Fig. 108) The competition entry by Hans Poelzig prevailed over designs by Paul Bonatz (Stuttgart), Richard Riemerschmid (Cologne), and Friedrich

40 Ludwig Kapeller, "Die neue Senderäume der Berliner Funkstunde," *Funk* 38 (1928): 269–270.

41 Walter Benjamin, "Reflections on Radio," in *The Work of Art in the Age of Its Technological Reproducibility and Other Writings on Media*, ed. Michael W. Jennings, Brigid Doherty, and Thomas Levin (Cambridge, MA: Harvard University Press, 2008), 391–392. The text, unpublished during Benjamin's lifetime, first appeared in Walter Benjamin, "Reflexionen zum Rundfunk," in *Gesammelte Schriften*, 2 vols. (Frankfurt am Main: Suhrkamp, 1980), 1506–1507. For more on Walter Benjamin and radio, see Jeffrey Mehlman, *Walter Benjamin for Children: An Essay on His Radio Years* (Chicago: University of Chicago Press, 1993); Jonathan Lutes, Lisa Harries Schumann, and Diana Reese, eds., *Radio Benjamin* (London: Verso, 2014).

Eugen Scholer (Stuttgart).[42] A perspective shows the 150 m main facade along the north side of Masurenallee, directly opposite the entrance to the exhibition grounds designed by Poelzig and Martin Wagner a few years earlier.[43] Clad in brick and purple ceramic, and crowned with a large sign bearing the legend "Reichsrundfunk" (Reich radio), the building has a rather monumental presence, reflecting the importance of the newly created RRG. (Fig. 106, 107) While the project was heralded in the popular press as Germany's "technical and organizational center for radio know-how," the preliminary sketches for the competition make it clear that Poelzig struggled with the organizational diagram.[44] The program for the new broadcasting house had to reconcile the slew of technical constraints that came with a completely new form of production with the administrative requirements of three different institutions. The plan sketches are color-coded: blue for Berliner Funk-Stunde, yellow or black for Deutsche Welle and the RRG. In one study, Poelzig tries to organize the fragmented plan of offices and small rooms along a ring with a double-loaded corridor. On another level, a larger portion of the plan is occupied by two broadcasting studios (*Senderäume*), one circular, the other oval. Tangential lines are used to relate the elongated facade towards Masurenallee with the circular shapes defining the broadcasting studios and the corners of the overall scheme. Interestingly, this rather awkward early iteration of the plan is actually composed of fragments of other building types. The broadcasting studios draw on precedents from theater or concert hall design, while the ring with administrative offices follows a logic similar to that of Poelzig's proposal for the extension of the Reichshauptbank in Berlin. (Fig. 109) A note on another sketch—"Isolierung der Senderäume" (isolation of the broadcasting studios)—points to the way Poelzig would eventually resolve the institutional and technological problem of the project.[45] (Fig. 110, 111) In essence, the winning scheme consists of two oval broadcasting studios that are protected from street noise by an undulating belt of administrative programs.

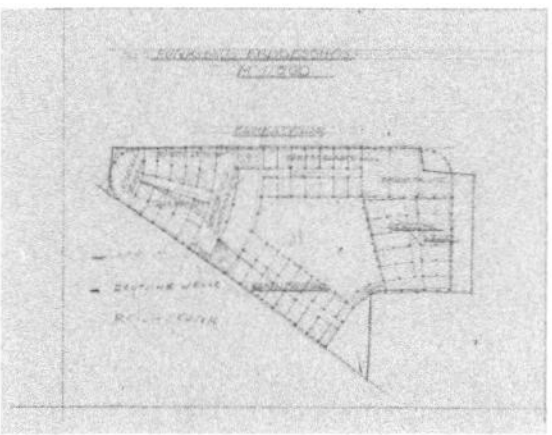

106 → P. 175

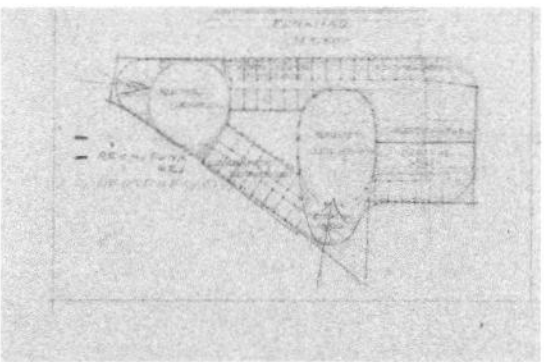

107 → P. 175

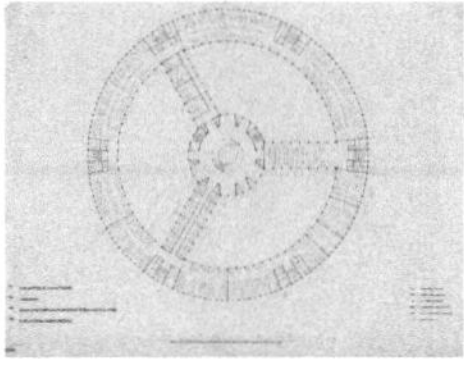

109 → P. 176

110 → P. 177

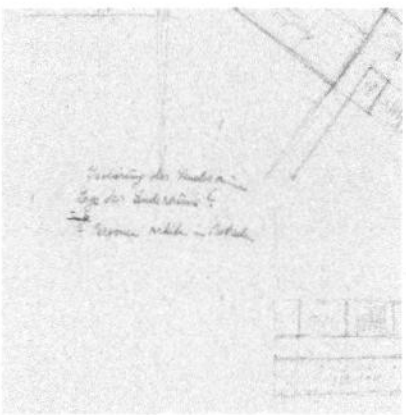

111 → P. 177

42 Fritz L. Büttner, *Das Haus des Rundfunks in Berlin* (Berlin: Haude & Spenersche, 1965), 17.

43 For more on the design of the trade fair grounds by Hans Poelzig and Martin Wagner, see chapter 1.

44 Menzel, "Blick in Deutschlands Rundfunk," 75. Author's translation.

45 Drawing in TU Berlin Architekturmuseum, Inv. Nr. HP 039,010.

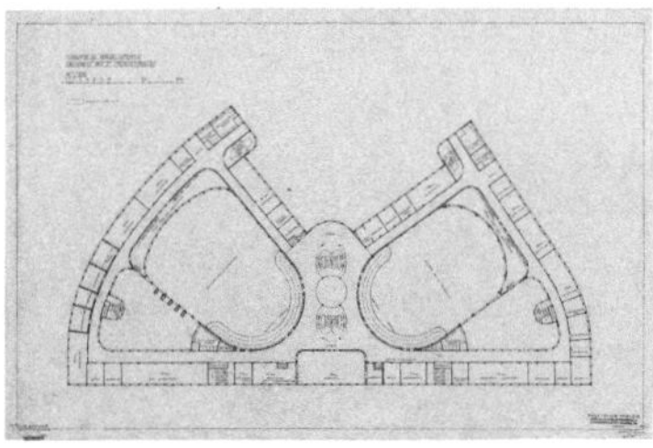

112 → P. 179

46 Theodor W. Adorno and Max Horkheimer, *Dialectic of Enlightenment* (Stanford: Stanford University Press, 2020), 95.

With this, Poelzig not only provided the RRG with an institutional image; he also, more importantly, produced a patently new building type. (Fig. 112)

The unexpected connection between Franz Ehrlich's early experience of working in Hans Poelzig's office in 1930 and the commission, twenty-two years later, to design a replacement for the Poelzig building traces a singular arc in both the life of the architect and the development of a particular form of architectural production. Poelzig's Haus des Rundfunks represented a new typology specifically conceived for organizing the ever-increasing production of media content. In that building, radio became mass media. But equally one could say that the *culture industry* found its *type* in the Haus des Rundfunks. Poelzig's proposal for the RRG could be seen as the first embodiment of the phenomenon whereby "the technical antithesis between few production centers and widely dispersed reception necessitates organization and planning by those in control"—as Theodor W. Adorno and Max Horkheimer famously framed the issue in the mid-1940s.[46] Adorno and Horkheimer's analysis saw this "organization and planning" reified almost exclusively in the safe, standardized "commodities" produced by the culture industry in America. However, in the Haus des Rundfunks, we can see how that form of organization and planning was first cemented—in the most literal sense of the word, in brick and mortar—not in postwar America, but in Weimar Germany, in an architectural form linking cultural and industrial production. Tracing the development and global dissemination of the typology of the broadcasting house not only reveals the technical mechanisms that brought together culture and industry but, more importantly, opens the door to rethinking the overall relationship between architectural form and modernity. The modern culture industry gave birth to a new form of organization and planning, which in turn demanded a new building type.

Though overlooked by most architectural and cultural critics at the time, the Haus des Rundfunks did not escape the notice of Adorno's contemporary, Siegfried Kracauer. In a column that appeared in the feuilleton of *Die Frankfurter Zeitung* on the occasion of its inauguration in 1931, he wrote:

> Yet Poelzig's building does not only make manifest the opacity inherent in the idea of radio: beyond that, it reveals an essential feature of the radiophonic enterprise. Not by coincidence is it so rigorous and so similar to the headquarters of an industrial authority. Through its entire stylistic posture, it marks with the utmost precision the commodity character of the intellectual activities that take place there. In the three rooms, these activities are produced like any other commodity, packaged and delivered to the consumers in their homes through wired or wireless paths. And it is precisely this double meaning of production and commodity that lends the new building the uncanny [*unheimlich*] character which is, at the same time, its most outstanding feature. Perhaps unintentionally, it makes visible the hidden property of radio: that of being a great enterprise which transforms the production of scholars, writers, and artists into products ready to be used.[47]

In responding to a genuinely modern program, the broadcasting house expanded the notion of functionality, particularly in relation to the emerging culture industry.[48] Articulating the relationship between commodity and cultural production, the nascent broadcasting house typology is characterized primarily by the organization of its plan. The main concert hall (*Grosser Sendesaal*) lies at its core, with the administrative spaces, smaller studios, and circulation arranged along its perimeter to act as a buffer against external noise. What Kracauer saw with mysterious precision in Poelzig's building was the "uncanny" relationship between the organization of the building and the content it produced, that is, between its form and its function. In programmatic terms, the Haus des

47 Siegfried Kracauer, "Sendestation: Das Haus," *Die Frankfurter Zeitung*, January 23, 1931. Author's translation.

48 For more on the role of commodities in relation to industrial production in interwar Germany, see Frederic J. Schwartz, "Commodity Signs: Peter Behrens, the AEG, and the Trademark," *Journal of Design History* 9, no. 3 (1996): 153–185. In one of the most influential modernist theories of "function," published a few years prior to the inauguration of the building, Adolf Behne addressed the relation between "creation" and "commodity." He conceptualized function (*Zweck*) in terms of the balance between the building as a "tool" and a "toy." See Adolf Behne, *The Modern Functional Building*, trans. Michael Robinson (Santa Monica: Getty Research Institute, 1996), 87–89. Original German edition: *Der Moderne Zweckbau* (Berlin: Drei Masken, 1926). For more on the role of electricity for the spatialization of industrial production in Germany at the time, see Tilo Amhoff, "The Electrification of the Factory or the Flexible Layout of Work(s)," in *Industries of Architecture*, ed. Katie Lloyd Thomas, Tilo Amhoff, and Nick Beech (London: Routledge, 2016), 259–270.

113 → P. 181

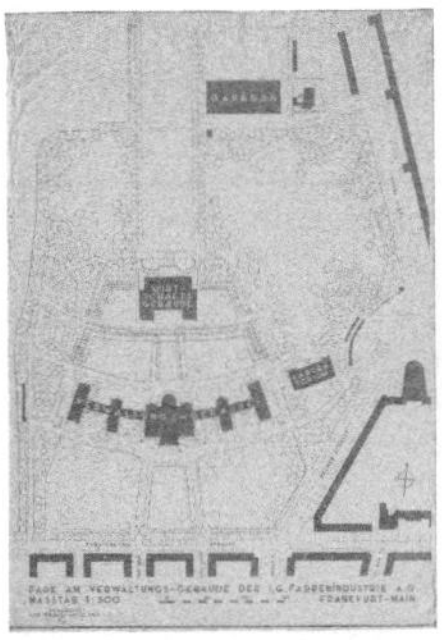

114 → P. 181

115 → P. 182

Rundfunks was simultaneously a factory and a theater, but in its internal logic and scale, it was specifically designed for the purpose of affecting a remote mass audience. In this sense, the building gives expression to a more complex notion of functionality latent in the new typology. The plan is extremely fragmented, prioritizing shorter circuitry paths over additional circulation. Mechanical ventilation and artificial light are used to isolate discrete spaces and keep them silent. In addition, internal surfaces are covered with various textures that produce the reverberation necessary for the sound waves to be picked up by the microphones. For Kracauer, the building was the embodiment of the agitated debate that had unfolded over the previous decade, as it became apparent that the boundaries between industrial and cultural production were being blurred, if not thoroughly redefined, by the emergence of mass media, and in particular of radio, which called into question both the traditional form and the social role of theater and music.[49] Considerable resources—intellectual and financial—were also devoted to radio's pursuit of a scientific and technological understanding of signal transmission through the electromagnetic spectrum. This concentration of resources is evident in the various research institutes, universities, and private and public companies that emerged in the service of the radiophonic enterprise in the early decades of the twentieth century.[50] Poelzig's architecture during those years reveals the same blurring of once separate domains, oscillating between cultural and industrial production.

(Fig. 113, 114, 115) By the early 1930s, Poelzig had gained significant recognition for his work in both the industrial and the cultural sectors. The chemical factory he designed for Milch & Co. in Luban (1911–1912) became a symbol of the contribution that architects could make to emerging industries, being described by Adolf Behne as "probably the best and most *sachlich*" of all "built" factories.[51] Further examples of Poelzig's prolific output as an industrial architect in the Weimar years are the complex he designed for the newly formed chemical conglomerate, IG Farben, in

49 See Rudolf Arnheim, *Radio: An Art of Sound* (London: Faber & Faber, 1936); Bertolt Brecht, "Der Rundfunk als Kommunikationsapparat," *Blätter des Hessischen Landestheaters*, no. 16 (July 1932). For the English translation, see Bertolt Brecht, "The Radio as an Apparatus of Communication," in *The Weimar Republic Sourcebook*, ed. Anton Kaes, Martin Jay, and Edward Dimendberg (Berkeley: University of California Press, 1994), 615–617. See also Theodor W. Adorno, "A Social Critique of Radio Music," in *Current of Music: Elements of a Radio Theory*, ed. Robert Hullot-Kentor (Frankfurt am Main: Suhrkamp, 2006), 201–217. For more on the format of the radio play, see Mark E. Cory, "Soundplay: The Polyphonous Tradition of German Radio Art," in *Wireless Imagination: Sound, Radio, and the Avant-Garde*, ed. Douglas Kahn and Gregory Whitehead (Cambridge, MA: MIT Press, 1992), 331–372. See also Bryan Gilliam, *Music and Performance during the Weimar Republic* (Cambridge: Cambridge University Press, 1994).

50 Roland Wittje, *The Age of Electroacoustics: Transforming Science and Sound* (Cambridge, MA: MIT Press, 2016).

51 Behne, *The Modern Functional Building*, 109.

Frankfurt (1928–1930), and the Cassirer cable works in Spandau, Berlin (1928). (Fig. 116, 117, 118) On the cultural side, Poelzig designed a series of cinemas during the 1920s for the emerging film industry. This new form of spectacle called for a new architecture, the picture palaces (*Lichtspielhäuser*) that Kracauer would describe as "optical fairylands ... events of the three-dimensional stage [that] blend imperceptibly into two-dimensional illusions."[52] Poelzig designed the Capitol am Zoo in Berlin as well as the Deli-Kino in Breslau (now Wroclaw). On the productive side of this cultural program, the Capitol am Zoo was not only a cinema but also the anchor for a larger development promoted by one of Berlin's most prominent real-estate magnates, Heinrich Mendelsohn (who was coincidentally behind the Friedrichstrasse competition now best known for the glass skyscraper entry submitted by Mies van der Rohe).[53] Even more dramatically, Poelzig placed this new form of metropolitan distraction at the heart of the urban transformation of the Berlin Mitte neighborhood, locating the Kino Babylon opposite the existing Volksbühne theater as part of an eight-block residential and commercial development. (Fig. 119, 120) But Poelzig was not only involved in housing the new experience of urban cinema; he was literally building for the films themselves. He collaborated with the director Paul Wegener, designing the sets for *The Golem: How He Came into the World* (1920) and the architecture for *Living Buddhas* (1925).[54] (Fig. 121, 122, 123) He also worked with theater producer, director, and entrepreneur Max Reinhardt on the renovation of the Grosses Schauspielhaus in Berlin in 1919. That project led to a further, unrealized collaboration on the design of the Festspielhaus for Salzburg theater festival, cofounded by Reinhardt with Richard Strauss and Hugo von Hofmannsthal. (Fig. 125, 126, 127) As he did in the film industry, Poelzig worked on both sides of the stage, designing, amongst others, the scenography for Shakespeare's *King Lear* (1923, dir. Bernhard Reich) and *Midsummer Night's Dream* (1926, dir. Gustav Hartung) as well as Mozart's *Don Giovanni* (1923, dir. Franz Ludwig Hoerth).[55]

116, 117, 118 → P. 182

119 → P. 184

120 → P. 184

121, 122, 123 → P. 185

125, 126, 127 → P. 188

52 Siegfried Kracauer, "Cult of Distraction, On Berlin's Picture Palaces," in *The Mass Ornament: Weimar Essays,* ed. Thomas Y. Levin (Cambridge, MA: Harvard University Press, 1995), 323–328, here 323.

53 For more on the role of Heinrich Mendelsohn in shaping Weimar Berlin, see Wolfgang Schäche, Daniel Ralf Schmitz, and David Pessier, eds., *Building Berlin: Developers Who Shaped the Emerging Metropolis* (Berlin: Jovis, 2019), 183–211.

54 Hans-Peter Reichmann, *Hans Poelzig: Bauten für den Film* (Frankfurt am Main: Deutsches Filmmuseum, 1997).

55 For more on Poelzig's work in theater, see *Der dramatische Raum: Hans Poelzig, Malerei, Theater, Film* (Krefeld: Museum Haus Lange, 1986). For illustrated volumes on cinema architecture in Berlin, see Heinz

By the time Poelzig won the competition to design the Haus des Rundfunks, the embryonic culture industry had already begun to undermine the traditional distinction between cultural and industrial production. Poelzig seems to have been well aware of the increasing interplay between the two spheres. Addressing the Federation of German Architects (Bund Deutscher Architekten) in 1931, just a few days after the inauguration of the new broadcasting house, he emphasized that "architecture is concerned with spatial design, with form for its own sake. Its laws are not on the technical plane, although this can serve architecture, and indeed it must influence its forms."[56] Though he enjoyed a reputation as a successful industrial architect, Poelzig maintained a critical distance from those architects who were attempting to make technology "visible" through architecture—for him, a lineage that ran from Peter Behrens to Walter Gropius. But in the same lecture, he also insisted "an artist cannot ignore technology, or craftsmanship, as the material vector of his creativity."[57]

For Poelzig, architecture and technology were two different things that nevertheless had to be addressed in tandem. Architecture had to have its own spatial purposes and logics, and yet use technology as its "material vector." With a further remark, related to his own experience on the Haus des Rundfunks, he made even more explicit the complex dialectical relationship between the possibility of an autonomous architecture and the most radical technology of his time: radio.[58] "Today, technology is pursuing its own course with its own achievements, using, for example, radio or wireless telegraphy to lose itself in the formless and elevate itself to technological magic."[59] The question then arose: if architecture was to rely on technology as its "material vector," how was it to work with radio, as the very technology that "magically" dissolved matter into the ether?

The paradox of radio—of an apparently immaterial medium being the material vector of an entire form of production—is precisely the design problem that

Frick, *Mein Gloria-Palast: Das Kino vom Kurfürstendamm* (Munich: Universitas Verlag, 1986), and Rolf-Peter Baacke, *Lichtspielhausarchitektur in Deutschland: Von der Schaubude bis zum Kinopalast* (Berlin: Frölich und Kaufmann, 1982).

56 Hans Poelzig, "Der Architekt," in *Hans Poelzig: Gesammelte Schriften und Werke,* ed. Julius Posener (Berlin: Gebr. Mann, 1970), 220–246, here 232. This and subsequent quotes from the work are the author's translation.

57 Poelzig, "Der Architekt," 232.

58 For more about the relationship between the notion of autonomy and architecture as it relates to external forces of technology and capital, see Pier Vittorio Aureli, *The Project of Autonomy: Politics and Architecture within and against Capitalism* (New York: Princeton Architectural Press, 2008).

59 Poelzig, "Der Architekt," 232.

Hans Poelzig was commissioned to address. The issue would occupy not only the architect, but a whole array of political, technical, and cultural actors. What the RRG required was an apparatus—part building, part machine—with the capacity to transduce mechanical inputs (sounds) into electronic outputs for broadcasting. This idea of *transduction* is illustrated in a paper by sound engineer Hans-Joachim von Braunmühl discussing the difference in the dynamic range between broadcast sounds and the experience in a concert hall. According to Braunmühl, the dynamic range in a live concert is 1:1000 (meaning that the loudest undistorted sound is one thousand times louder than the quietest discernible sound), but for broadcasting, the range is only 1:100: "On the top [loudest] range, the limit is set by the transmission devices, particularly the radio transmitter, which could experience distortions and interference on certain frequencies. On the lower [quieter] range, the limit is set by the recording facilities and space of reception, where ambient sounds could either be captured by the microphone or be louder than those emitted by the speaker."[60]

By pointing to the artificial construction of "natural" sounds for the emerging culture industry, Braunmühl's analysis illustrates a singular trait of this new form of production, namely, the indissoluble link between electronic technical media such as microphones and amplifiers, and the mechanical architectural elements of walls, floors, and ceilings. Within this new broadcasting facility, the sound waves emitted by actors' voices and musical instruments were to be modulated by the geometric and material conditions of the wall surfaces, and then modulated again by electronic technical media. However, this act of transduction was determined not only by the technical constraints described by Braunmühl, but defined in equal measure by the administrative and organizational challenges that came with the conjoining of industrial and cultural production. This new facility, in other words, had to resolve the technical problems presented by sound and

60 Hans-Joachim von Braunmühl, "Uber die Intensitätsverhaltnisse von natürlichen Klangbildern mit besonderer Berücksichtigung der Rundfunksendung," *Zeitschrift für Technische Physik* 14 (1933): 68–70, here 68. Author's translation.

124 → P. 187

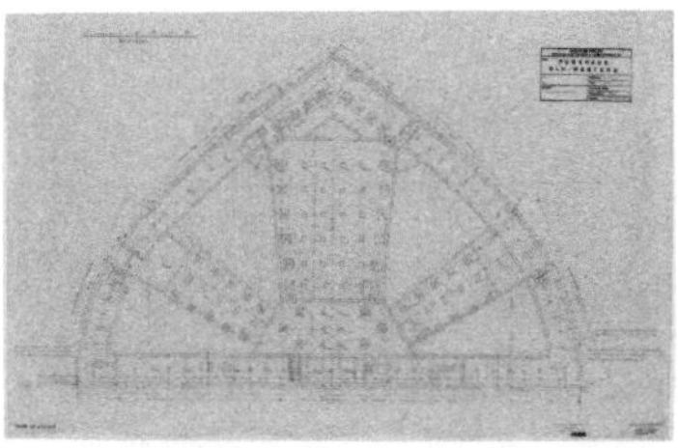

128 → P. 190

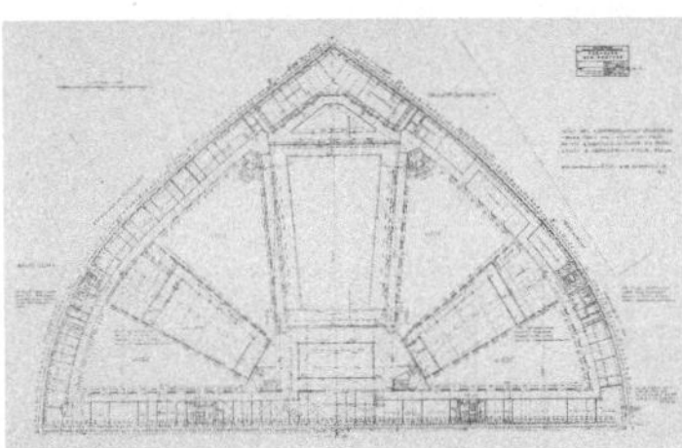

129 → P. 191

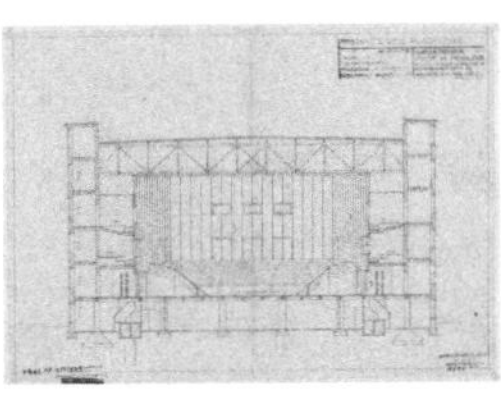

130 → P. 193

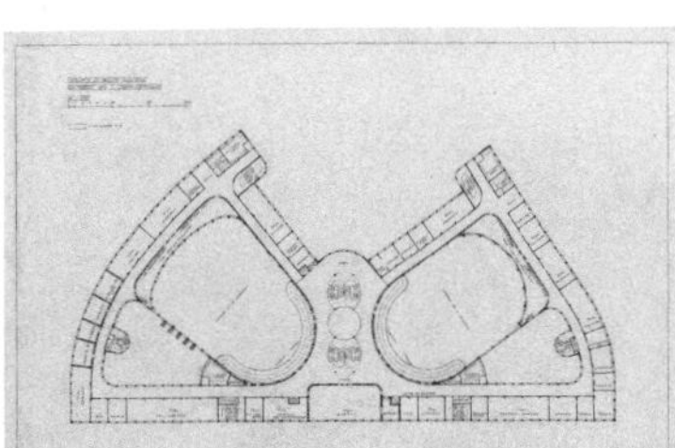

112 → P. 179

electroacoustics for many types of performances taking place simultaneously. Given the different institutions housed in the same building, countless meetings and administrative tasks had to be orchestrated in parallel. Technicians going in and out of the rooms had to contend with both noisy mechanical ventilation systems and the sanctity of those silent spaces. And all of this had to be bundled together within a coherent spatial form that gave an urban identity to a national public organization. (Fig. 124) The complexity of the program that the building had to accommodate is apparent from an image showing a presentation of the project in the newly inaugurated trade fair building across the street. We see a photographic collage of significant global events tethered to a model of the Haus des Rundfunks at the rear of the space: “the whole world is passed through the filter of the culture industry.”[61] It is worth noting that the visibility of these developments was considered less important than their audibility. Wars, political events, protests, theater and music performances—all were communicated to the masses through the medium of radio, filtered by the massive material infrastructure of the building before being carried through the air and transmitted throughout the territory.

(Fig. 128, 130) Outwardly coherent and unified, Poelzig’s design in fact consists of four adjacent, semi-independent buildings integrated by means of a square grid. The three trapezoidal spaces at the center of the ground plan are recording studios (*Sendesäle*). (Fig. 129) Each rests on independent foundations and is insulated from street noise by the perimeter belt of administrative spaces. Four courtyards bring natural light and ventilation to the rooms at the core of the project. Each has double-layered walls providing space for mechanical heating and ventilation as well as for the circulation of the audience and staff. It is interesting to compare these construction documents with Poelzig’s initial competition scheme. (Fig. 112) Whereas the early proposal aspired to an almost baroque unified idea of totality, that vision would

61 Adorno and Horkheimer, *Dialectic of Enlightenment*, 99.

be fragmented by the demands of the electroacoustic technology. Ultimately, what was built had a feigned sense of wholeness—dictated by the institutional requirements—that veiled an operative disjunction of programmatic pieces mandated by the unruly nature of wires and waves.

131 → P. 194

132 → P. 194

The description of the Haus des Rundfunks by the RRG's chief acoustic engineer, Walter Schäffer, makes apparent this oscillation between formal totality and functional fragmentation, between institutional requirements and technical mandates. The first point that Schäffer highlights is the binding relationship between the spatial qualities (*Raumeigenschaften*) and a specific audible condition (*Hörsamkeit*). Studies conducted by the American engineer Wallace Sabine at the turn of the century had laid the basis for the modern science of electroacoustics, establishing that reverberation time could be measured and prescribed by the relationship between the volume of the room and the materiality of the surfaces.[62] Schäffer, however, placed a much greater emphasis on the architectural implications of this principle, clarifying "When building a recording room for radio, one has to consider that the task is theoretically different from that of constructing a concert hall for a live audience. In the latter case, it is desirable to provide every seat in the auditorium with an equally good sound quality, whereas for radio, one only need provide the desired sound quality in the few spots where microphones are located."[63]

The word *Klangbilder* (sound images) recurs frequently in Schäffer's description of the Haus des Rundfunks, as if the qualities of the space had the potential to trigger specific sonic images in the minds of distant listeners. (Fig. 131, 132) In order to achieve the desired sound images, Schäffer proposes that one of the recording studios at the core of the building should be equipped with a system of "door-like foldable panels, clad on the one side with wood and on the other with sound-absorbing material."[64] Acknowledging the disparity between the scientific

62 See Wallace C. Sabine, *Collected Papers on Acoustics* (Cambridge, MA: Harvard University Press, 1922). See also Emily Thompson, *The Soundscape of Modernity: Architectural Acoustics and the Culture of Listening in America, 1900–1933* (Cambridge, MA: MIT Press, 2002).

63 Walter Schäffer, "Die neuen Rundfunkaufnahmeräume im Haus des Rundfunks und ihre Akustik," January 23, 1931, in Archiv Universität der Künste Berlin (UdK-Archiv), Bestand 1b (Rundfunkversuchsstelle) / 12, Bl. 142–151, here 147.

64 Schäffer, "Die neuen Rundfunkaufnahmeräume," 147.

133 → P. 194

134 → P. 194

135, 136 → P. 195

137 → P. 195

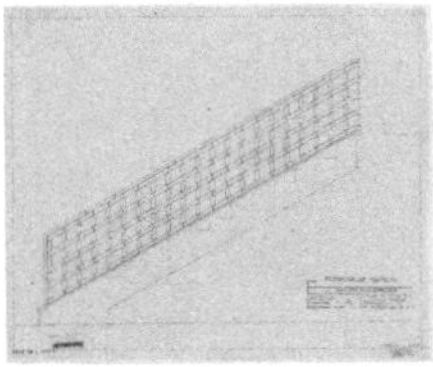

138 → P. 196

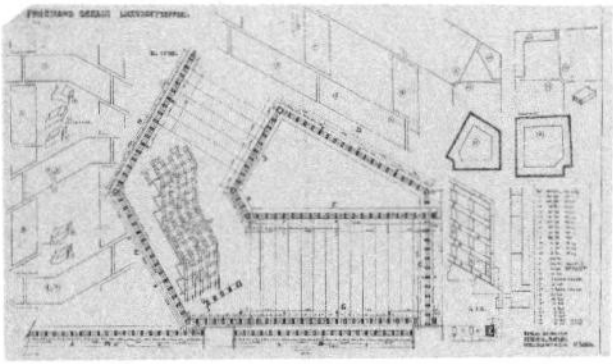

139, 140 → P. 196

predictability of sound characteristics within a controlled laboratory setting and the often variable results experienced in the radiophonic industry, the proposal aims to make it possible to adjust the spatial qualities according to the specifications of a particular microphone, or the number of musicians or listeners present in the room, or even the temperature of the interior at any given moment.[65] All of the recording studios were wired to control rooms where the mechanical sonic waves were received as discrete electrical signals. In the entrails of this new productive topology, wires would carry signals from silent room to silent room, while meandering corridors orchestrated the parallel circulation of people going in and out of the same rooms. It was here, in the design of the overall plan and the shaping of the wall surfaces, that the techniques of architecture and sound engineering would meet.[66] Poelzig's plan was specific enough to organize the various activities and related institutions, yet at the same time open enough to allow the emergent science of electroacoustics to determine the technical aspects of the project.

With the exception of the plan, however, everything in this building conforms to one of the most common modernist design tropes: the square grid. (Fig. 133, 134, 135, 136, 137) The main facade towards the street takes the form of a 150 m-long grid of squares. Three larger square bays at the center of the facade give access to the building. Inside, the central courtyard is built as a three-dimensional grid. The structure of columns and slabs forms a double square, while the balustrade made of yellow ceramic tiles forms an even denser version of the same grid. To the rear, we see square windows also subdivided into squares. The same grid of squares divides the lamps hanging from the skylight in the courtyard, but it also has a unifying quality, as seemingly the only organizational device capable of reconciling the noisy flow of the human occupants of the building with the silent flow of sounds through electrical wires. (Fig. 138, 139, 140) The staircases climbing up through the building are also gridded, as are the

65 Walter Schäffer, "Raumakustik für Rundfunkdarbietungen," *Funktechnische Monatshefte* 2 (1932): 51–54.

66 For a recent collective study on the notion of "design techniques," see Zeynep Çelik Alexander and John May, *Design Technics: Archaeologies of Architectural Practice* (Minneapolis: University of Minnesota Press, 2020).

skylights above them. (Fig. 141, 142, 143) To sum up, the building has a gridded facade, gridded windows, gridded interior courtyard facades, and gridded entrances to the main recording rooms as well. The surfaces inside the recording studios also comprise grids, and still another grid of squares marks the wall with the pipe organ. Thus the visual imprint of the grid remains consistent, even though the pivoting panels Schäffer designed for the recording studios modify the space in acoustic terms.[67]

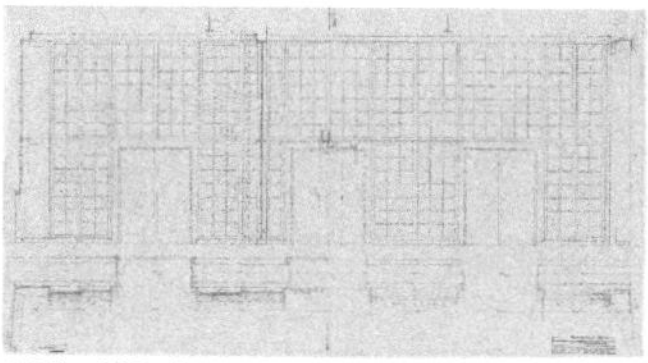

141, 142 → P. 197

143 → P. 198

The point here is that Poelzig recognized there were only two areas in which an architect working for the culture industry could intervene in the design. One was the organization of the plan, which gave definition to the overall scheme, the *Grossform*. The other was the gridded surface of the walls, given that the form and size of the rooms were already determined by acoustic considerations. If, in Poelzig's phrase, the "material vector" of this building was technology, and the technology at work here was the apparent immateriality of radio, the grid was the chosen mechanism for linking the realms of the virtual and the actual—that is, for connecting bricks and sounds, tiles and wires, wood surfaces and microphones. With this, it becomes clear that Poelzig not only created a new building typology, but acted within radically different techno-political conditions of possibility, conditions that demanded a different relationship with architectural form on the part of the architect.[68] Typologically, what the Haus des Rundfunks established for this new mode of production was a built manifestation of the co-generative principles of design derived not from an accumulated and anonymous knowledge of "culture" but rather from political and technological guidelines dictated by the embryonic state of the emerging culture industry.[69] Responding to this new productive regime, Poelzig's proposal was truly sensitive to the zeitgeist, in recognizing the limited agency of the architect but at the same time identifying the precise sites of action that were still possible within those limits.

67 For a recent study that problematizes the relationship between abstraction and modern capitalist modes of production, see Pier Vittorio Aureli, *Architecture and Abstraction* (Cambridge, MA: MIT Press, 2023).

68 For a comprehensive synopsis of the transformation of the idea of type since the eighteenth century, see the entry on type in Adrian Forty, *Words and Buildings: A Vocabulary of Modern Architecture* (London: Thames and Hudson, 2004), 304–311.

69 An attempt to insert the Haus des Rundfunks within the discourse of type and typology requires further clarification. Though rejected as a term in German architectural debate for most of the twentieth century, ideas of type permeated the discourse of the Werkbund. For more, see Frederic J. Schwartz, "The Type," in *The Werkbund: Design Theory and Mass Culture before the First World War* (New Haven: Yale University Press, 1996), 121–146; Frederic J. Schwartz, "Commodity Signs: Peter Behrens, the AEG, and the Trademark," *Journal of Design History* 9, no. 3 (1996): 153–185; and Hermann Muthesius and Henry van de Velde, "Werkbund Theses and Antitheses" [1914], in *Programs and Manifestoes on Twentieth-Century Architecture*, ed. Ulrich Conrads (Cambridge, MA: MIT Press, 1971), 28–31. For more on the reception and echoes of the German debate on type and standardization, see Alla Vronskaya, "Process: Organicist Aesthetics of Soviet Standardization," in *Architecture of Life: Soviet Modernism and the Human Sciences* (Minneapolis: University of Minnesota Press, 2022), 107–139.

144 → P. 200

145 → P. 201

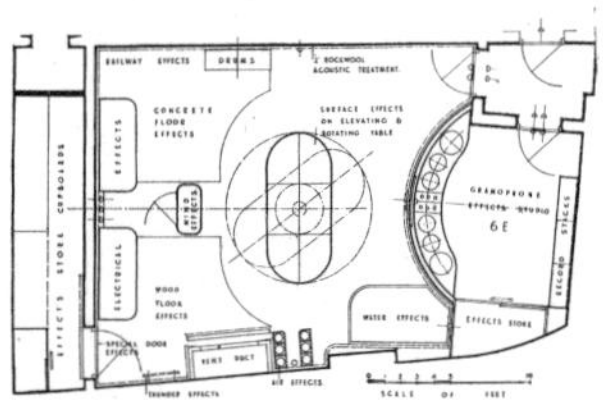

147 → P. 202

148 → P. 202

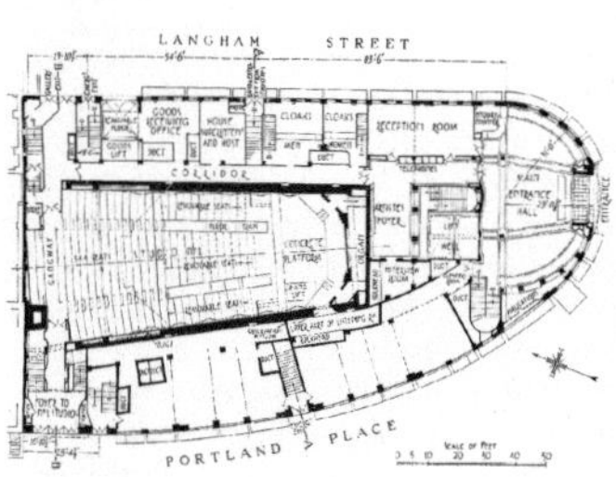

146 → P. 201

The spatial and organizational arrangement proposed by Poelzig would circulate around the world almost as seamlessly as radio waves themselves. In 1932, the BBC (British Broadcasting Corporation) inaugurated its first purpose-built home for broadcasting in Langham Place, central London. (Fig. 144) Adrian Forty has described how "the most noticeable effect of early radio was to create a sense of unreality by introducing the sounds and voices of the outside world into the home in a way that had never happened before."[70] The BBC's new headquarters carefully orchestrated this notion of a modern fiction or "unreality." Located next to All Souls' Church (designed by John Nash), with its antennas doubling as spires, it asserted that religion no longer had the monopoly on connecting immaterially with the beyond. Architecture, in other words, could bridge the virtual and the actual by dissolving bodies into sounds and disseminating them through electromagnetic waves.

(Fig. 145) (Fig. 147, 148) The scheme by George Val Myer incorporated a number of independent rooms by different architects. The one designed by Edward Maufe is chapel-like, mimicking the sonic and visual presence of religious space. Dorothy Warren Trotter designed the Greek-Revival lounges, Serge Chermayeff the band-room, Wells Coates the dramatic control room as well as the special effects studio. All serve to illustrate the fragmented totality associated with this new mode of production. *Architectural Review* dedicated an entire issue to the new building, proudly celebrating the fact that "the enlightenment which the BBC employs in its programmes, it has also employed in its new building."[71] An editorial referred to the precedent of Poelzig's building, noting "the Rundfunkhaus is impressive and its planning expressive."[72] (Fig. 146) The typological arrangement of the BBC building clearly relates to the organization of the Haus des Rundfunks. All the studios are located in its central heavy masonry tower built, like the adjacent church, in Portland stone. As at the Haus des Rundfunks, the production-related functions are

70 Adrian Forty, "Wireless Style: Symbolic Design and the English Radio Cabinet, 1928–1933," *Architectural Association Quarterly* 4 (1972): 22–31, here 27.

71 Editorial, "The New Tower of London," *Architectural Review (*August 1932): 43.

72 Editorial, "English and Foreign Precedents," *Architectural Review (*August 1932): 44.

buffered from street noise—sealed off acoustically and programmatically—by administrative spaces.

In Britain, as in Germany, the state was reluctant to put the new mass medium of radio into the hands of private corporations. The BBC's purpose, as defined by the towering figure of its first director general, John Reith, was to *inform, educate, entertain*—to be an instrument of social and cultural change. At the same time, however, the technology of radio and the collection of fragmented worlds within the building not only represented but directly enabled the invisible, yet audible, omnipresence of Britain in its colonies.[73]

151 → P. 204

152 → P. 205

(Fig. 151, 152) Another iteration of the emerging typology of the broadcasting house appeared on the other side of the Atlantic, shaped by a rather different system of production. In 1932, NBC (National Broadcasting Company) opened its new studios in the recently completed Rockefeller Center in midtown Manhattan, at the base of the RCA (Radio Corporation of America) tower. As discussed earlier, in the United States there was a *laissez-faire* corporate approach to the development of the mass medium of radio. RCA was founded in 1919 by General Electric and the Marconi Wireless Telegraph Company, and NBC was the entertainment subsidiary of General Electric. By 1932, RCA had become a large corporation under the leadership of the eccentric and charismatic David Sarnoff, who had ties to manufacturers of electronic equipment as well as companies related to content production. In contrast to the BBC in Britain and the RRG in Germany, the radiophonic enterprise in the United States had few, if any, links to the federal government. However, by law, eighty percent of the stock of RCA had to be owned by US individuals or corporations, making it a *de facto* national monopoly, and members of the US Navy served on its advisory committee.[74]

The story of how RCA and NBC came to be located in Rockefeller Center is symptomatic of the cultural

73 For more on the social and political ambitions of the BBC and their relationship to architecture, see Shundana Yusaf, *Broadcasting Buildings: Architecture on the Wireless, 1927–1945* (Cambridge, MA: MIT Press, 2014). On the presence of British radio in the African colonies, see Brian Larkin, *Signal and Noise: Media, Infrastructure, and Urban Culture in Nigeria* (Durham: Duke University Press, 2008), 48–72. See also Simon J. Potter, "The Lonely Listener in the Bush: BBC Empire Service Programmes and Audiences," in *Wireless Internationalism & Distant Listening: Britain, Propaganda, and the Invention of Global Radio, 1920–1939* (Oxford: Oxford University Press, 2020), 171–200; and David Clayton, "British Colonial Broadcasting in the 1940s," in *The Wireless World: Global Histories of International Radio Broadcasting*, ed. Simon J. Potter et al. (Oxford: Oxford University Press, 2022), 93–98.

74 For more on the origin of RCA and its influence as a powerbroker in the first half of the twentieth century in the United States, see Thomas S. W. Lewis, *Empire of the Air: The Men Who Made Radio* (New York: Harper Collins, 1991).

shift that accompanied the development of radio as a mass medium. The first promoter of the project, the financier, philanthropist, and director of the Metropolitan Opera Company, Otto Kahn, had originally earmarked the site for a new opera house, an idea John D. Rockefeller Jr. embraced when he joined the project in 1928, having helped to acquire the site on a leasehold from Columbia University. Proposals for the new opera house were put forward by the architects Joseph Urban and Benjamin Wistar Morris, among others, but none was taken further. By the time of the Wall Street Crash in 1929, the Metropolitan Opera had withdrawn from the project. Amidst a global financial collapse, Rockefeller found himself without a primary tenant for one of the most significant real-estate investments ever made. It was only weeks, however, before the matter was resolved. By the beginning of 1930, RCA, NBC, and Radio-Keith-Orpheum had been signed up as the main tenants. Corporate capital had replaced cultural philanthropy. Mass media had ousted highbrow values from the prime location. Live music performance had given way to electroacoustics and remote transmission. The culture industry had landed in the center of New York City.[75]

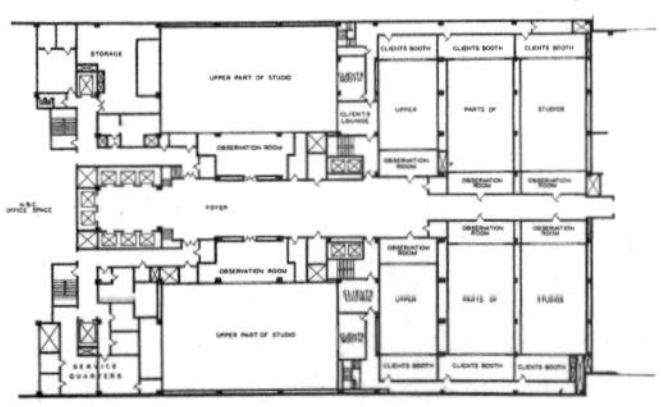

149 → P. 203

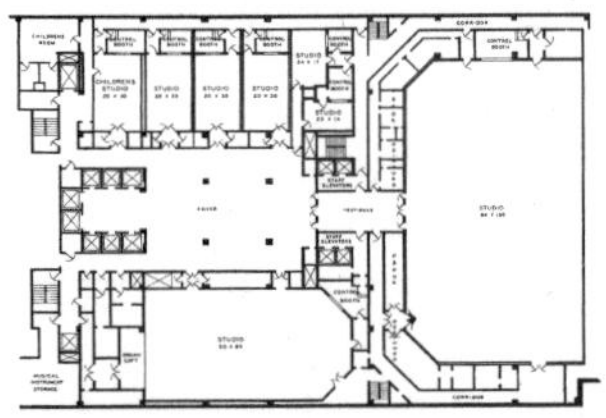

150 → P. 203

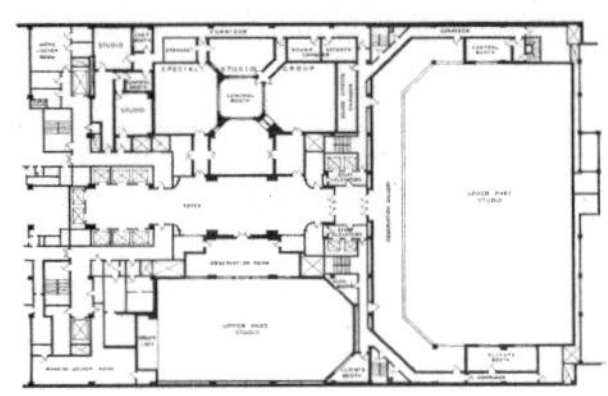

154 → P. 206

The advent of radio brought a paradigm shift in the performance and perception of music—one encapsulated in Adorno's observation that even if "today music is considered ethereal and sublime, it actually functions as a commodity."[76] Adorno was talking not just about the content of the music, but about its ties to corporate and financial structures. This connection is made explicit in the plans and sections of the NBC studios, where the new facilities are entwined with the RCA tower, which operated as the material structure of the overall corporate organization. (Fig. 149, 150, 154) NBC occupied "the ten-story extrusion of the site in the early skyscraper tradition of sheer multiplication of surface."[77] The four steel columns legible at the center of the plan correspond to the loadbearing structure of the tower; thus, the plan is organized with offices on the left-

75 For more on the history of Rockefeller Center, see Robert A. M. Stern, Gregory Gilmartin, and Thomas Mellins, *New York 1930: Architecture and Urbanism Between the Two World Wars* (New York: Rizzoli, 1987), 617–670, and Giorgio Ciucci, Francesco Dal Co, Mario Manieri-Elia, and Manfredo Tafuri, *The American City: From the Civil War to the New Deal* (Cambridge, MA: MIT Press, 1979). For the multiple variations of the project with and without the Metropolitan Opera House as a central tenant, see the text and illustration of Raymond Hood, "The Design of Rockefeller City," *Architectural Forum* 56, no. 1 (January 1932): 2–9. For an iconographic interpretation of the sculptures and paintings at the entrance of the RCA tower, see Eugene Clute, "The Story of Rockefeller Center X and XI: The Allied Arts," *Architectural Forum* 57 (October 1932): 353–358. See also Thompson, *The Soundscape of Modernity*, 295–324.

76 Adorno, "A Social Critique of Radio," 201–217, here 208. For a critical account of NBC and its effects on music, see Theodor W. Adorno, "Analytical Study of the NBC Music Appreciation Hour," in *Current of Music: Elements of a Radio Theory*, ed. Robert Hullot-Kentor (Frankfurt am Main: Suhrkamp, 2006), 247–318.

77 Rem Koolhaas, *Delirious New York: A Retroactive Manifesto for Manhattan* (New York: Monacelli Press, 1994), 199.

hand side and studios on the right. The studios are placed "to avoid interference with this massive steel supporting the weight of the tower," at the precise point where the required large spans are structurally feasible.[78] (Fig. 153) In the longitudinal section, floors five and seven contain double-height studios, while the ninth floor houses a four-story recording space. The central control room is sandwiched between the two main studio floors, connected by wire to the twenty-seven recording spaces above and below. Recall that in the Haus des Rundfunks in Berlin, this spatial organization occurred horizontally, on the same level. In New York, the combination of the prime midtown location and the corporate ethos of NBC meant that type was reorganized vertically, in section. This presented new challenges for the architecture of radio. Just as microphones, wires, and mixing boards channeled the signals in perfect synchronicity in this new form of production, with sound traveling up and down the structure via the electric wires connecting the studios with the control room, so elevators and corridors channeled the flow of people in a precise choreography of activities among three different kinds of users—technical staff, artists or performers, and tourists eager for a glimpse of the new technological wonder.

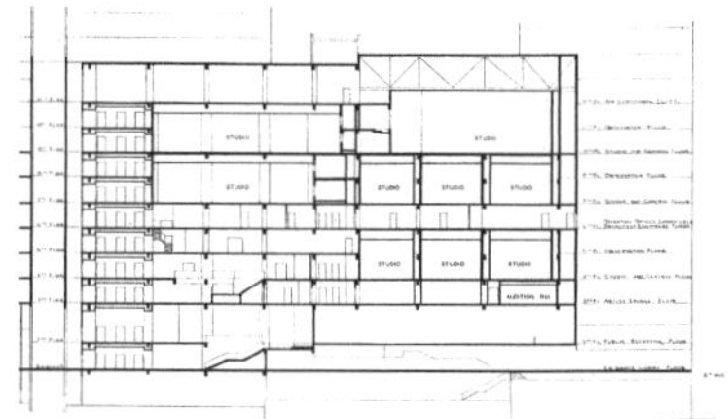

153 → P. 206

The "single electronic arena" of the NBC studios, as Rem Koolhaas defined it, was not the "first architecture that can be broadcast,"[79] but was, nevertheless, a pioneering vertical translation of an earlier typology. In this case, steel served as the actual material translator from an industry organized primarily in plan to one organized in section. Both the Germans and the British had been careful to avoid the use of steel framework in their new broadcasting houses, as it transmits sound and mechanical vibrations. But at the NBC studios, which were integrated into one of the most ambitious real-estate ventures ever undertaken, the structure *had* to be made out of steel. NBC's chief engineer, O. B. Hanson, would lament "the problem is forced upon us and we cannot avoid it."[80]

78 O. B. Hanson, "The Plan and Construction of the National Broadcasting Company Studios," *Architectural Forum* (August 1932): 153–160, here 153. For a more in-depth description of the technical aspects of this project, see O. B. Hanson, "Planning the NBC Studios for Radio City," *Proceedings of the Institute of Radio Engineers* 20, no. 8 (August 1932): 1296–1309; for Hanson's general take on broadcasting studios, see O. B. Hanson and R. M. Morris, "The Design and Construction of Broadcast Studios," *Proceedings of the Institute of Radio Engineers* 19 (January 1931): 17–34.

79 Koolhaas, *Delirious New York*, 199.

80 Hanson, "The Plan and Construction," 153–160, here 159.

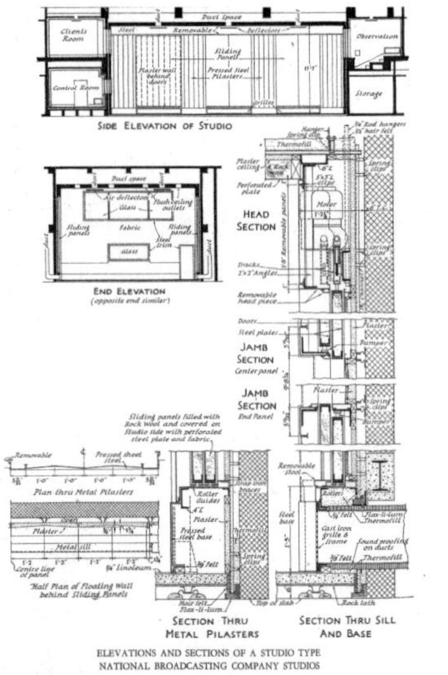

155 → P. 207

156 → P. 207

(Fig. 155) To avoid the transmission of sounds through the steel structure of the building, each recording studio was suspended, with a few centimeters separating the interior spaces from walls, ceilings, and floors. The detailed wall sections show that each room within the complex was a levitating capsule, allowing multiple sonic realities to coexist within the same building.[81] Analyzing Rockefeller Center in *Space, Time and Architecture*, architectural historian Sigfried Giedion observed that "a certain feeling of hovering, of suspension, emanates from the surfaces" of the walls.[82] Giedion would depict that "feeling" in a famous collage that breaks up the facades of the building into multiple fragments, superficially mimicking the most common of the cubist formal tropes. (Fig. 156) In relation to the NBC studios, however, we could say that Giedion's observation was both right and wrong. *Yes*, the Rockefeller Center is one of the first manifestations (after Berlin) of the coexistence of multiple, differentiated environments, and *yes*, the walls of RCA are effectively "suspended," but Giedion was arguably looking at the *wrong* walls and the *wrong* kind of fragmentation. The kaleidoscope of adjacent and coexisting realities was actually unfolding within the studios, made possible by the levitating walls seen in the section. The modern architectural space was being transformed from within the tower by the entangled relations between architectural and electroacoustic elements, and corporate capital.

The typological continuity of the buildings made for radio in Berlin, London, and New York was taken to another level with the end of the Weimar Republic and the rise to power of the Nazis. Between 1934 and 1940, the number of Reich stations (*Reichssendern*) mushroomed throughout Germany. More than fifteen of these stations were built within a mere five years, all of them following the typological guidelines defined by Poelzig.[83] On one level, the Haus des Rundfunks provided a spatial template for relating the mechanical and electroacoustic architectural elements, making this combination replicable in different places and

81 The other relevant means of sound contamination within the building came by way of the emerging technology of air conditioning. See Wulf Böer, "Air-Conditioning, Architecture, and Modernism: On the History of the Controlled Environment, 1911–1952" (Dr. Sc. diss., ETH Zurich, 2019).

82 Sigfried Giedion, *Space, Time and Architecture: The Growth of a New Tradition* (Cambridge, MA: Harvard University Press, 1962), 750.

83 For more on the building documentation of the transmission stations and broadcasting houses designed by the RRG, see Deutsches Rundfunkarchiv Frankfurt, A48 RRG Bauzeichnungen, 388–430. See also chapter 1.

according to different modes of organization. On another level, the type was flexible enough for subsequent transformation, which is exactly what happened when Adolf Hitler and Joseph Goebbels assumed control of the centralized organization of the mass media.

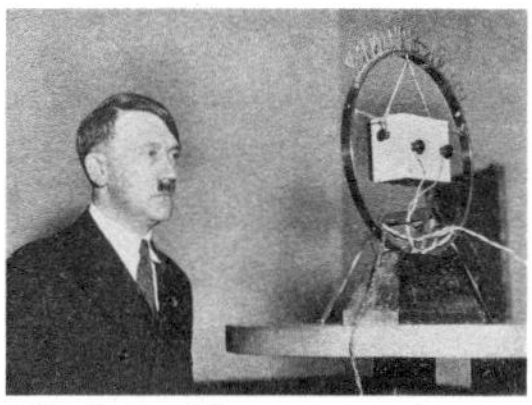

157 → P. 208

(Fig. 157) Adolf Hitler gave his first radio address as Chancellor from the Haus des Rundfunks on February 1, 1933. A photograph shows him standing in front of the Berliner Funkstunde's iconic microphone, which was suspended by flexible chords from a peripheral ring. In capturing Hitler's voice and channeling it directly into domestic interiors, the broadcast set in motion the notorious process of *Gleichschaltung*, which was designed to turn Germany into a single-party state, synchronized through both an electric and an electromagnetic medium. There was no possibility, at the RRG, for dissent.[84] Its serving chairman, the radio pioneer Hans Bredow, resigned on the day the Nazis seized power, telling his colleagues that he was quitting because "the personal and political fight waged against me has ultimately reached the radio and, with it, all of you."[85] There would be a radical change in both the content of the radio broadcasts and the cadre of employees producing it. The RRG's new production director, Eugen Hadamovsky, jubilantly reported, "On January 30, 1933, *the people* broke into the broadcasting house. Within a few weeks, the theater and literary hordes had evaporated before their eyes."[86] For Hadamovsky, the center of German culture was now the *Volk*, rather than the cultural elite.

This shift in direction required material changes to the recently inaugurated broadcasting house. The archives of the RRG's building department contain hundreds of plans and drawings documenting alterations to the Haus des Rundfunks, beginning within a month of the National Socialists taking it over. The kitchen and the dining hall were expanded and acquired new furnishings bearing Nazi iconography; the control room was modified; the seating in the main concert hall was redistributed while the stage was redesigned,

84 For more on the relationship between radio and totalitarianism in Germany, see Horst J. P. Bergmeier and Rainer E. Lotz, *Hitler's Airwaves: The Inside Story of Nazi Radio Broadcasting and Propaganda Swing: Jazz, Swing, and Nazi Radio Propaganda* (New Haven: Yale University Press, 1997), and Josef Kopperschmidt, *Hitler der Redner* (Munich: Brill I Fink, 2003).

85 For the full text of Hans Bredow's speech, see Deutsches Rundfunkarchiv Frankfurt, Akte Bredows Lebensweg, Bd. 1933, Bl. 12–17. Author's translation. Previously referenced in Büttner, *Das Haus des Rundfunks*, 53.

86 Eugen Hadamovsky, *Dein Rundfunk: Das Rundfunkbuch für alle Volksgenossen* (Munich: Zentralverlag der NSDA, 1934), 47. Author's translation.

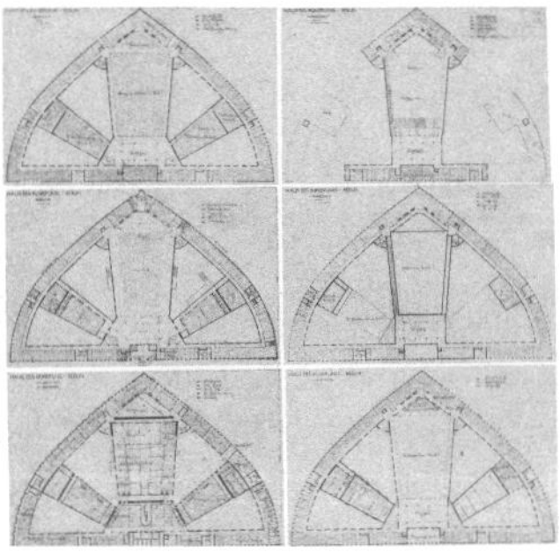

158 → P. 208

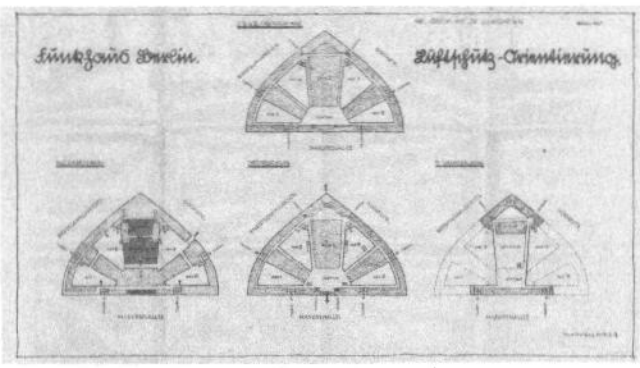

159 → P. 209

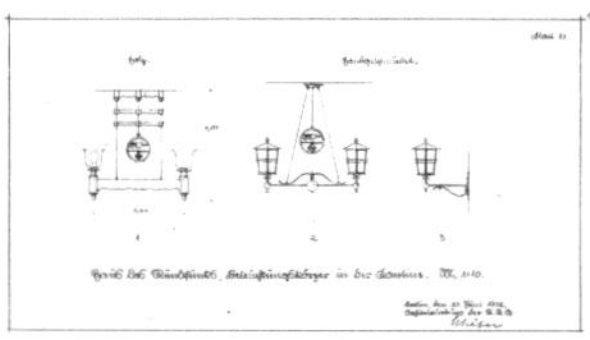

160 → P. 209

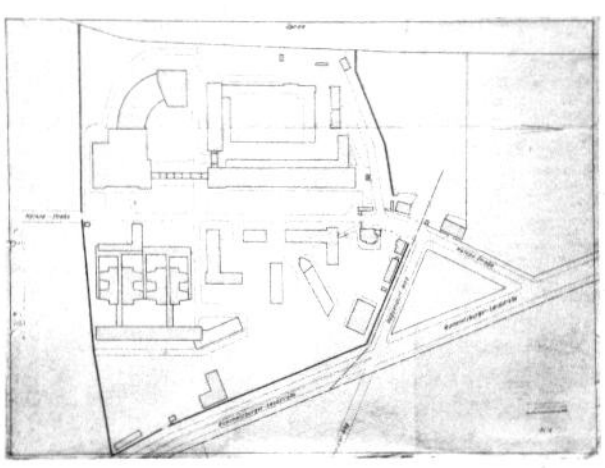

161 → P. 211

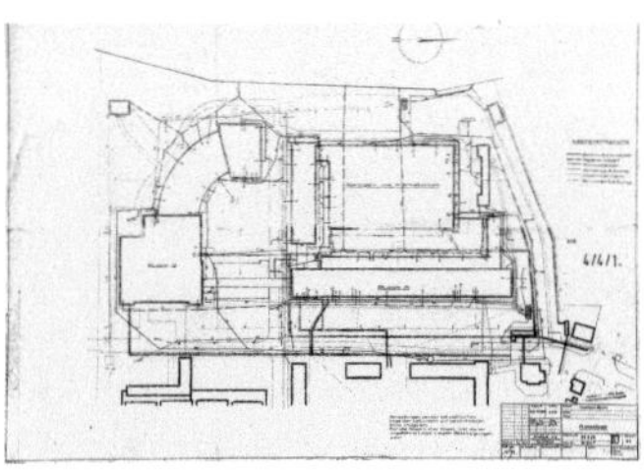

162 → P. 213

requiring work on the foundations for that portion of the building; offices were reorganized; television was soon incorporated as a new form of production;[87] the interior climate was regulated with new machines and technologies; and the system of circulation was reconfigured to allow the basement to be converted into an air-raid shelter. (Fig. 158, 159, 160)

More than just minor programmatic reforms, the transformations of Poelzig's building speak to the increasing effort to centralize the architecture of radio during the Nazi regime. If the formation of the RRG in 1925 had allowed the state to gain control of semi-autonomous local stations, that project of centralization gained considerable momentum after 1933, when the *Gleichschaltung* of the Reich was orchestrated from the Haus des Rundfunks, as the central node connecting the radio-activities—administrative, technical, and programmatic—of all of Germany.

III

As discussed in the previous chapter, media historian Friedrich Kittler argued that the postwar audiovisual media industry was materially shaped by the appropriation of electronic technical media developed during World War II. Kittler was referring specifically to the way in which both military and civilian sectors took over and transformed devices such as tape recorders, mixing boards, new kinds of microphone, and amplifiers.[88] But the new building typologies developed during the interwar years were also appropriated, modified, and re-used for new purposes in the Cold War period. By typologies I mean the immaterial sets of organizational ideas linked to buildings such as the broadcasting house, which were understood as accumulated knowledge worth applying to new domains. It is this process of "typology transfer" which gave facilities like the Funkhaus on Nalepastrasse a fundamental operative role under the new political and technological regime of the GDR. (Fig. 161, 162, 163, 167) The GDR began

87 For more on the invention of television in Germany, see Friedrich Kittler, *Optical Media: Berlin Lectures 1999* (Cambridge: Polity Press, 2010), 207–224.

88 Friedrich Kittler, "Unconditional Surrender," in *The Truth of the Technological World*, ed. Hans Ulrich Gumbrecht (Stanford: Stanford University Press, 2013), 195–208, here 203.

transmissions from the Funkhaus on Nalepastrasse after it was expelled from the Haus des Rundfunks in 1952. In the two broadcasting houses, completed twenty years apart, we can trace not only the technological development of radio but also the way the production and dissemination of information through electromagnetic waves became an industry in itself. The floor plan of Poelzig's Haus des Rundfunks delineates a *Grossform*, a single grand formal gesture organizing the various functions and activities demanded by the embryonic culture industry. By incorporating an array of rooms—all with different sizes, materiality, and lighting and acoustic conditions—into a cohesive form, Poelzig effectively defined the broadcasting house type, developing a formal and programmatic arrangement that would be replicated and modified in different locations around the globe. However, when we look at the plan of the East German Funkhaus designed by Franz Ehrlich, that unity no longer exists. In place of a single freestanding building, there is an ensemble of different buildings, each serving a particular function defined by a specific formal and spatial character.

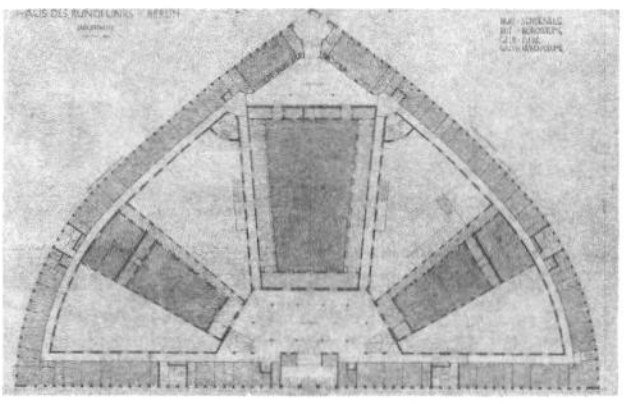

163 → P. 214

167 → P. 217

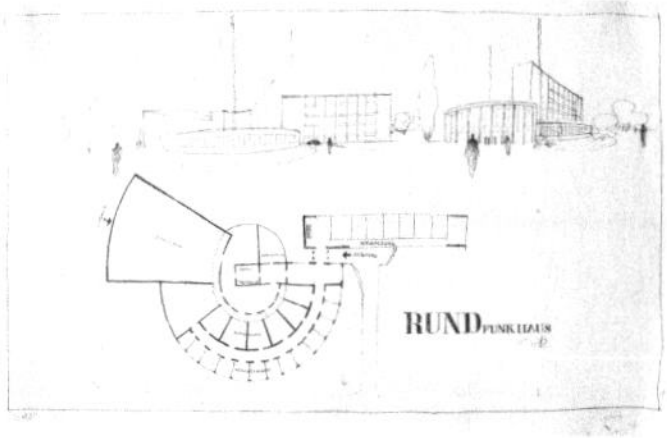

164 → P. 216

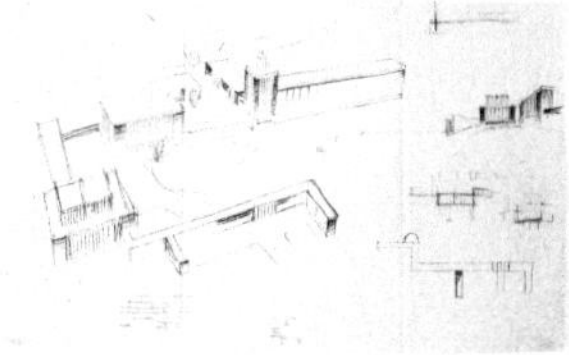

165 → P. 216

(Fig. 164) Ehrlich's scheme didn't start out that way. Initially, he had conceived of a unified object in the spirit of the original broadcasting house. An early sketch shows him building on the centripetal scheme envisioned by Poelzig, with an internal ring of recording studios lapped by an outer ring of spaces for the editorial team. A larger portion of the circular arrangement was taken up by the main concert hall, while the administrative program was accommodated in an independent pavilion. (Fig. 165) In a later plan sketch, however, the Poelzig typology is dismembered, atomized. In this new iteration, there is a complex of "blocks," each accommodating only a fragment of the industrial process that turns mechanical sounds into electrical signals. Each of the buildings that make up the ensemble mediate a host of relationships between people and machines, between mechanical sounds and electroacoustic devices, between political interests and cultural sensibilities. The new political

168 → P. 217

169 → P. 217

166 → P. 217

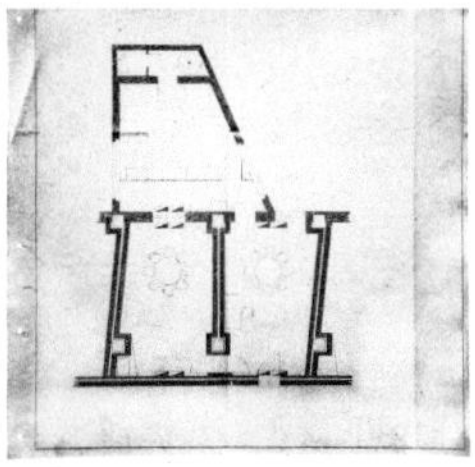
170 → P. 218

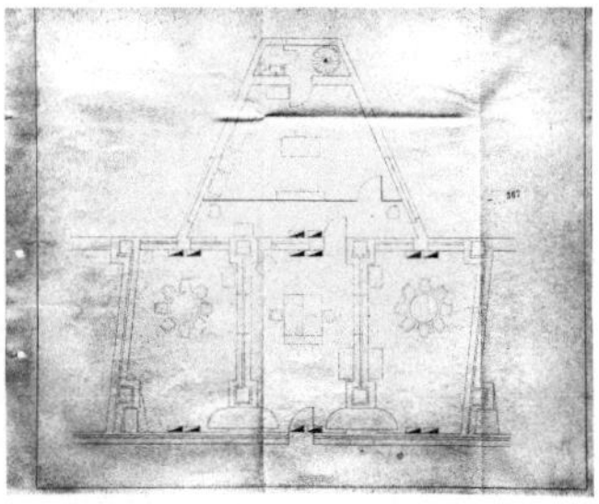
171 → P. 218

and technological context of the Cold War years had forced Ehrlich to drastically expand the typology of the broadcasting house, transforming a single unified building almost into a city in its own right.

By consolidating all of media institutions around a centralized complex, the GDR was not only redefining the scale of the broadcasting house but was, more fundamentally, designing a new spatial device for disseminating political ideas far beyond its borders. A closer look at the design process of the Funkhaus on Nalepastrasse reveals how the production and dissemination of information was transformed by the combination of the political conditions intrinsic to the Cold War and the technological innovations that emerged out of that context. Far from depending on the design vision of a specific architect, the changes in the typology of the broadcasting house were driven from within, by the forces exercised by electronic technical media, by the spatial configurations they enhanced or forbade, and by the national and transnational scale of the Cold War information conflict. In this sense, the broadcasting type, while providing material evidence of those historical transformations, also served as an agent of historical change, reshaping the ways of producing, transmitting, and receiving information.

In a signed document dated May 1951, Franz Ehrlich described the existing conditions of the Funkhaus site, which had previously been used as a wood manufacturing facility. Entering from the east side, through Nalepastrasse, visitors encountered a reinforced concrete structure, clad in brick and dating from 1942, which had once contained wood workshops and production facilities. Four stories high, 125 m long and 18 m wide, the building had a small six-story tower, 20 m by 20 m, at its southern end. The first operation on the site was to renovate this building and convert it into Block A, which would host most of the administrative program of the new facility. To accommodate the large number of working spaces required by the centralized infrastructure, the proposal added one story to the longer portion of the existing building, and three extra levels to the tower. (Fig. 168, 169) Looking at the building today, one can still see the slight variations in materiality between the original structure and its 1952 extension.

(Fig. 166) As the existing building did not have a basement, all the technical equipment was installed on the ground level, while the floor above was used exclusively for special technical rooms. The new facility organized these in clusters, with small groups of editors and program producers, to avoid a parallel arrangement of walls. (Fig. 170, 171, 172, 173) Ehrlich designed furniture and light features for these rooms as well. At the end of the building, the renovated tower housed the new administrative offices as well as the vertical circulation for the whole block. With these modifications, a building conceived for the wood industry was transformed into more than five hundred independent rooms dedicated to radiophonic production. Looking closer, it is interesting to see the particular spatiality of the proposed rooms—to see, for example, how views are enabled while sounds are not, or how the circulation of people is limited by layered walls while the flow of information is enhanced and articulated by the cables that wire the rooms together.[89] This was the only existing structure that was retained. Other smaller buildings on the site were demolished as they were unsuited to the required spaces of the radio program.

Perpendicular to Block A was Block C, containing a canteen for the workers at ground level, and an auditorium with more than five hundred seats on the level above. (Fig. 174) Though independent, the building derived its material and formal vocabulary from the existing building, albeit at a smaller scale. (Fig. 175, 176) None of its rooms were used directly for any kind of radio production, but served instead for social gatherings and activities. On the north side of Block C was Block D, an impressive industrial hall of more than 2,500 m^2 covered by eight concrete half-vaults. (Fig. 177, 178) The massive hall sheltered all

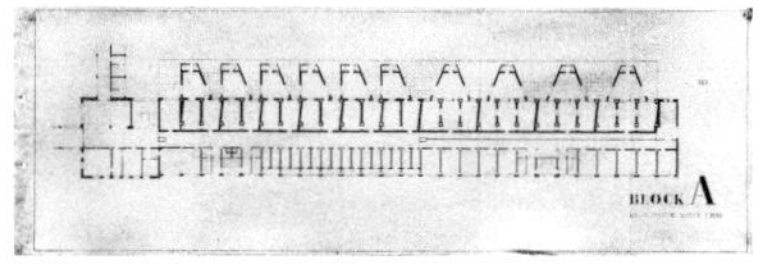

172 → P. 218

173 → P. 218

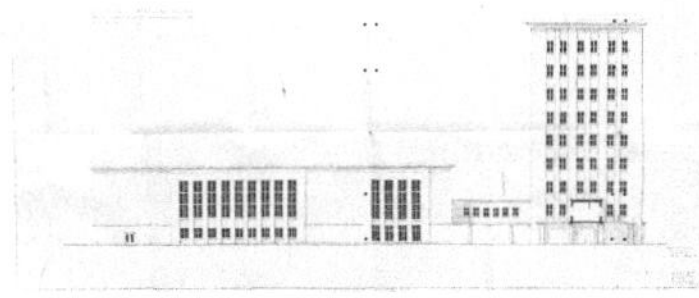

174 → P. 219

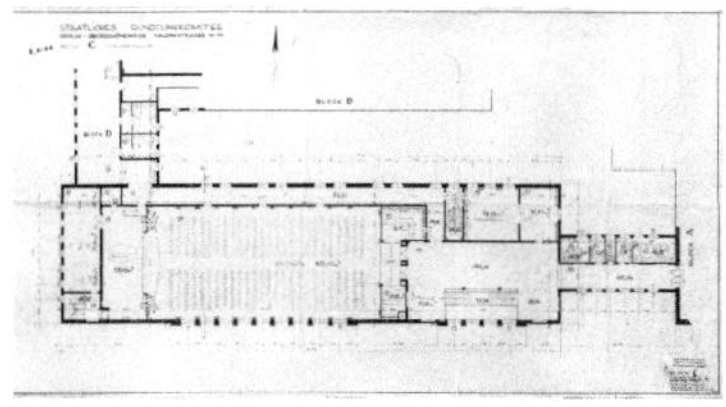

175 → P. 219

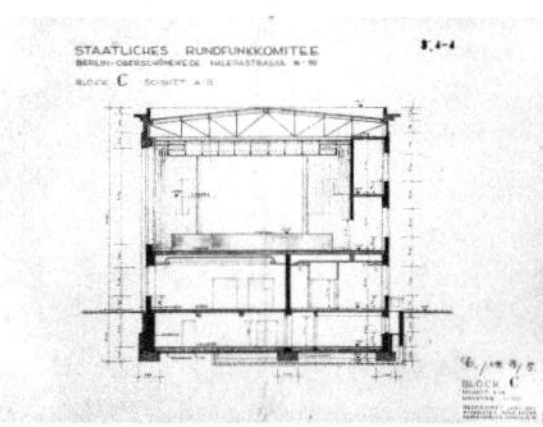

176 → P. 219

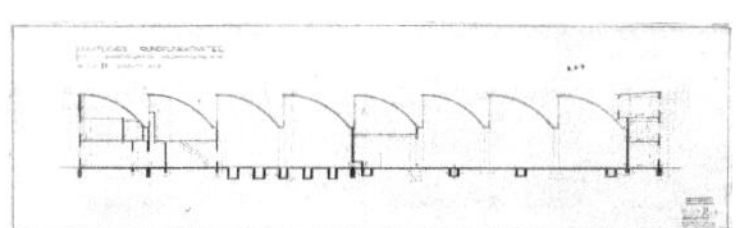

177 → P. 220

178 → P. 221

89 For the entire description of the early design of the project by Franz Ehrlich, including estimated costs and early design drawings, see "Vorprojektierungen 1951–1952 Verwaltungsgebäaude HV Bau-Industrie Nalepastrasse VVB Industrie-Entwurf, June 7, 1951," in Deutsches Rundfunkarchiv Babelsberg, F 088-01-02/0005.

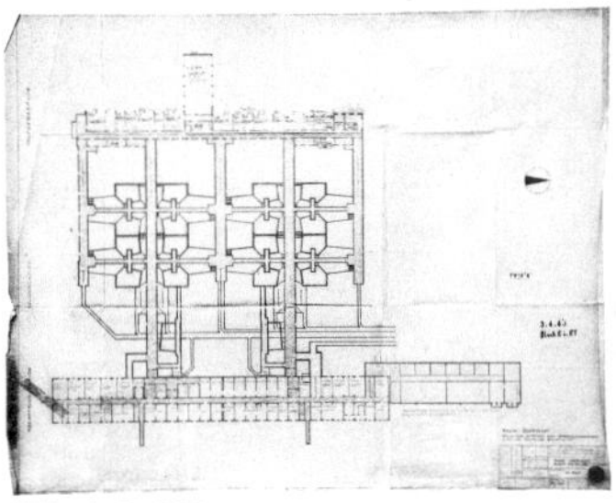

179 → P. 222

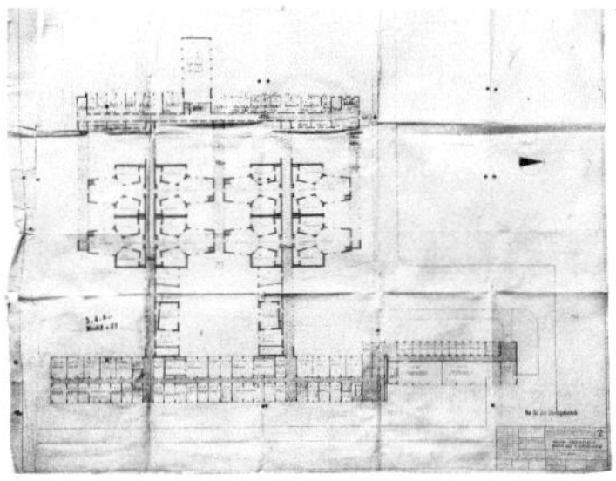

180 → P. 224

90 For all of the requisite technical details for Block D operations, see "Vorbereitungsraum für Ü-Stelle," in Bundesarchiv, DM 301/11254.

91 For more on Radio Berlin International see Odermann, *Wellen mit tausend Klängen*. The problem of broadcasting to African nations took on relevance at both political and technical levels in the GDR during the 1960s. For more on reports from GDR technicians attending international conferences on the subject, see H. Götze and F. Täumer, "Problem der Planung von Lang- und Mittelwellenrundfunksendern für die afrikanische Rundfunkzone," *Technische Mitteilungen des RFZ* 9, no. 1 (1965): 49–53. For more on the diplomatic relations established between the GDR and India through RBI, see Anadita Bajpai, *Cordial Cold War: Cultural Actors in India and the German Democratic Republic* (New Delhi: Sage, 2021), 63–110.

the mobile infrastructure for radio production. It was where the outside broadcast vans were stored, cleaned, tested, and repaired. Painted in blue, these vehicles crisscrossed GDR territory, reporting and transmitting news directly from local venues. The massive size of the hall indicates the scale of the territorial expansion of media production effected by these "studios on wheels."[90]

Since the new Funkhaus would accommodate thousands of workers, it was conceived as a city-within-the-city. And like a piece of the city, it was planned and built over time, adjusting to ever-changing requirements. In 1959, Radio Berlin International (RBI) was set up to create an international presence for the GDR—one that would counterbalance the voice of West Germany broadcast through Deutsche Welle—with programs in German, Swedish, Italian, English, French, Arabic, Danish, Spanish, Portuguese, and Swahili.[91] The ambition to disseminate "socialist solidarity" globally had an impact on the production facilities on Nalepastrasse, as work immediately began on a new block to host studios and production facilities for the international programs. The nearby Nauen station, which had given birth to wireless telegraphy in Germany at the dawn of the century, took on a new role, transmitting RBI's international programs.

(Fig. 179) In the new Block E, the extension of the program was organized in two parts. On the east side, a single five-story bar housed the expanded editorial team. It was equipped with offices for program and content production, as well as the technical capacity for storing and processing tapes recorded elsewhere in the Funkhaus. On the west side, closer to the rest of the complex, was an irregular mat-building in which circulation was interwoven with courtyards and rooms of different shapes and sizes. This part of the block contained new facilities for recording mainly speech-based programs, political commentaries, and news. (Fig. 180) It was organized in eight clusters of rooms mirrored along four different axes of symmetry. Each of the clusters was made up of a control room (*Regieraum*) and a preparation room (*Vorbereitungsraum*), flanked on either side by a recording studio for talk-based programs (*Sprecherraum*). The five corridors serving the eight clusters were differentiated: three gave access to the control rooms; the other two to the recording studios. The topology expressed in the plan—a spatial matrix of rooms connected without circulation, combined with the corridors linking the various clusters—speaks to the mechanisms by which information was manufactured in this part of the Funkhaus. One can imagine the corridors serving the recording studios in constant use, with people coming and going between the editorial offices, collecting scripts before they recorded the programs. Once they were finished, the magnetic tapes carrying the recorded sounds would be carried from the control room along the separate corridors that led to the editing suites in the other part of the building.

This topology facilitated two independent flows—one of people, the other of tapes encoded with information—all moving with the orchestrated precision of choreographed production. Overlapping voices speaking in many different languages filled these rooms around the clock, ready to be broadcast around the world. “It is Bendix, Metal, Lucky, and Honeywell who are producing electronic equipment, toys, and textiles in Mexico and are getting rich on low production costs and super exploitation”; “Es ist zehn Uhr Weltzeit. Hier ist Radio Berlin International, die Stimme der Deutschen Demokratischen Republik. Wir senden täglich zu dieser Zeit das Programm in deutscher Sprache für Hörer im Nahen Osten, Nordwestafrika, Südostasien und Europa”; “Expressions of surprise and even amazement dominated in the West German media at the beginning of this year when the neo-Nazi republicans received seven and a half percent in the West Berlin elections, giving them eleven seats in the parliament”; “This will not last forever. The measures taken by Iraq and Chile against the foreign monopolies are just two

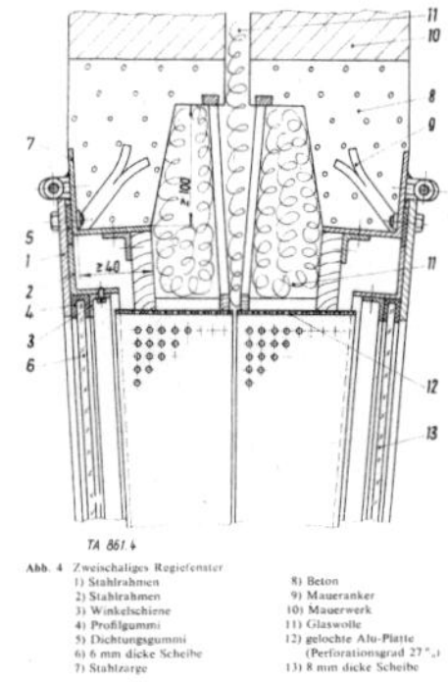

181 → P. 226

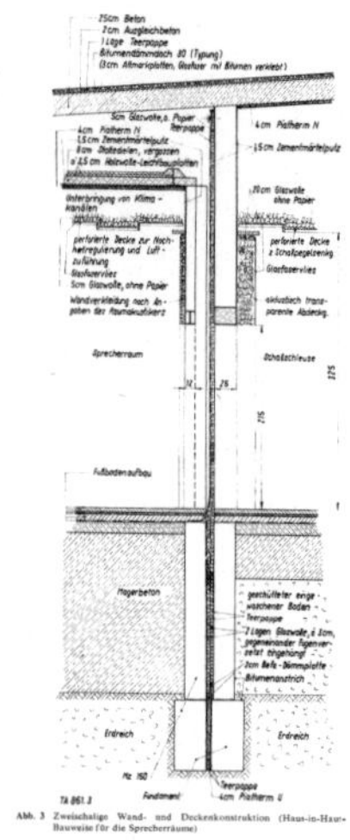

182 → P. 226

183 → P. 227

examples showing the future trend." (Fig. 181, 182) Multilayered dividing walls allowed this polyglot cacophony to take place with a maximum of spatial density and a minimum of sonic interference. The section of the walls shows how the transmission of sounds was avoided by changing the medium of propagation many times through the use of different materials. The construction of the roof and ceiling eliminated external noise that might interfere with production, and in particular the sound of planes from nearby Schönefeld airport. Mechanical ventilation was enhanced without transmitting sounds, and most importantly, cables wove in and out of the walls.[92] (Fig. 183, 184) These walls also articulated an unusual working environment for postwar Germany, with women sitting at futuristic-looking tables managing the magnetic tapes on one side, and brown-skinned men talking on the other side. (Fig. 185, 186)

After Block E was completed, the company VEB Neontechnik was commissioned to design a sign that would communicate the sum of institutional, administrative, and infrastructural transformations instigated by the radiophonic enterprise at a national level. The neon sign, 5 m high and 60 m long, would dominate the roof of the rather abstract and rationally gridded facade of the editorial section of Block E. Placed centrally, over the symmetrical axis of the building, were the words "Deutscher Demokratischer Rundfunk," flanked on one side by the logos of Berliner Rundfunk and Radio DDR, and on the other by the logos of Deutschlandsender and Radio Berlin International. If the multiple transformations in the typology of the broadcasting house had effectively dismembered its program, obscuring the overarching institutional coherence of radiophonic production, the overscaled signs were intended to restore the suppressed unified whole of the project. The neon lights atop the building captured in one grand gesture the sum of the reforms initiated in 1952 by the State Radio Committee, and the four logos joined by the phrase "Deutscher Demokratischer Rundfunk" would mark the institutional organization of radio in the GDR until the fall of the Berlin Wall.[93]

92 For more technical details on the construction of Block E of the Funkhaus on Nalepastrasse, see Gisela Herzog, "Der Studioneubau des Funkhauses Berlin," *Technische Mitteilungen des RFZ* 10, no. 4 (1966): 141–148.

93 For the original documents, see "Neonbeleuchtung Block E," July 25, 1968, in Deutsches Rundfunkarchiv Babelsberg, F 088-01-02/0003.

Radio Berlin International's global aspirations, along with Block E's architecture, would be tested by an event that took place 12,000 km from Berlin. At 09:10 on September 11, 1973, the socialist president of Chile, Salvador Allende, addressed the Chilean people through radio, announcing the fall of his government following a violent US-sponsored coup d'état.[94] Sharp, historically conscious, and poetic as always, Allende's last words on the radio still haunt Chile's collective subconscious. The president had used the only architecture of radio available to him after the bombardment of the transmission towers of Radio Portales and Radio Corporación in what the military called "Operación Silencio." He had picked up "the Iron"—the magnetic telephone that directly connected the presidential office in La Moneda with the broadcasting studio of Radio Magallanes, not yet seized by the new regime—and asked to be put on air. The station's director, Guillermo Reves Santis, responded immediately, and within seconds Allende was giving his last address.[95] The media condition that made possible this historic event was a precarious chain: a telephone line pressed against a microphone, the microphone inside a studio linked to a tape recorder, which was in turn wired to a transmission station located in Renca, on the northwest outskirts of Santiago. With a prophetic tone, Allende famously set down in tape—and not in stone—that "history is ours, and it's made by the people."[96] His consciousness of being at a "historical crossroads" flashes against a sonic background of bullets and fire on his end of the line, and the sounds of people running and screaming on the other. Beyond Allende's words, everything that was *real* in that moment got stored on tape. The president recognized he would "pay with my life for my loyalty to the people." Four hours later he was dead, and everyone who worked for the few radio stations loyal to the socialist regime had either been killed or forced into hiding.

From the day of the coup, the entire infrastructure for producing and transmitting information in Chile was controlled by, or associated with, the military junta.

184 → P. 227

185 → P. 228

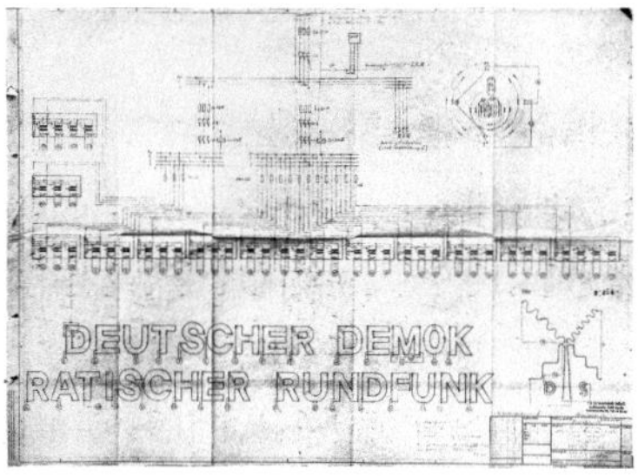

186 → P. 229

94 To read the testimonies of those involved in the heroic transmission of Allende's last address, see José Miguel Varas, "La verdadera historia del rescate del ultimo discurso de Salvador Allende," https://www.ciperchile.cl/2008/06/26/la-verdadera-historia-del-rescate-del-ultimo-discurso-de-salvador-allende/. See also Ascanio Cavallo and Margarita Serrano, "Las 24 horas que estremecieron a Chile," *La Tercera Reportajes,* August 24, 2023, 8–23, and Ascanio Cavallo and Margarita Serrano, *Golpe 11 de Septiembre 1973: Las 24 horas más dramáticas del siglo 20* (Santiago: uqbar, 2013). I would like to thank Agustina Labarca for pointing me to these references.

95 Salvador Allende Gossens, "Discurso del Presidente Salvador Allende en la radio Magallanes," September 1, 1973, in Biblioteca Nacional de Chile, MC0043425, Compact Disc, 6'17'. Author's translation. The address is widely available online.

96 "Discurso del Presidente Salvador Allende en la radio Magallanes."

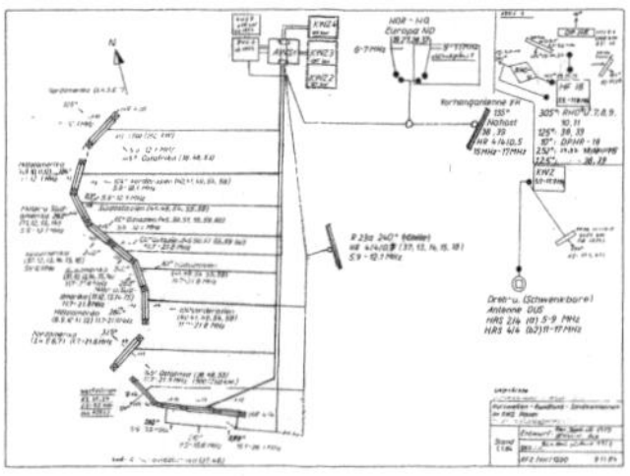

187 → P. 230

It took only a day, however, for East Berlin media to propose countermeasures. A memo addressed to the head of the GDR's Foreign Office for Transport and Telecommunications, General Feist, reported: "As of September 15, 1973, three 100 kW transmitters broadcast programs to Chile at 22:45–24:00 for the first time."[97] It went on, "in agreement with the State Radio Committee, transmission times to the target area Chile will be extended." Just one day after that, "as of September 16, 1973, a total of five 100 kW transmitters broadcast programs to Chile from 00:00–04:00. Thus, transmissions to Chile have increased from 1.5 hours to 3 hours per transmitter (15 hours in total)."[98] A plan was put forward for immediate and long-term infrastructural reforms to "increase effectiveness in the target area of Chile." In principle, the report suggested, "there is the possibility of using all five shortwave transmitters exclusively for Chile (multi-frequency operation). With the existing antennas, the whole of Chile can be covered (transmission range 218°–271°)." As a consequence of this infrastructural maneuver, however, "all other programs broadcast by these transmitters in other directions and languages had to be cancelled." In the longer term, the report raised the possibility of deploying the much more powerful 500 kW shortwave transmitter in Nauen for transmissions to Chile, a move that would require an additional antenna to supplement the existing ones, which were all directed towards Europe and the Middle East.[99] (Fig. 187) Nauen would be adapted to extend its range to Chile. A plan shows a snake of antennas connected to the various transmitters, depicting the material basis for the apparently immaterial global reach that RBI aspired to: North America, Central and South America, and Southeast Asia each have their own dedicated antennas. Separated from the "snake" is the antenna built specifically to counteract the operation intended to silence a country some 12,000 km away, across the Andes.[100]

On September 15, 1973—that is, two days after the coup against Allende's democratically elected

97 "Wirksamkeit der Sendungen von RBI im Zielgebiet Chile," September 20, 1973, in Bundesarchiv, DY30, 39093. This and subsequent quotes are translated by the author.

98 "Wirksamkeit der Sendungen von RBI."

99 "Wirksamkeit der Sendungen von RBI."

100 For more about the origins of Nauen in the early twentieth century and its impact in global networks of communication ever since, see chapter 1.

government—RBI broadcast the first edition of *Chile al Día*. The new Spanish-language program would go out every day, without interruption, for the next sixteen years. Pretty much everyone who had the slightest relevance for the cultural and political situation in Chile passed through the Funkhaus on Nalepastrasse to address their fellow countrymen. The infrastructural changes to the historically important Nauen transmission station allowed for a direct connection between the interior spaces of Block E of the East Berlin broadcasting house and the interior spaces of people in Chile, bringing the voices of the thousands who were in exile to those who stayed at home.[101] RBI, together with programs from Moscow aimed at a broader swath of the Chilean population, were the only sources for accessing information not aligned with the military regime.[102] During the sixteen long years of dictatorship, many in Chile had a secret daily listening ritual. Once the children were in bed and silence finally filled the home, transistors were tuned to the right frequency. The otherworldly opening refrain of the Inti-Illimani song "Altura" would be followed by the message "Radio Berlin International presents *Chile al Día*, news, chronicles, commentaries, and interviews by and about Chile. Chilean, you are not alone."[103] In one program, the Argentinian writer Julio Cortázar recalled his meeting with Allende in Chile and reminded the audience of the importance of the continuing struggle in Latin America.[104] In another, the exiled Joan Jara, the widow of the musician Victor Jara, who was murdered right after the coup, tearfully recounted how much she missed Chile.[105] The activist Gladys Marín reported from the 1975 Women's Congress in Berlin and sent emotional messages to all the woman held in prisons or concentration camps in Chile.[106] Armando Uribe, a former diplomat, gave a firsthand account of the hostility towards Allende's government that he had witnessed in the US.[107] And there were many, many more in this vein, until the final transmission of the program, on March 11, 1990. *Chile al Día* went out on a high, celebrating the inauguration of Patricio Aylwin, the country's first

101 The reception of the program was intensely surveyed by Pinochet's secret police (DINA). For more on the DINA report on Radio Berlin International, see "Intercepción Radio Berlin International," February 26, 1976, in Museo de la Memoria y de los Derechos Humanos, 00000057-000001-000029.

102 The battle to control the infrastructure of information was ubiquitous during the Cold War, and far from being the exception, Chile was an emblematic case where the socialist government tried to implement an early cybernetic infrastructure to counteract the information sabotage undertaken by ITT after Allende's election. For a complete historical account, see Eden Medina, *Cybernetic Revolutionaries: Technology and Politics in Allende's Chile* (Cambridge, MA: MIT Press, 2014). For an earlier theoretical analysis of "Cybersyn," see Claus Pias, "Der Auftrag. Kybernetik und Revolution in Chile," in *Politiken der Medien*, ed. Daniel Gethmann and Markus Stauff (Zurich; Berlin: Diaphanes, 2005), 131–154. For more about the role of radio during and after the coup in Chile, see Luis Zaragoza, *Voces en las Sombras: Una Historia de las Radios Clandestinas* (Madrid: Ediciones Cátedra: 2016), 333–343, and José Miguel Varas, *Escucha Chile: Radio Moscú* (Santiago: LOM Ediciones, 2012).

103 Opening line of the radio program *Chile al Día* extracted from "Isabel Allende auf einer Konferenz in Helsinki," July 28, 1973, in Deutsches Rundfunkarchiv Babelsberg, Chile-Programm, 2044023, digitized audio file. Author's translation.

104 "Rede von Julio Cortázar bei der Eröffnung der Neruda-Schule im Rahmen der Solidaritätskonferenz mit der chilenischen Kultur in Torun," May 4, 1979, in Deutsches Rundfunkarchiv Babelsberg, Chile-Programm, 2043958, digitized audio file.

105 "Interview mit Joan Jara," November 4, 1977, in Deutsches Rundfunkarchiv Babelsberg, Chile-Programm, 2044014, digitized audio file.

106 "Interview mit Gladys Marín, 1975–76," in Deutsches Rundfunkarchiv Babelsberg, Chile-Programm, 2044042, digitized audio file.

107 "Aussage von Armando Uribe vor der Internationalen Kommission zur Untersuchung der Verbrechen der Militärjunta in Chile im März 1974 in Helsinki, April 1974," in Deutsches Rundfunkarchiv Babelsberg, Chile-Programm, 2043843, digitized audio file. For the complete written record of the programs broadcast from East Berlin to Chile, see Deutsches Rundfunkarchiv Schriftgutbestand Hörfunk: Radio Berlin International / Redaktion Lateinamerika / Chile.

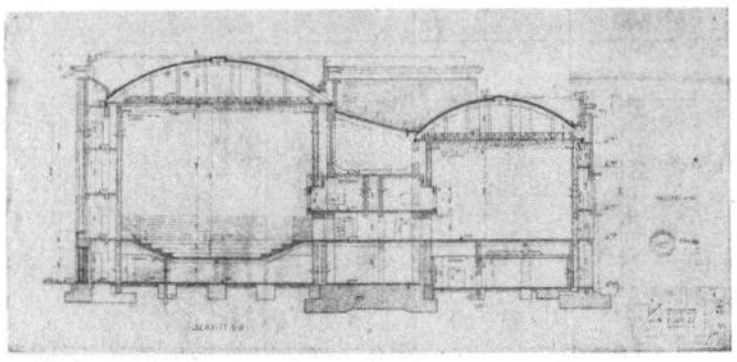
189 → P. 234

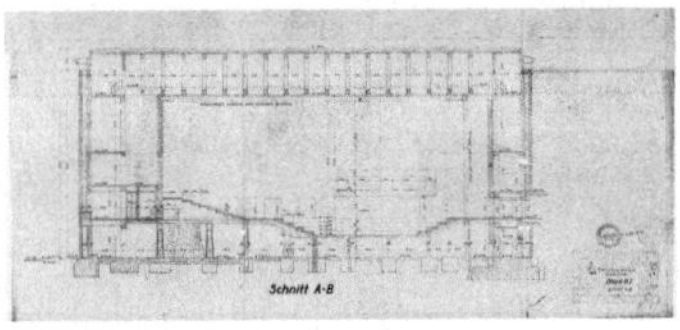
190 → P. 235

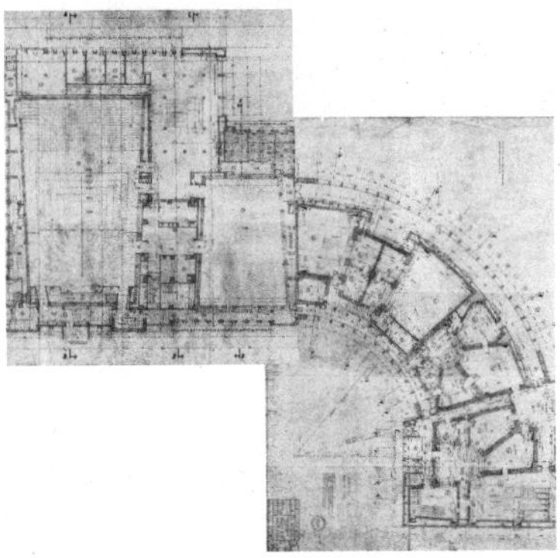
192 → P. 237

democratically elected president since Allende. There was a celebratory note, too, when Jorge Alvarado, *Chile al Día*'s anchor for the sixteen years of its existence, was joined by others involved in the making of the program in a roundtable discussion of its history and achievements.[108]

To the south of Block E, the international section of the Funkhaus on Nalepastrasse, was Block B, arguably the most complex, sophisticated, and interesting part of the complex, with an interlocking system of recording studios dedicated to the production of music and radio plays. As we have seen, in the production process of radiophonic content, the task of recording is defined by both program and technology. Recording is the spatial phenomenon through which the world of vibrating instruments and voices—of people speaking and performing, of bodies interacting in space—leaves the realm of the mechanical and enters the realm of the electrical as mediated by buildings and microphones. It is *in* and *through* the walls of these studios that the mechanical vibration of the air—as the transmitting medium of sound—is transduced into electrical signals. The studios here perform as the envelope of that transduction process. At the same time, they function as the means for orchestrating that transduction in accordance with the requirements of this particular form of production.

Here, it is important to reiterate the urban and territorial ramifications of the transmission, reception, and blocking of radio waves in Berlin.[109] At the scale of the country, the precise location of transmission stations defined and expanded the limits of dissemination of socialist space through radio. At the scale of the city, the location of the jamming transmitters effectively formed a pre-emptive scrambling wall, turning signals into noise and preserving the sovereignty of the GDR's airwaves. At the scale of the building, finally, the aim was to build walls that would articulate a new and recognizable sonic content. In simple and concrete terms, the sonic signature of the broadcast socialist space was

108 "Rundtischgespräch zum Abschluss des Chile-Programms von RBI," March 11, 1990, in Deutsches Rundfunkarchiv Babelsberg, Chile-Programm, 2042934, digitized audio file. Author's translation. To listen to this and other episodes of *Chile al Día* broadcast by Radio Berlin International from Berlin to Chile, go to the following link https://soundandscience.net/contributor-essays/chile-al-dia-and-the-funkahus-in-the-nalepastrasse/ or use the QR Code placed below.

109 For the territorial implications of the transmission and reception of radio signals, see chapter 1.

shaped and constructed by the material composition and geometric disposition of the walls in this part of the Funkhaus.[110] In this respect, the particular organization of walls in Franz Ehrlich's design adheres to a precise and heterogeneous set of programmatic, geometric, and performative requirements.

188 → P. 233

In the original publication of the Funkhaus in a 1956 issue of *Deutsche Architektur* dedicated to industrial architecture, Block B was presented alone, as an independent building. In the text, Ehrlich describes it as "one external house and eight internal houses."[111] The program is located above ground, to avoid the complexity and expense of constructing basement levels in an area close to the River Spree. (Fig. 189, 190, 192) The external shell is of loadbearing concrete, partly clad in brick. Large concrete beams with a variable section of up to 5 m allow a free span from one side to the other. The studios along with the other "internal houses" are structurally independent from the external shell. All are constructed using a sophisticated system of multilayered double walls. Wires and technical facilities occupy the technical *poché* between them, so the spaces are interconnected not by traditional means of circulation, but by electricity. Despite—or precisely because of—the wireless nature of radio, everything within the Funkhaus had to be wired.

On the floor plan of Block B, looking from right to left, the first "house" is a cluster of five rooms surrounding a six-sided control room tailored to the performance and recording of radio plays. (Fig. 188) Each side of the control room connects with a separate room through an acoustically sealed window. Within the control room, the mixing board and the tape recorder form the interface between the visually connected but sonically sealed spaces. They could be described as "space machines," gathering all of the discrete sounds produced in the different rooms and routing them by wire to the central control space.[112] Thus, the walls defining the cluster perform as material and spatial extensions of the montage-based

110 The notion of broadcasting socialist space is introduced in chapter 1, where I discuss the dialectical opposition between socialist and capitalist space proposed by Henri Lefebvre in *The Production of Space*, 1991.

111 Franz Ehrlich, "Aufnahme- und Studiogebäude des Staatlichen Rundfunkkomitees," *Deutsche Architektur* 9 (1956): 399–409, here 400.

112 The notion of "Raummaschine" was used by Ehrlich himself to describe his collaboration with Gropius on the Total Theater project for Piscator in 1927. See Ehrlich, "Aufnahme- und Studiogebäude," 399–409. While in the context of the theater, the notion refers to the mechanical adaptability of the stage designed to blur the boundary with the audience and enable simultaneous performances, in the context of the Funkhaus and its electronic media apparatus it can be understood as enabling the coexistence of various sonic spaces and connecting the recording studios with a much broader audience. For more on the models and original drawings of the project, see "Total Theater for Erwin Piscator, Berlin, 1927," in Harvard Art Museums, HUAM169343. See also Peter Walsh and Evelyn Rosenthal, *Walter Gropius' Total Theater Design of 1927: A Variable Theater-Instrument,* exh. cat. (Cambridge, MA: Harvard University Art Museum, 1992).

191 → P. 236

193 → P. 238

195 → P. 238

194 → P. 238

capacities of the mixing board and the tape recorder, mediating between sonic signals and electronic machines.

Invented during World War II by German engineer Walter Weber under the supervision of Hans-Joachim von Braunmühl, the magnetic tape recorder revolutionized media production in two different ways. Previously, in the absence of a means of recording high-quality sounds for longer periods, radio had essentially to be transmitted live. Poelzig's Haus des Rundfunks, for example, was organized around this principle. With the arrival of the tape recorder, the production of content could be uncoupled from transmission, making simultaneous productions both possible and desirable. On another level, the tape recorder allowed different sonic signatures to be edited and mixed, turning the production of sound into a veritable process of sonic montage. The architecture of the Funkhaus, and of Block B in particular, should be viewed through the lens of this technological invention. Combined with the mixing board in the control room, the tape recorder allowed the independent and fragmentary production of sounds to be mixed, combined, and articulated. Walls, wires, and windows all took on a different spatial agency following the introduction of the tape recorder, with the device decisively transforming the overall typology of the broadcasting house from within.[113]

Continuing with a reading of the plan of Block B, a corridor separates the cluster just analyzed from another that follows similar morphological and programmatic principles. The adjacent trapezoidal space, Saal IV, was designed for the recording of dance music, with a reverberation time of 1.2 seconds.[114] (Fig. 191, 193) The four trapezoidal rooms that constitute the fourth "house" were relatively small and used for solo performances: each room was acoustically and visually independent from the others. Here, the surfaces of the walls were covered with prisms painted with fleurs-de-lis. The ornamental pattern was not only intended to please the eye but

113 For a more in-depth reading of the invention of the tape recorder in the context of the RRG during World War II, see John T. Mullin, "Creating the Craft of Tape Recording," *High Fidelity Magazine* 26, no. 4 (1976): 62–67; Peter Hammar and Don Ososke, "The Birth of the German Magnetophon Tape Recorder 1928–1945," *db Magazine* 16, no. 3 (1982): 47–52. See also "Bericht Trautweins über Vorführung des Magnetophons der AEG, 1935," in Archiv Universität der Künste Berlin (UdK-Archiv), Bestand 1b (Rundfunkversuchsstelle) / 9. According to historian Stephen Lovell, tape recording existed in the Soviet Union as early as the late 1920s. See Stephen Lovell, *Russia in the Microphone Age* (Oxford: Oxford University Press, 2015), 6.

114 Reverberation time, usually measured in seconds, is the duration between the emission of a sound and the decay in its intensity below the level of human perception, in a given room. For an early study on reverberation time, see Sabine, *Collected Papers on Acoustics*.

also, and perhaps more importantly, lent the sonic signal a specific character. Returning to the plan, the large trapezoidal room of Saal III was designed for the recording of chamber music, with a reverberation time, again, of 1.2 seconds. All the music-recording rooms were trapezoidal to avoid the acoustic problems associated with parallel walls. Slight variations in angles, combined with the varied materiality of each wall, gave individual rooms unique sonic signatures. (Fig. 195) The adjacent large trapezoidal room of Saal II was designed for small orchestras, with a reverberation time of 1.4 seconds. At the far end of the complex spatial distribution was Saal I, the largest recording studio in the world, able to accommodate an entire symphony orchestra, with a reverberation time of 1.8 seconds. Finally, a pair of control rooms—one facing Saal II, the other facing Saal I—separate and mediate the two major spaces in the building.[115] (Fig. 194)

Access to Saal I was through a separate entrance, leading to a colonnade of black stone columns resting on a reflective black marble floor. From there, one ascended a stair that arrived at a wood-paneled foyer offering spectacular views of the River Spree. (Fig. 196, 197) This impressive entrance sequence, however, was not designed for a large audience drawn from the public. (Fig. 198, 199) In previous iterations of the broadcasting house typology built in the 1930s in Berlin, London, and New York, the plan and program were essentially structured around a conventional concert hall, with a stage and space for a seated audience. By contrast, the highly refined concert hall in the Funkhaus had seats for a full symphony orchestra, but no place for a live audience. Here, the broadcasting apparatus substituted for the collective space for listeners. (Fig. 201) Saal I was a room with no direct precedents, a space full of modernity but not modernism. Beneath its strange, undulating ceiling it resembled a classicist structure with cylindrical columns. The backdrop to the 10 m-high hall was a large pipe organ of the kind usually found in a church, only now there was a music

196 → P. 239

197 → P. 239

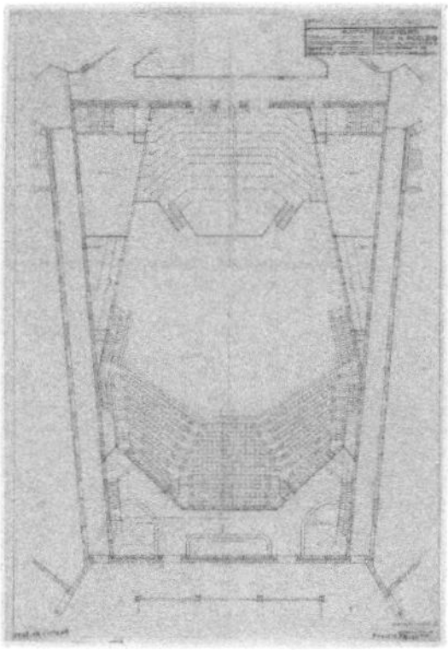
198 → P. 240

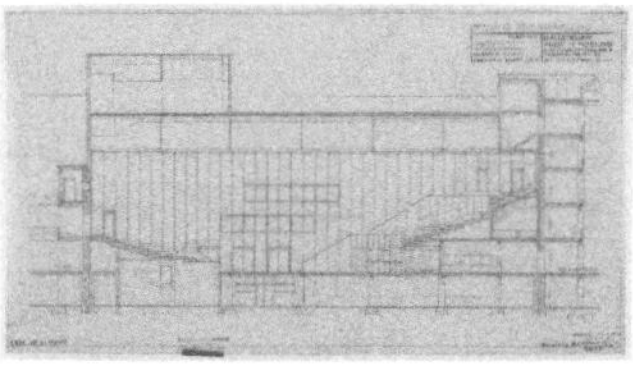
199 → P. 241

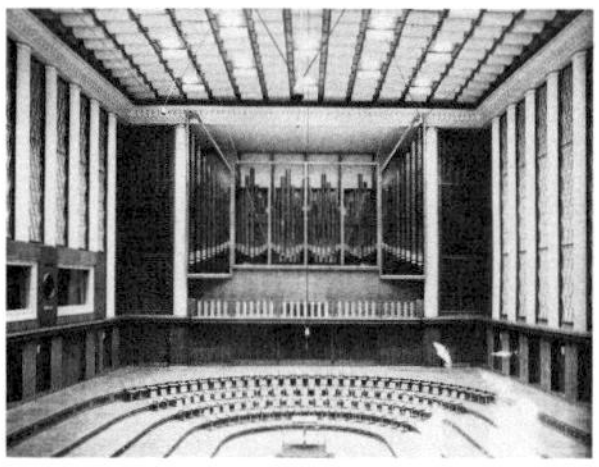
201 → P. 245

115 In February 1955, a major fire broke out in Saal I of the Funkhaus. GDR authorities pointed the finger at American spies. Saal I was quickly reconstructed in time for the inauguration of the building in 1956. For more, see Jan Eik, "Tod einer Legende—Der Funkhausbrand vom 16. Februar 1955," in *Besondere Vorkommnisse: Politische Affären und Attentate in der DDR* (Berlin: Das Neue Berlin, 2006), 9–64.

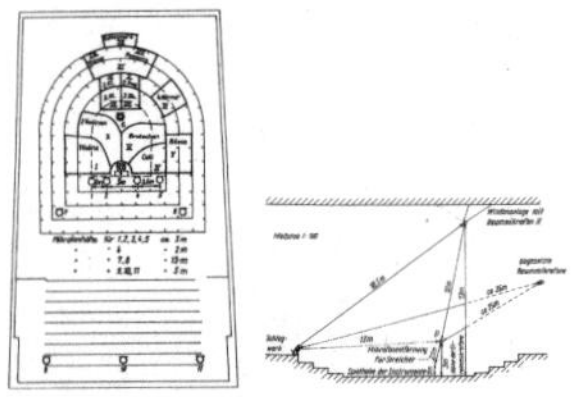

200 → P. 243

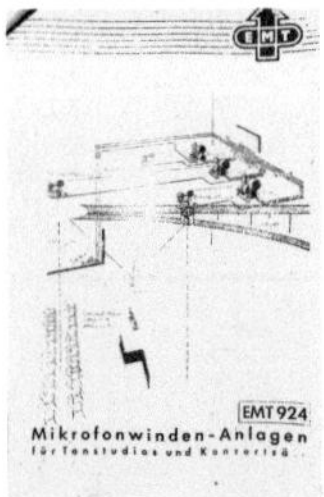

203 → P. 247

202 → P. 246

206 → P. 249

director in the pulpit rather than a priest. Suspended from the ceiling, and shining like little stars, were the innovative Neumann CMV 551 condenser microphones.[116] (Fig. 200, 203) The microphones were precisely positioned at different heights and distances and controlled by motorized winches. It could be argued that this was a hall for manufacturing, with the utmost refinement, nothing more than the vibration of air—air engineered as the medium that modulates and propagates sound. The ceiling, together with the fake, half-cylindrical columns, acted as sonic diffusers, while the wooden gratings on either side of the organ were bass traps to eliminate low frequencies. For the musicians in the orchestra, playing in the hall was like being "in a tub ... floating in space," listening only to each other.[117]

Through its materials and proportions, as well as the mechanical and electrical devices that populated it, Saal I embodied an alternative form of monumentality to the symbolic grandeur of Karl-Marx-Allee or the Soviet War Memorial in Treptower Park. Forged in response to the war in the ether, it represented the aesthetic and technological consolidation of a utopian dream to broadcast the most refined version of socialist space into the homes of people around the world.[118]

(Fig. 202, 206) Taken as a whole, the Funkhaus complex contains interesting moments that speak to a profound shift in the approach to the relationship between users and information. In Poelzig's Haus des Rundfunks, the circulation of sounds through electric wires was subordinated to accommodating the activities of people. Within its unified form, a centralized hall welcomed the audience; here, users were granted a comprehensive grasp of the whole of the building, while vertical and horizontal circulation connected rooms according to the overall hierarchy of the scheme. The Funkhaus reverses this hierarchy. People circulate throughout the facility—for example, along the two elevated routes that connect Block A with Blocks B and C—in isolation from all the

116 For more on the design and costs of the organ in Saal I, see "Planung zum Bau der Orgel für den Grossen Sendesaal im Block B des Funkhauses Nalepastrasse, 1952–1955," in Deutsches Rundfunkarchiv Babelsberg, F 088-01-03/0001.

117 Ehrlich, "Aufnahme- und Studiogebäude," 399–409.

118 For more on the design of the movable microphone system in Saal 1, see "Auftragserteilung zur Montage von 4 Stck. mtor. Mikrofonwindenanlagen i. Saal I des Blockes B in der Nalepastrasse," November 26, 1956, in Bundesarchiv, DM 301/11250. For a more general understanding of stereophonic recording in the context of the GDR, see Klaus Wagner and Wolfgang Hoeg, *Stereofonie-Aufnahmetechnik* (Berlin: VEB Technik, 1970), 80–86. See also Alexander Raschkowitsch, "Neue Musik- und Hörspielstudios," *Radio und Fernsehen* 5 (1955): 204–207. See also Gerhard Steinke, "Akustisch-musikalische Grenzprobleme in der Rundfunk- und Fernsehtechnik," *Technische Mitteilungen des RFZ* 11 (1967): 149–159.

associated programs, raised above the ground to avoid any disruption to the operations underway around them. It is almost as if the users were another type of signal circulating around the Funkhaus, another set of variables among the semi-autonomous electroacoustic components that, in their co-functioning, make up a larger *ensemble of technical forms*. These components are wired together, literally by cables, and metaphorically by circulation elements that look—and act—as is they too were to perform as such links.

The differences, and continuity, between the Funkhaus on Nalepastrasse and the broadcasting houses built in Berlin, London, and New York some twenty years before offer a lens through which to examine the changing role and nature of building types. Writing in the late 1970s, a period of renewed interest in typology, Anthony Vidler proposed that from the mid-eighteenth century onwards, the production of architecture had been informed by two distinct typologies. The first was grounded in the rational order of nature, the second, responding to the need to confront mass production, referred especially to the machine. Both aimed to make universal claims about the relationship between form and function. Both typologies, Vidler wrote, "were firm in their belief that rational science, and later technological production, embodied the most progressive 'forms' of the age." He then described an emerging typology of a "radically different order" embodied in the work of the Italian Rationalists. Closing the Hegelian triad for Vidler, this third typology was an "ontology of the city" rooted in the accumulated experience of urban culture, and political in the sense that it placed "its faith in the public nature of all architecture."[119] Looking at the genealogy of the broadcasting house, it could be argued that we are confronted with a *fourth typology*. This new form of production—building on earlier models of both industrial and cultural production—brings to the fore a different rationale for conceiving buildings, one that certainly has form and function as central elements, but is equally informed by the agency of electronic technical media

119 Vidler, "The Third Typology," 1–4. See also Forty, *Words and Buildings*, 304–311.

and politics. If a disciplinary discourse around type and typology ever resurfaces in architecture, it will need to ponder the agency of electronic technical media and politics as central agents for the production of architectural form.

This idea can be illustrated by means of three electroacoustic innovations that radically redefined the typology of the broadcasting house and, with it, the media landscape of postwar Germany. I have already discussed how the magnetic tape recorder introduced a new temporality in the process of recording and broadcasting: while uncoupling the production of radio content from the process of transmission, it simultaneously transformed recording techniques by expanding the possibilities for combining different sounds and spaces. Since the invention of the magnetic tape recorder, each sound that is broadcast is the outcome of a process of montage—a process enhanced in the Funkhaus by the form and nature of the floor plan. The microphone, in turn, transformed the mechanical phenomenon of sound into electric signals. Concert halls designed for live audiences gave way to rooms for modulating and broadcasting sound. The third transformative innovation, developed during World War II, was FM radio transmission.[120] The use of frequency modulations (FM) instead of amplitude modulations (AM) had an impact on the organization of the territory. As discussed in the previous chapter, FM's much smaller radius of transmission meant that more transmitters were required to cover the full expanse of the sovereign territory, giving a new twist to the problem of jamming signals. But the use of FM to disseminate information and sounds also had an impact at the scale of the buildings designed for content production. As FM carries a much purer signal, without interference or distortion, the design of the Funkhaus had likewise to become more refined, to ensure the signals captured by microphones and tape recorders were broadcast with a similar quality.[121]

An analysis of the impact of electronic technical media on the origin and definition of the broadcasting

120 For an in-depth reading of the impact of FM radio, see Wolfgang Hagen, *Das Radio: Zur Geschichte und Theorie des Hörfunks—Deutschland/USA* (Munich: Wilhelm Fink, 2005), 339–365. See also Lewis, *Empire of the Air,* 247–278.

121 It is worth noting that the development of the Funkhaus was undertaken in parallel with the design of the "Zwillinger" in Köpenick, the GDR's first FM transmitter, with a power of 150 kW. For the declassified design protocols for the Köpenick transmission station, see "Stand der Arbeiten am Objekt 'Z3' Stichtag, September 4, 1951," in Deutsches Rundfunkarchiv Babelsberg, F 210-00-00/0007. For the adaptation of the transmission station in Nauen—originally designed by Hermann Muthesius in 1920—to the FM shortwave range, see K. E. Muller, "Eine neuartige Kurzwellenrichtantenne," *Technische Mitteilungen RFZ* 9, no. 1 (1965): 1–6.

house casts new light on the relationship between ideas of *type* and modernity in architecture. In a broader sense, too, it offers an oblique and alternative approach to thinking about the relationship between sound, technology, and space in modernity. One of the most influential studies about this intersection is Emily Thompson's *Soundscape of Modernity* (2002), which studies the experience of modernity in North America in relation to the process of decoupling *sound* from *space*. According to Thompson, the early work developed by engineer Wallace Sabine initiated a progressive trajectory whereby reverberation—the mechanical phenomenon linking sound and space—was equated with undesirable noise: "the sound of space," she writes, "was effectively eliminated from the modern sound as reverberation came to be considered an impediment, a noise that only interfered with the successful transmission and reception of the desired sound signal."[122] For Thompson, the modern soundscape in America became equivalent to a space purified from reverberation, and therefore, noise.

As profound as Thompson's study is, the underlying principles of its progressive narrative are called into question by an analysis of the relationship between sound and space that developed during the global Cold War, particularly in Berlin. The plan and sections of the Funkhaus provide material evidence of the renewed relevance of linking sound and space. Given the competition over the reception of signals in Berlin, where the channels of communication were less than "pure," a specific relationship between sound and space had to be established in order to make the signals distinguishable.[123] In addition, as the jamming maps drawn at the scale of the city attest, noise itself took on a new significance. A GDR citizen listening to RIAS at home might suddenly find the music of the Rolling Stones drowned out by engine noise—a deliberate counter-signal transmitted by the East German state, since radio transmissions were considered an intrusion into the sovereignty of acoustic space. In Berlin, at both the scale of buildings and the urban scale, the enduring relationship

122 Thompson, *The Soundscape of Modernity*, 234.

123 For another perspective on the reinsertion of reverberation as a technological and aesthetic trope after World War II, see Jonathan Stern, "Space within Space: Artificial Reverb and the Detachable Echo," *Grey Room* 60 (Summer 2015): 111–127.

204 → P. 247

205 → P. 247

207 → P. 250

208 → P. 251

209 → P. 251

between sound and space was hardly eliminated. On the contrary, under the sway of Cold War logic, reverberation and noise once again became relevant agents in the construction of acoustic space. (Fig. 204, 205, 207) As an illustration of this idea, we can trace the path that sound, as raw material, followed within the entire complex of the Funkhaus, as it was transformed—almost as in a production line—from the realm of mechanics into electronic signals ready to be broadcast via electromagnetic waves. We might start with the bus terminal in Block D, which was covered by eight concrete half-vaults and served over five thousand employees daily, before moving onto the recording rooms in Block B, which had a massive staircase, finished in raw concrete, carpet, and wood, that led nowhere, its sole purpose being to serve as a resonating device for recording footsteps under different acoustic conditions. From there, we could go to the studio next door, where all the flooring was exchangeable, again using a wide range of materials to mimic or construct sounds. Next stop: Saal I, where sound waves bounced off the geometric patterns of the internal surfaces of the room before reaching the Neumann microphones, designed to sense those waves with greater precision. From there, the sound was channeled through wires to the control and listening rooms where the tape machines were located, followed by the mixing board and mastering studios (Fig. 208, 209) and then the switch room (*Schaltraum*), where the signals were orchestrated and distributed to the various transmission stations—an entirely different typology that deployed large diesel engines, heavy machinery, and a 250 m tower to transform electric impulses into radio-electromagnetic waves (Fig. 210, 212) and deliver them to domestic radio receivers and speakers, through which they would reach the ears of the inhabitants of East and West Germany and, in some cases, spread across the entire planet. (Fig. 211)

Traditional readings of Berlin during the Cold War are usually centered on a city, or territory, defined by the Berlin Wall, and preoccupied with how the

Wall disrupted the flow of physical bodies across the Iron Curtain. The argument presented here is that the space of Berlin, and to a certain extent the world, was also strongly defined by the effects of other walls and their attendant electronic media apparatus. Both electronic technical media and buildings such as the Funkhaus on Nalepastrasse were designed to saturate the acoustic space with noises or signals loaded with a new and recognizable media content that served to block reception of the signals emanating from the West, which were just as ideologically charged. The radical pursuit of lost audiences through aesthetically differentiated media content led to a refinement of the relation of sounds and delicately designed walls and surfaces, all the way down to the micro-temporal level of reverberation time. Time, associated with the extended electroacoustic apparatus, defined the plan of the Funkhaus and, in so doing, temporally redefined the acoustic space of the city.

210 → P. 252

212 → P. 253

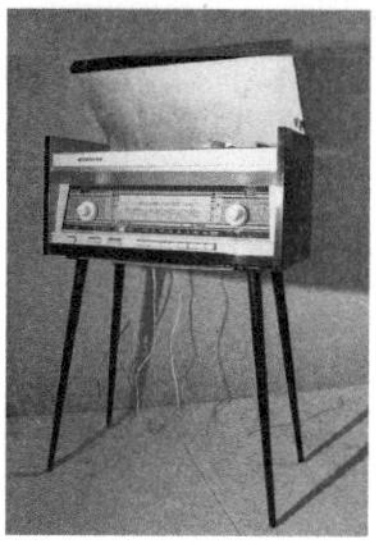

211 → P. 253

The Funkhaus contains different types of walls, performing at various levels. On the one hand, they divide and create heterogeneity through fragmentation, allowing for the coexistence of juxtaposed realities through a calibrated balancing of hermetic seals and permeability. On the other hand, the inner surfaces of the walls are constantly vibrating; they reflect and then radiate every sonic impulse that has been emitted, being the very medium through which sonic and spatial qualities are combined into electric signals. As has been argued, this building is not the outcome of architectural ideas alone, nor is it an autonomous box that houses and shelters a program totally independent from its architecture. Rather, the building appears here as the facilitator of interdependent spatial dispositions and electroacoustic devices. There is also a close interdependence of electroacoustic devices and architectural mechanical elements. At the same time, however, the functionality of the Funkhaus—that of creating a specific kind of space that can be recorded and broadcast—is performed and achieved in combination with other machines. More specifically, the Funkhaus constitutes an *ensemble of technical forms*, while at the same time being part of an even larger technical ensemble.[124] Thus, the architecture of the Funkhaus helps us to reframe the question of the wall within the environmental history of the war against information.

124 For more on the notion of ensemble of technical forms, see Gilbert Simondon, *On the Mode of Existence of Technical Objects*, trans. Cecile Malaspina and John Rogove (Minneapolis: Univocal, 2017), 63–83.

PART 3
THE LABORATORY
SOURCE MATERIAL

213

VEB FILMFABRIK WOLFEN · DEUTSCHE DEMOKRATISCHE REPUBLIK
ORWO
RFZ Labor für Akustisch-Musikalische Grenzprobleme
Band Nr. E 119 A
38
Komponist: Frederik Rzewski
Titel: " Zoologischer Garten "
- Einzelspuren auf 2 Kanälen -
Mitwirkende: Teil I :Spur 1 - links
Spur 2 - rechts
Spur 3/4 (Mono)
Teil:II:Spur 1 - links
Spur 2 - rechts
Spur 3/4 (Mono)
TeilIII:Spur 1 - links
Spur 2 - rechts
Spur 3 -(Mono)(
Dirigent:
Aufnahmeort: 14.8. - 14.12.65 RFZ - Studio
Aufnahmedatum:
Anlage: Schallm. 3
Herkunft: Urband
Tonmeister: -
Mikrofontyp:
Toningenieur: Bechstein
Dauer: Ca. 40 '00
Bandtyp: PER 555
Charge: -
Bemerkungen: Teil IV : Spur 1/2 - links
Spur 3 - rechts
Archiviert: Is.
Datum: 1.8.68
Abspielvermerk:

214

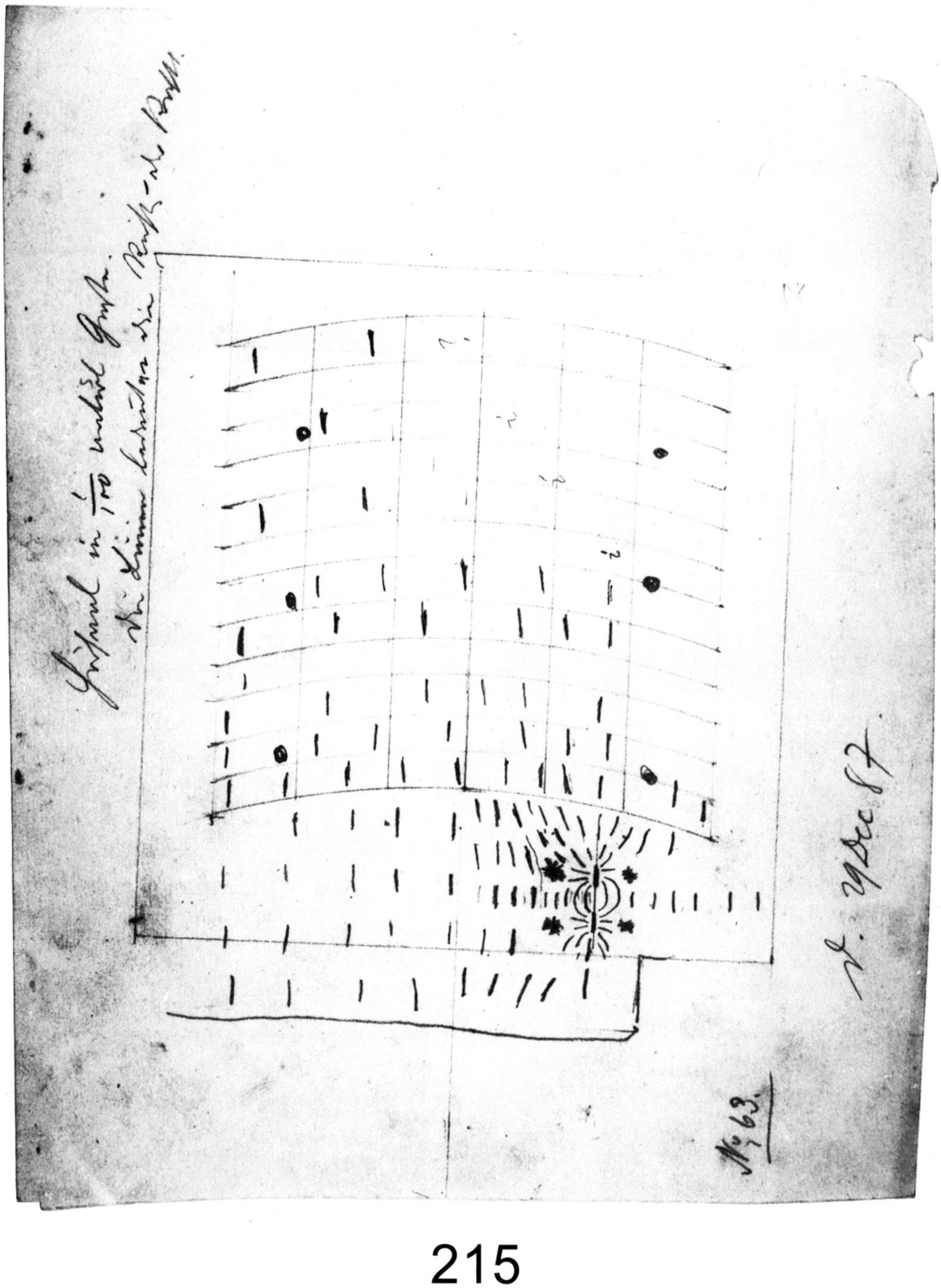

215

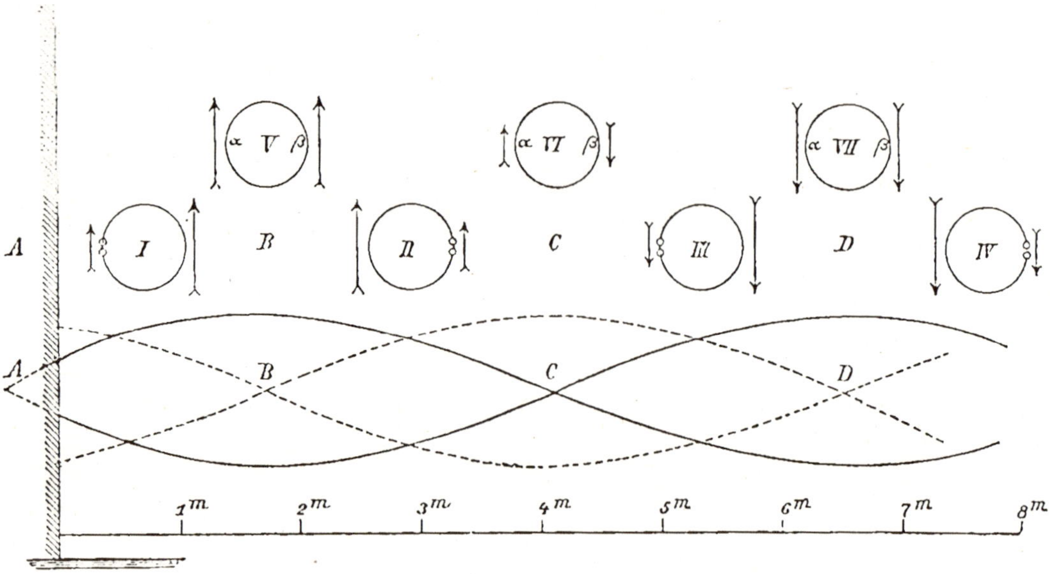
α V β
α VI β
α VII β
A
I
B
II
C
III
D
IV
A
B
C
D
1m
2m
3m
4m
5m
6m
7m
8m

216

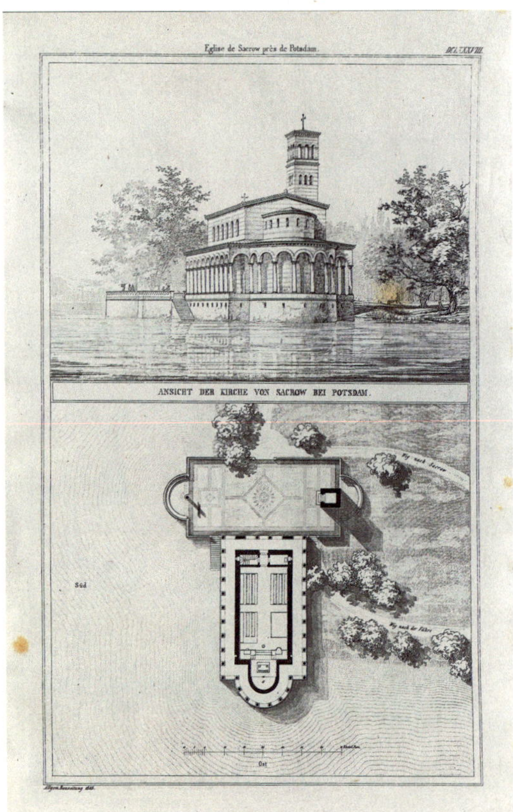
Eglise de Sacrow près de Potsdam.
ANSICHT DER KIRCHE VON SACROW BEI POTSDAM.
Süd

217

218

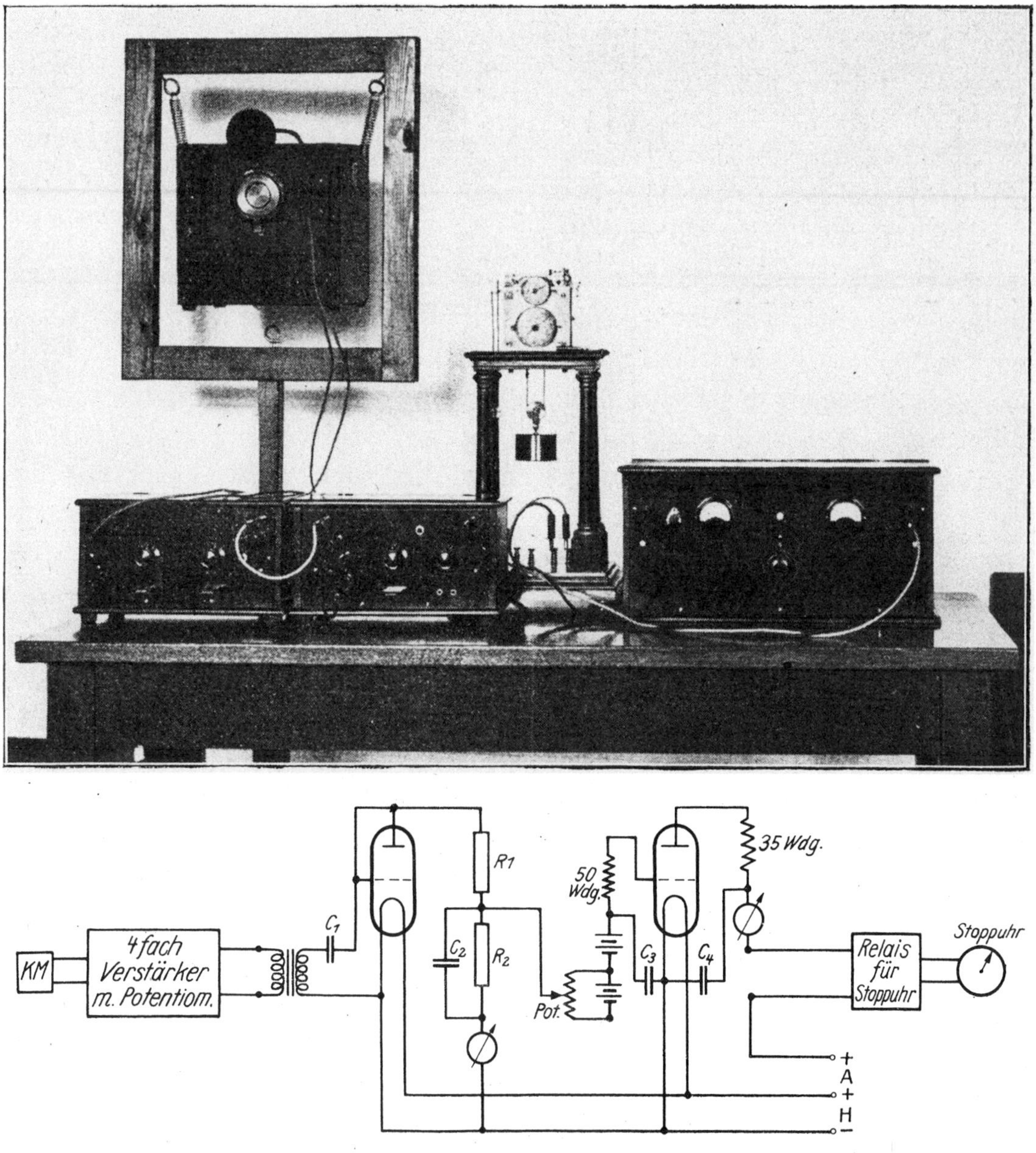
KM
4fach Verstärker m. Potentiom.
C_1
R1
C_2
R_2
Pot.
50 Wdg.
C_3
C_4
35 Wdg.
Relais für Stoppuhr
Stoppuhr
+
A
+
H
−

219

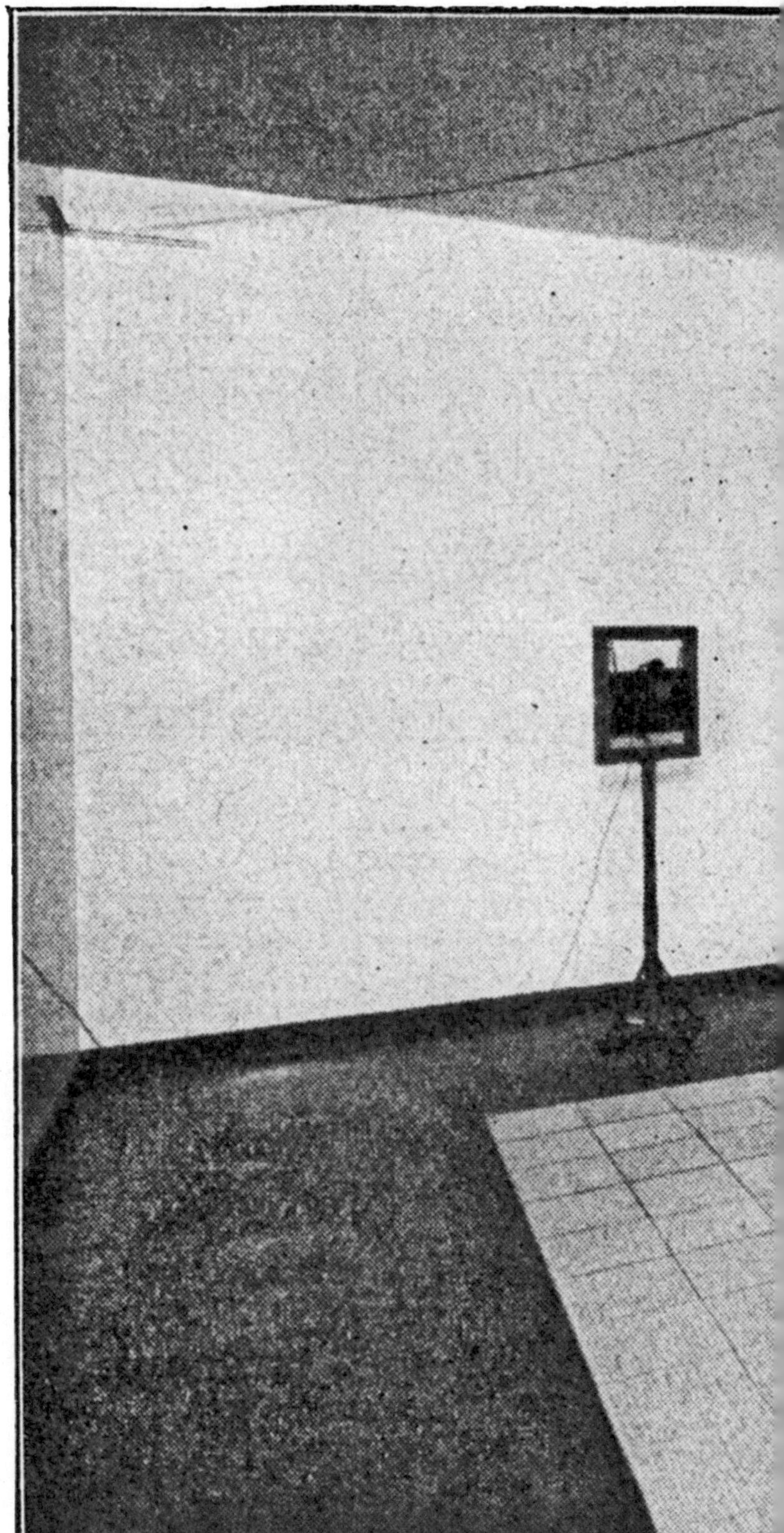

220

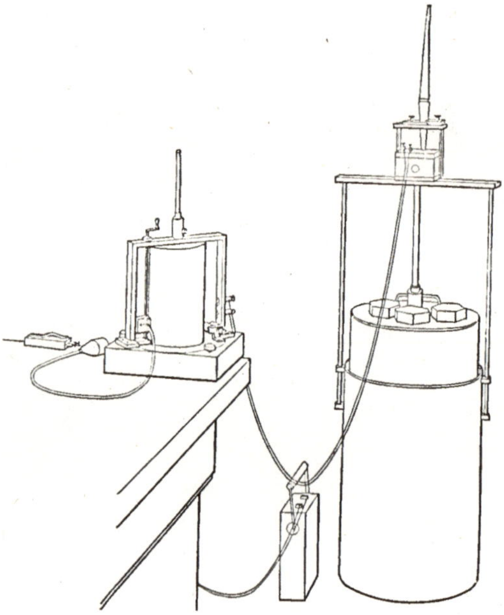

221

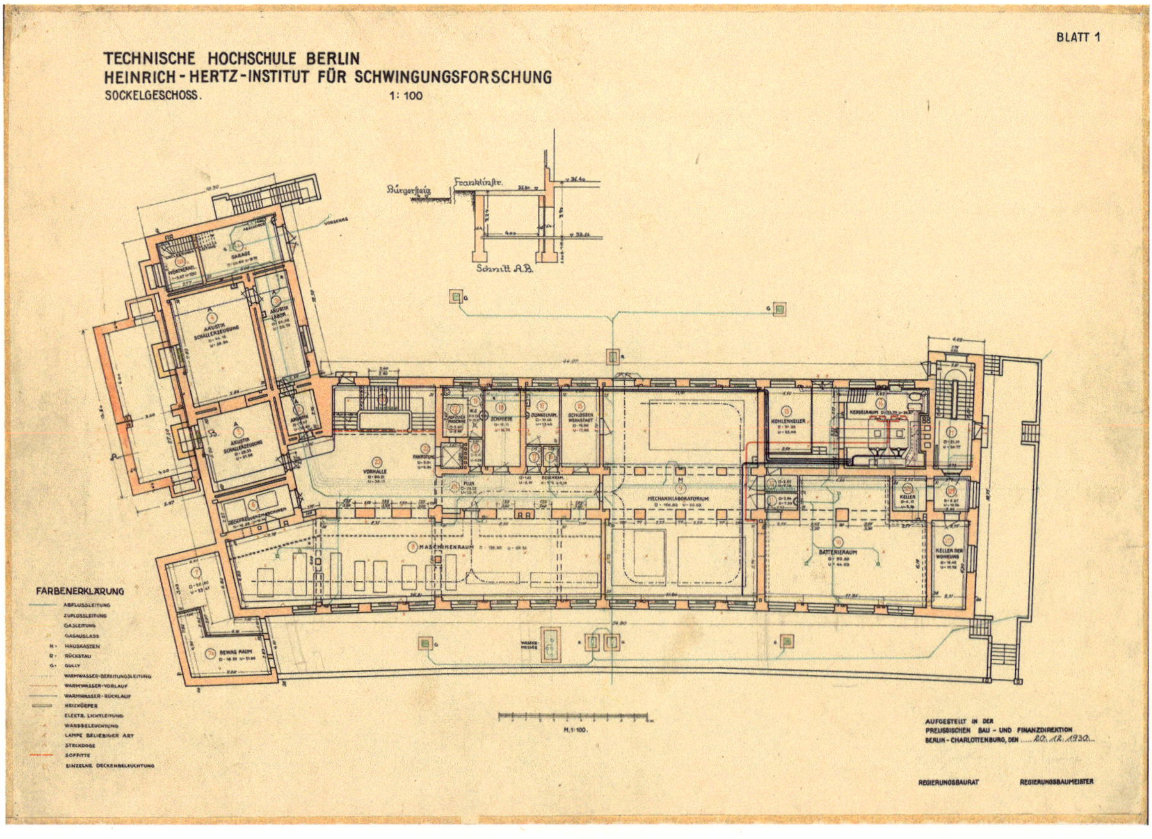
BLATT 1
TECHNISCHE HOCHSCHULE BERLIN
HEINRICH-HERTZ-INSTITUT FÜR SCHWINGUNGSFORSCHUNG
SOCKELGESCHOSS.
1: 100
Schnitt A.B.
FARBENERKLÄRUNG
M. 1:100.
AUFGESTELLT IN DER
PREUSSISCHEN BAU- UND FINANZDIREKTION
BERLIN-CHARLOTTENBURG, DEN 20. 12. 1930.
REGIERUNGSBAURAT
REGIERUNGSBAUMEISTER

222

223

224

225

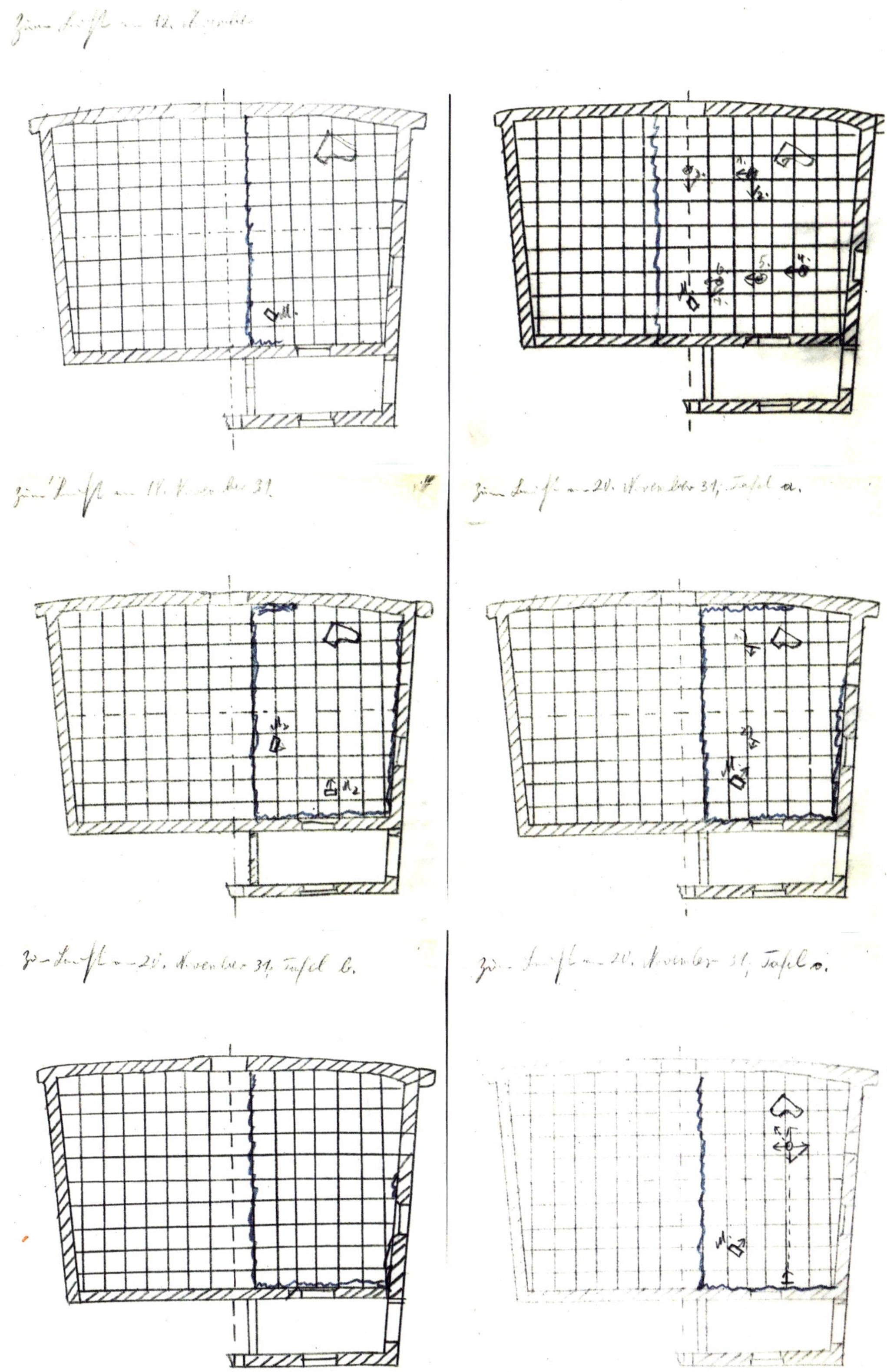
20. November 31, Tafel a.
20. November 31, Tafel b.
20. November 31, Tafel c.

Fig. 213 Telefunken Magnetophon 15a tape recorder in the archives of the Akademie der Künste. Photograph by the author.

Fig. 214 Image of magnetic tape. Photograph by the author. A recording of Frederic Rzewski's composition, "Zoologischer Garten," is available online at https://s9.imslp.org/files/imglnks/usimg/0/0c/IMSLP153834-WIMA.568d-Zoologischer-Garten.mp3.

Fig. 215 Sketch of the laboratory built by Heinrich Hertz inside the auditorium of the University of Karlsruhe. The drawing depicts the position of the improvised antenna, the grid of seats in the hall, and the form of propagation of the electromagnetic waves. Source: KIT Archiv. 28010 Allgemeine Fotosammlung, I / 3150, Versuchsprotokoll von Heinrich Hertz (Prof. für Physik, *1857, +1894) vom 23.12.1887.

Fig. 216 Diagram of electromagnetic wave oscillation depicting the reflecting character of the back wall of the room. It is worth noting that the drawing is a mixture of a graph and a section of the room. Source: Heinrich Hertz, "Electromagnetic Waves in Air and Their Reflection," in *Electric Waves*, trans. D. E. Jones (London: Macmillan & Co., 1893), 128.

Fig. 217 Perspective and plan of the Heilandskirche designed by Ludwig Persius in Potsdam-Sacrow, Germany. Source: *Abbildungen zur Allgemeinen Bauzeitung* 10 (1845): Tafel 688.

Fig. 218 Heilandskirche designed by Ludwig Persius in Potsdam-Sacrow, Germany. Photograph by the author.

Fig. 219 Photograph and circuit diagram of the machine designed by Erwin Meyer to automatically measure reverberation time. Source: Erwin Meyer, "Ein Neues Automatisches Verfahren der Nachhallmessung," *Zeitschrift für Technische Physik*, no. 7 (1930): 256.

Fig. 220 Interior view of the acoustic laboratory where Erwin Meyer tested the new procedure for automatically measuring reverberation time. Source: Erwin Meyer, "Ein Neues Automatisches Verfahren der Nachhallmessung," *Zeitschrift für Technische Physik*, no. 7 (1930): 258.

Fig. 221 The device developed by Wallace Sabine consisted of a chronograph battery and an air reservoir surmounted by the electro-pneumatic valve and organ pipe. Source: Wallace C. Sabine, *Collected Papers on Acoustics* (Cambridge, MA: Harvard University Press, 1922), 15.

Fig. 222 Heinrich Hertz Institute, plan of level 0. Architect: Otto Weissgerber. Source: TU Berlin Architekturmuseum, Inv. Nr. 20323.

Fig. 223 Heinrich Hertz Institute, exterior view. Source: TU Berlin Architekturmuseum, Inv. Nr. F 8108.

Fig. 224 Two harp players rehearse in front of a microphone in a classroom in the Rundfunkversuchsstelle (Radio Research Laboratory). Source: Archiv Universität der Künste Berlin (UdK-Archiv), Bestand 1b (Rundfunkversuchsstelle) / Fotogr. Nr. 70.

Fig. 225 A piano teacher, probably Dr. Erich Fischer, in front of a microphone at the Rundfunkversuchsstelle. Source: Archiv Universität der Künste Berlin (UdK-Archiv). Bestand 1b (Rundfunkversuchsstelle) / Fotogr. Nr. 72.

Fig. 226 Haus des Rundfunks, series of planimetric drawings of Saal II. The drawings depict the experiments conducted by the composer Hermann Scherchen. Source: Akademie der Künste. Berlin (AdK), Hermann-Scherchen-Archiv, Scherchen 371.

227

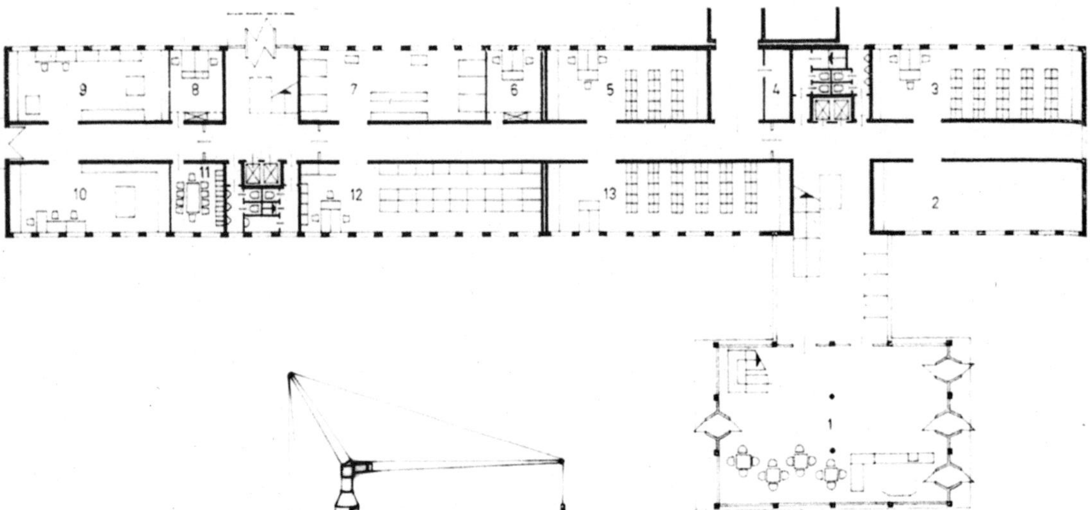

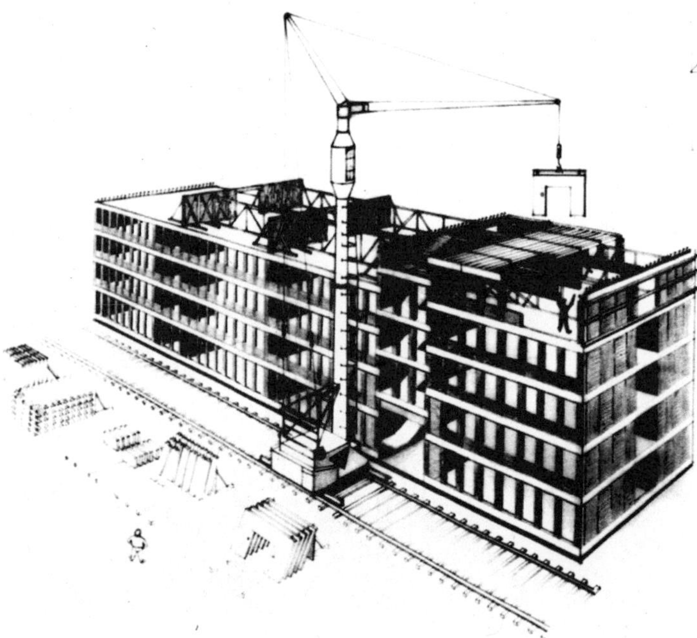

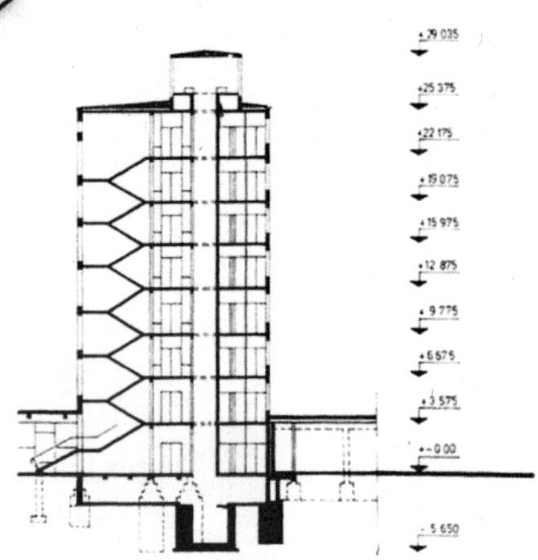

Grundriss EG Eingangs- und Hauptgebäude
Herstellung der Fertigteile auf der Baustelle
Hauptgebäude, Schnitt

229

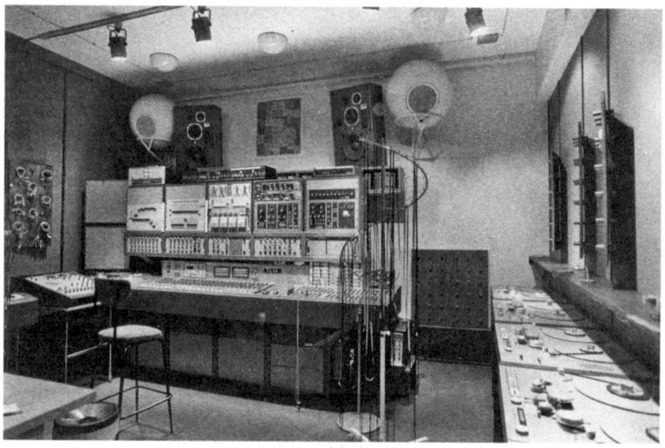

230

231

232

233

234

235

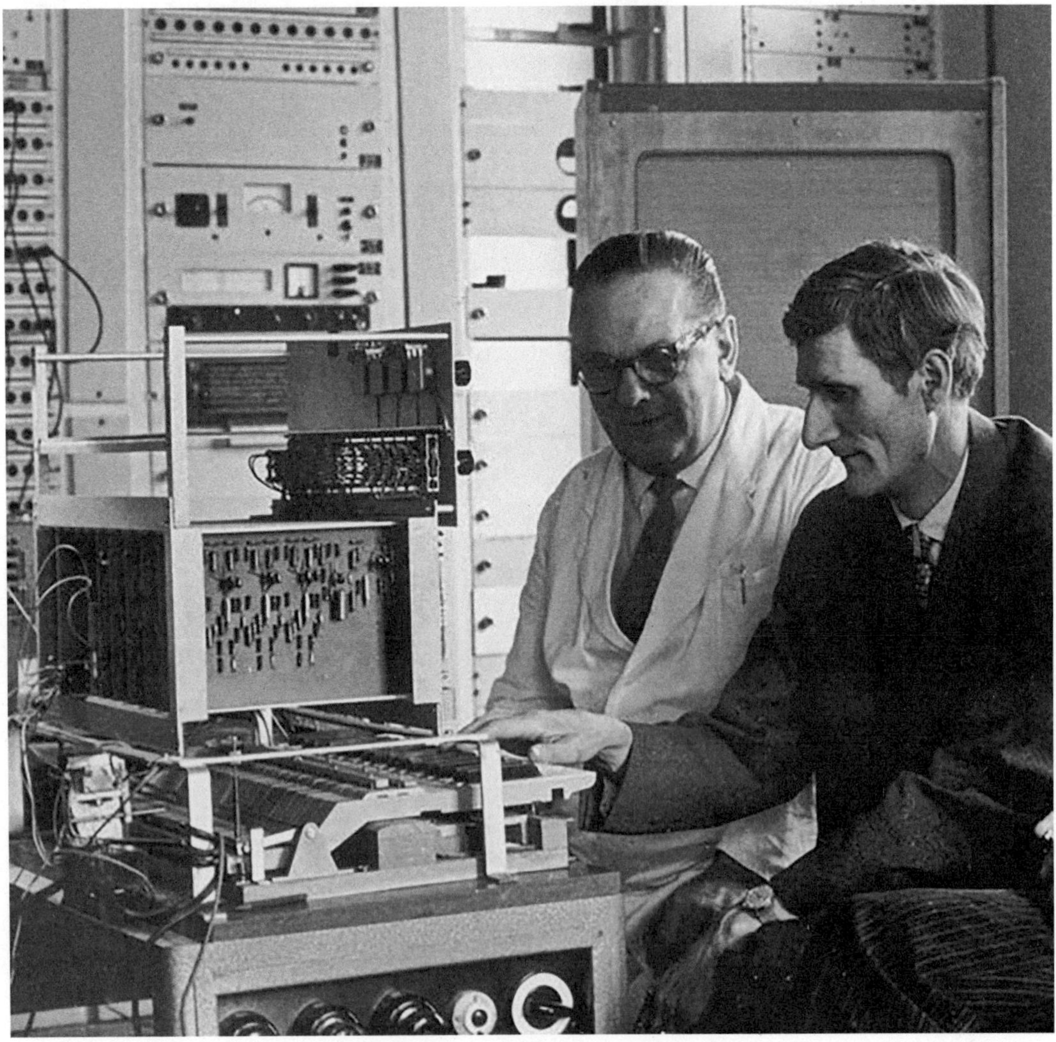

236

HELIRADIO

ELEKTROAKUSTISCHE STUDIOTECHNIK

radio mit ratio

Subharchord 2

Experimentelle Musik mit dem subharmonischen Klangerzeuger

Dipl.-Ing. G. Steinke,
Rundfunk- und Fernsehtechnisches Zentralamt der Deutschen Post, Berlin-Adlershof (RFZ)

Mit der Erzeugung von elektronischen Klängen und Geräuschen zur Verwendung als Gestaltungsmittel bei Funk, Film und Fernsehen befassen sich in den letzten Jahren immer mehr spezielle Studios in den verschiedenen Ländern. Bekannte Studios zur Klang- und Geräuscherzeugung bzw. für sogenannte elektronische oder experimentelle Musik befinden sich in Illinois, Köln, Mailand, Moskau, München, Paris, Rom, Tokio, Toronto, Utrecht, Warschau und Wien, um nur die wichtigsten zu nennen.

Diese Studios mit den unterschiedlichsten Zielen und Kompositionsrichtungen sind entstanden, nachdem sich im Konzertsaal bisher kein elektronisches Instrument so recht hatte behaupten können, außerdem konnten infolge von Unzulänglichkeiten des gegenwärtigen elektronischen Musikinstrumentariums die Bedürfnisse an elektronischen Klängen für die Untermalung von Filmen, Hör- und Fernsehspielen nicht erfüllt werden.

Elektronische Kleininstrumente - z. B. monophon oder polyphon spielbare Imitatoren herkömmlicher Musikinstrumente bzw. -klänge, elektronische Orgeln, Harmonien u. ä. - haben sich in der Unterhaltungsmusik einen gewissen Anwendungsbereich erobern können. Elektronische Orgeln fanden sogar Einzug in Kirchen, Opernhäuser und Konzertsäle, aber nur wenige sind bekannt geworden, die durch Beherrschung der Nachbildung von Ausgleichsvorgängen einer Pfeifenorgel (z. B. Barockorgel) wirklich ebenbürtig sind (wie z. B. die von E. SCHREIBER im VEB WERK FÜR FERNSEHELEKTRONIK, Berlin Oberschöneweide, entwickelte Toccata-Orgel). Diese inzwischen sehr bekannten und gewohnten Klänge entsprechen jedoch nicht den Vorstellungen moderner Komponisten, die wegen des Fehlens geeigneter neuer Instrumente zur elektronischen Synthese aus Sinus- bzw. Rechteckschwingungen mit Filtern und Verzerrern usw. übergingen, um neuartige Klänge statt der konventionellen zu erzeugen. Diesen Komponisten ist letzten Endes die Einrichtung spezieller Studios zu verdanken, denn die Erzeugung von Klängen mit elektronischen Mitteln war trotz aller Handfertigkeit in einem Konzertsaal nicht mehr zu erreichen. Der letzte Virtuose in

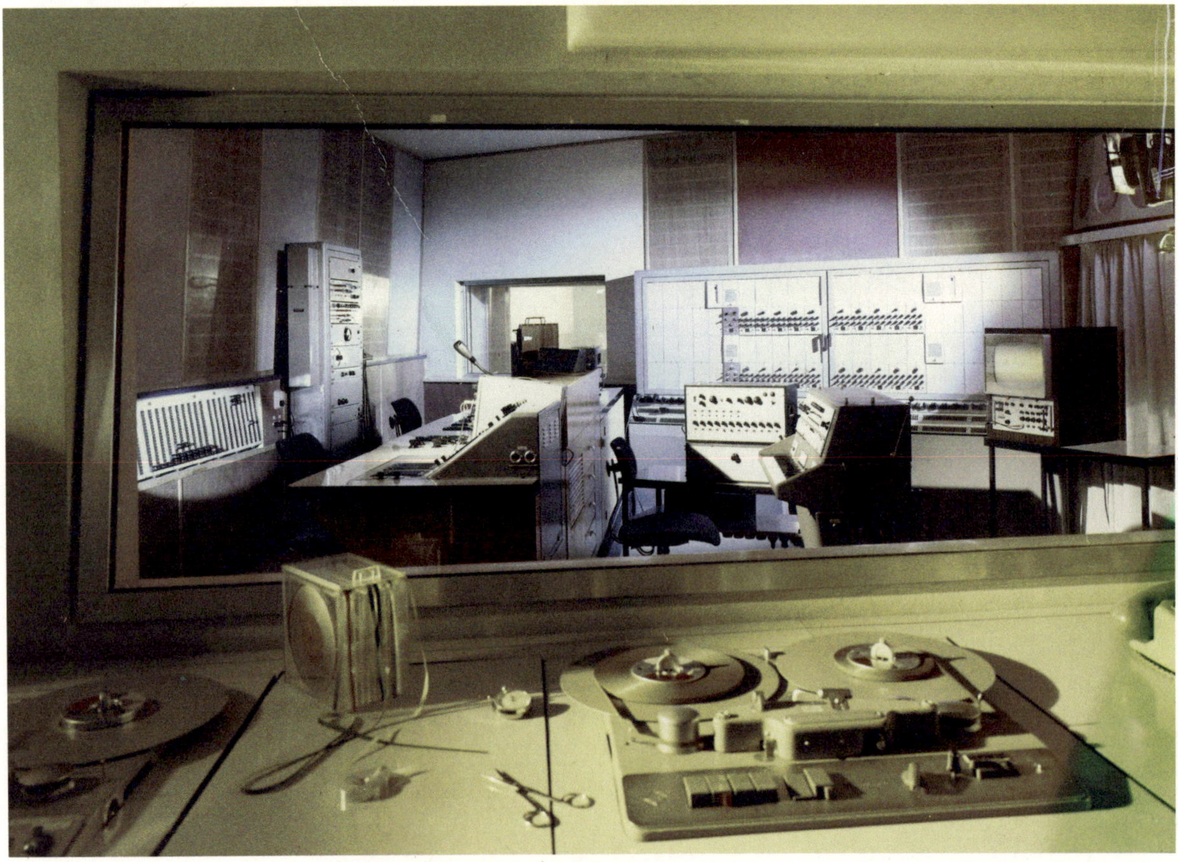

238

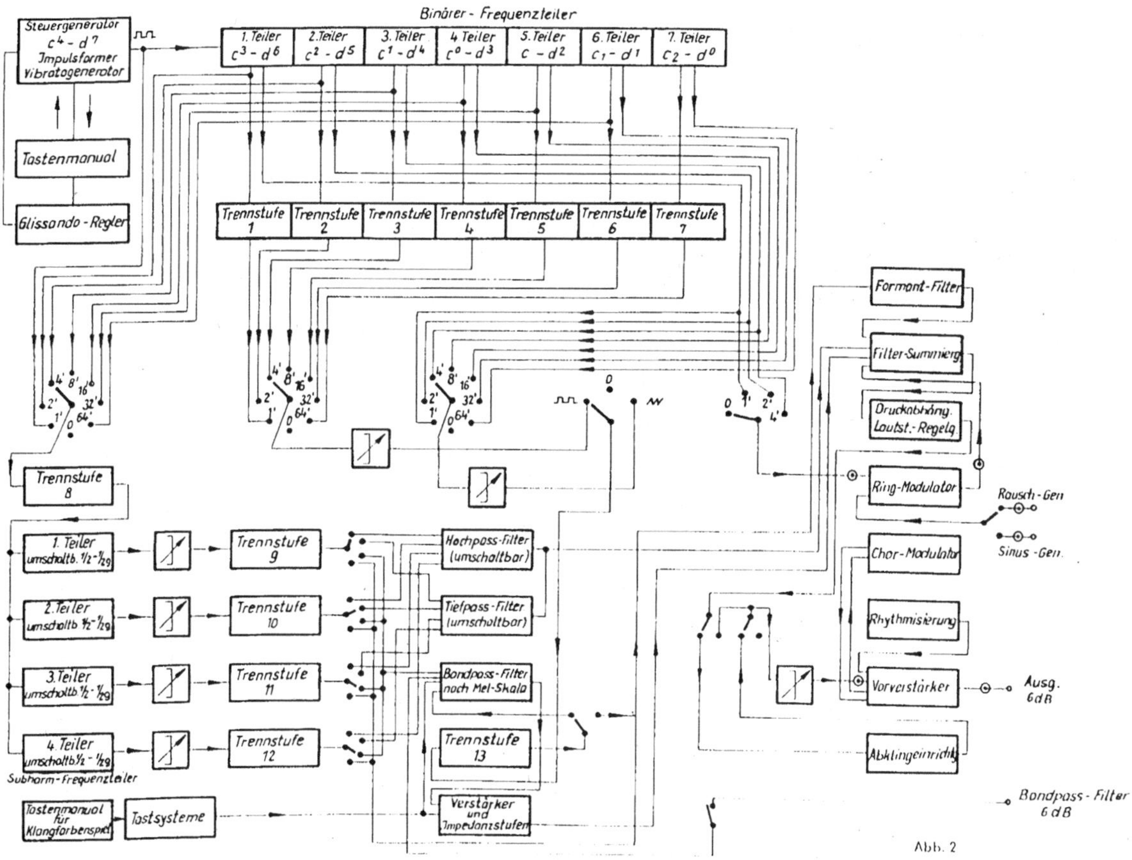
Binärer - Frequenzteiler
Steuergenerator
Impulsformer
Vibratogenerator
1. Teiler
2. Teiler
3. Teiler
4. Teiler
5. Teiler
6. Teiler
7. Teiler
Tastenmanual
Glissando - Regler
Trennstufe 1
Trennstufe 2
Trennstufe 3
Trennstufe 4
Trennstufe 5
Trennstufe 6
Trennstufe 7
Trennstufe 8
Formant-Filter
Filter-Summierg.
Druckabhäng. Lautst.-Regelg.
Ring-Modulator
Rausch - Gen
Sinus - Gen.
Chor-Modulator
Rhythmisierung
Vorverstärker
Ausg. 6dB
Abklingeinricht.
1. Teiler umschaltb. 1/2 - 1/29
2. Teiler umschaltb. 1/2 - 1/29
3. Teiler umschaltb. 1/2 - 1/29
4. Teiler umschaltb. 1/2 - 1/29
Subharm - Frequenzteiler
Trennstufe 9
Trennstufe 10
Trennstufe 11
Trennstufe 12
Trennstufe 13
Hochpass-Filter (umschaltbar)
Tiefpass-Filter (umschaltbar)
Bandpass-Filter nach Mel-Skala
Verstärker und Impedanzstufen
Tastenmanual für Klangfarbenspiel
Tastsysteme
Bandpass - Filter 6 dB

Abb. 2

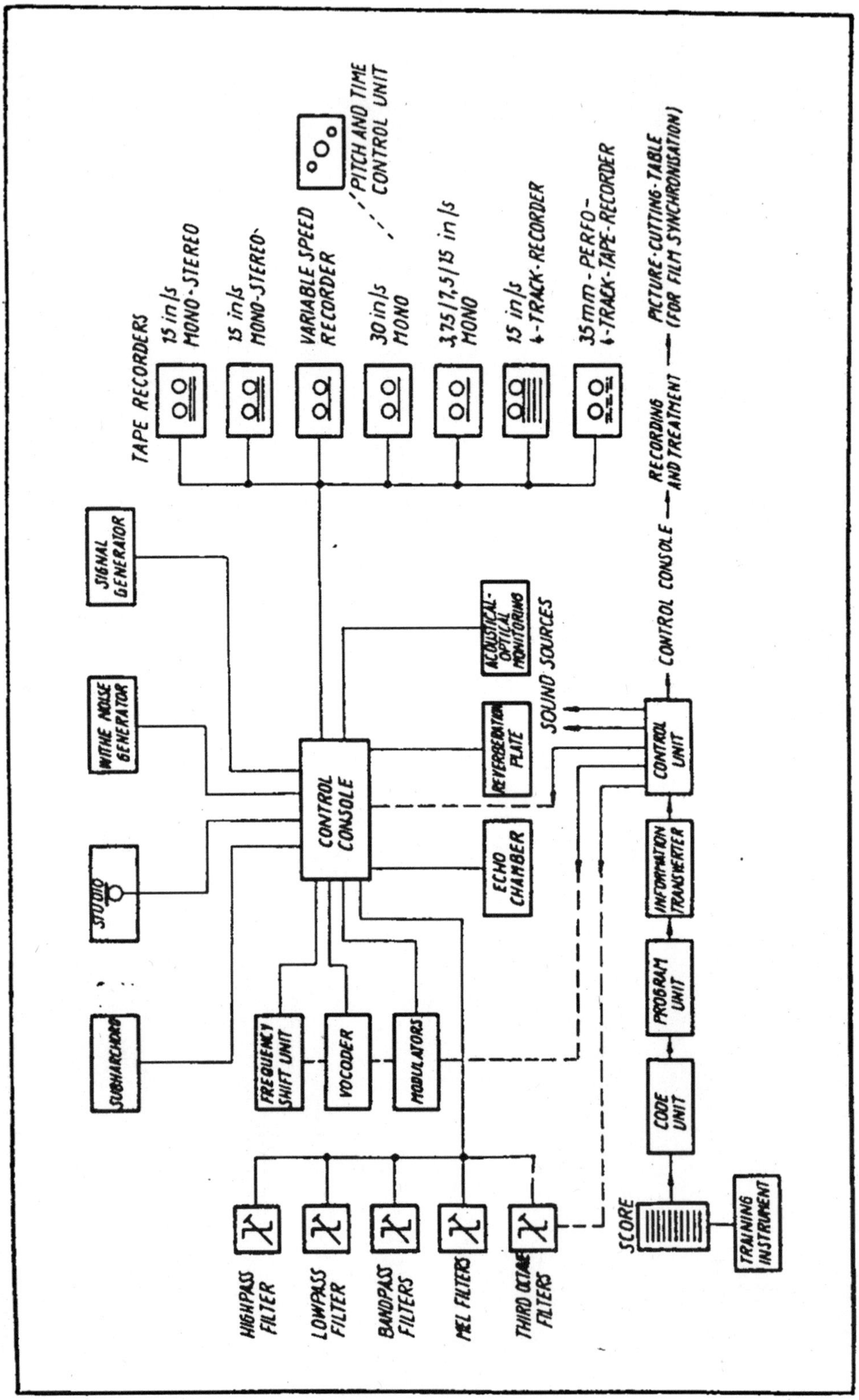
TAPE RECORDERS
15 in/s MONO-STEREO
15 in/s MONO-STEREO
VARIABLE SPEED RECORDER
PITCH AND TIME CONTROL UNIT
30 in/s MONO
3,75/7,5/15 in/s MONO
15 in/s 4-TRACK-RECORDER
35mm-PERFO-4-TRACK-TAPE-RECORDER
PICTURE-CUTTING-TABLE (FOR FILM SYNCHRONISATION)
RECORDING AND TREATMENT
CONTROL CONSOLE
SIGNAL GENERATOR
WITHE NOISE GENERATOR
STUDIO
SUBHARCHORD
CONTROL CONSOLE
ACOUSTICAL-OPTICAL MONITORING
REVERBERATION PLATE
ECHO CHAMBER
SOUND SOURCES
CONTROL UNIT
INFORMATION TRANSFERTER
PROGRAM UNIT
CODE UNIT
SCORE
TRAINING INSTRUMENT
FREQUENCY SHIFT UNIT
VOCODER
MODULATORS
HIGHPASS FILTER
LOWPASS FILTER
BANDPASS FILTERS
MEL FILTERS
THIRD OCTAVE FILTERS

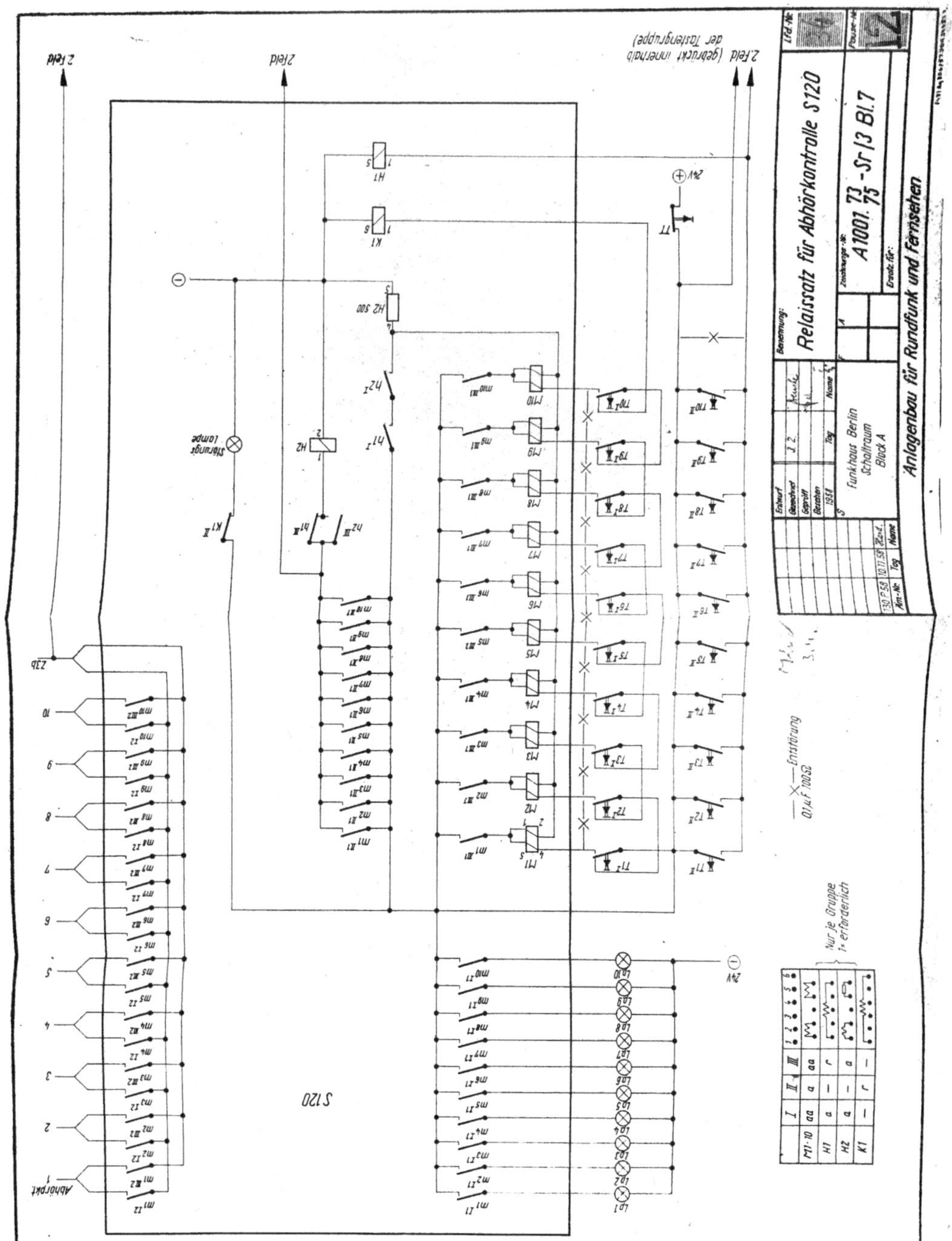
Relaissatz für Abhörkontrolle S120
A1001.73/75-Sr/3 Bl.7
Funkhaus Berlin
Schaltraum
Block A
Anlagenbau für Rundfunk und Fernsehen
S120
Störungslampe
Entstörung 0,1µF 100Ω
Nur je Gruppe 1× erforderlich
2.Feld (gedrückt innerhalb der Tastengruppe)
Diese Unterlage ist unser Eigentum, Mißbrauch, Vervielfältigung oder Mitteilung an Dritte wird strafrechtlich verfolgt.

KP	Stückliste Nr.	Blatt
3	F. 91802	2 / 2

PRINZIPSCHEMA EINER ANLAGE MIT 3 WINDEN

(Eingezeichnet jedoch nur die Stromläufe für 1 Winde)

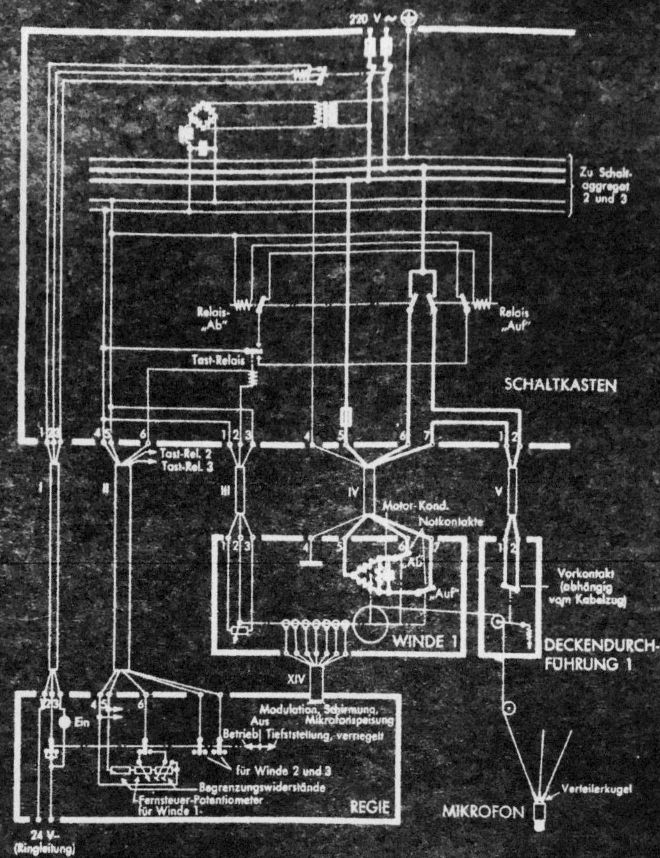

LISTE DER KABELLEITUNGEN FÜR EINE ANLAGE MIT 3 WINDEN

(ergänzend siehe Seite 1)

Nr.	Zweck		Adernzahl	Spannung	Strom
I	Einschaltung und Signal		3	24 V –	0,2 A
II	Fernsteuerung		5*	60 V –	0,1 A
III	Fernsteuerung	Winde 1	3	60 V –	0,1 A
IV	Motor	„ „	4	220 V ~	1 A
V	Deckendurchführung	„ „	2	220 V ~	1 A
VI	Fernsteuerung	Winde 2	3	60 V –	0,1 A
VII	Motor	„ „	4	220 V ~	1 A
VIII	Deckendurchführung	„ „	2	220 V ~	1 A
IX	Fernsteuerung	Winde 3	3	60 V –	0,1 A
X	Motor	„ „	4	220 V ~	1 A
XI	Deckendurchführung	„ „	2	220 V ~	1 A
XII	Anschluß Netz 220 V ~		3	220 V ~	4 A
XIII	Anschluß Ringleitung 24 V –		2	24 V –	0,5 A
XIV	Mikrofonanschluß und Schirmung		6 + Schirm	–	–

*Die vorgesehenen Kabeleinführungen gestatten auch die Verwendung von 2 Kabeln zu 3 Adern.

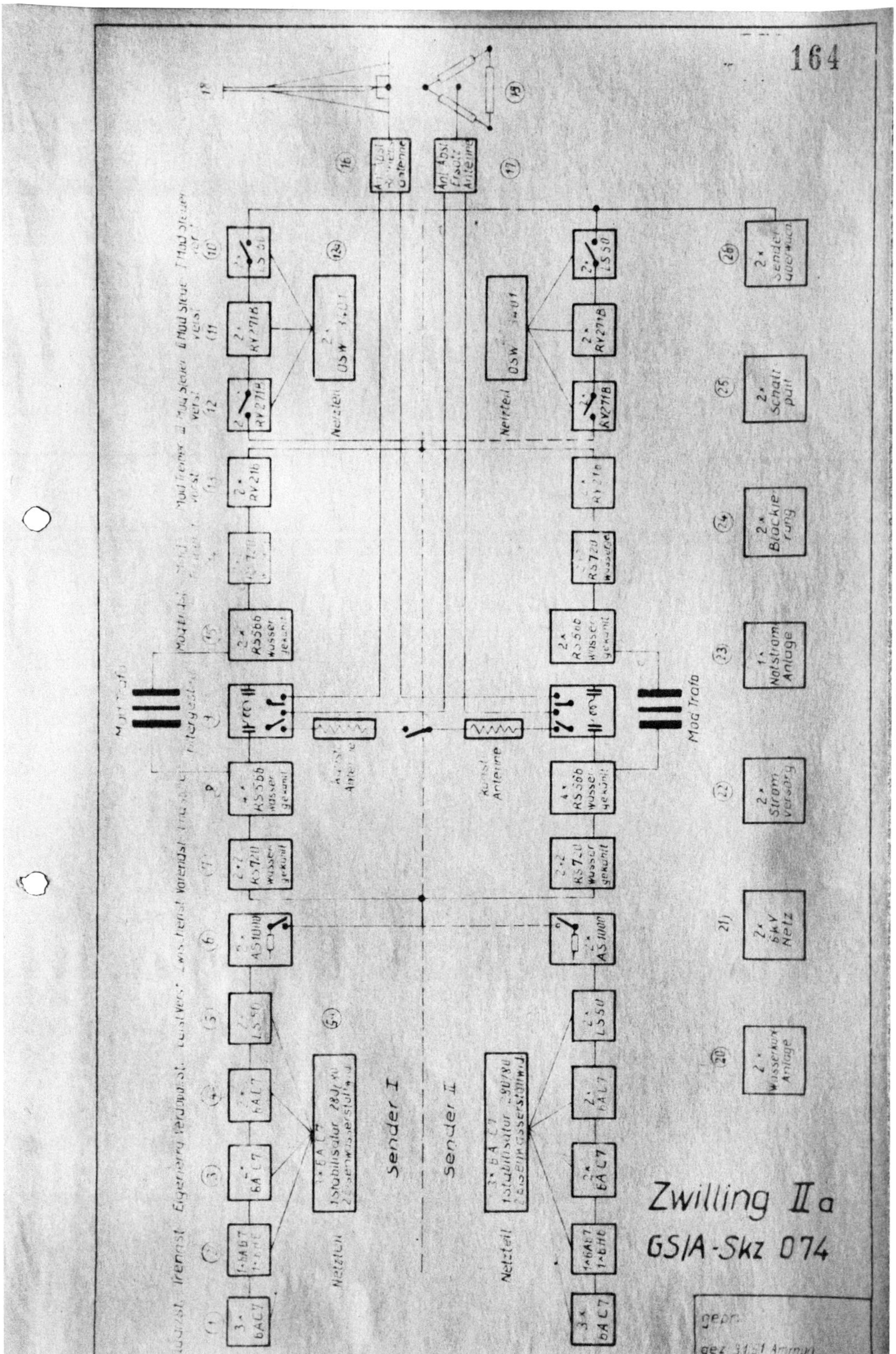
164
Zwilling IIa
GS/A-Skz 074
Sender I
Sender II
Mod Trafo
Kunstl. Antenne
Netzteil

Fig. 227 Rundfunk- und Fernsehtechnisches Zentralamt (RFZ), Adlershof, exterior view of the building designed by Peter Flier. Source: Peter Flier, "Rundfunk- und Fernsehtechnisches Zentralamt Berlin-Adlershof," *Deutsche Architektur* 11 (1964), 662.

Fig. 228 RFZ, plan and axonometric view of the prefabricated construction process of the building. Source: Angelika Petruschat, *Peter Flier Architekt* (Berlin: form+zweck, 2010), 50.

Fig. 229 Studio für elektronische Musik, Cologne, interior view. Source: Nordwestdeutschen Rundfunk, *Sonderheft über elektronische Musik* (Cologne: Hausmitteilungen des Nordwestdeutschen Rundfunks, 1954).

Fig. 230 Groupe de Recherches Musicales (GRM), Paris, interior view of studio showing the EMI mixing desk. Source: INA Groupe de Recherches Musicales.

Fig. 231 Centro sperimentale elettroacustico, Gravesano, interior view. Source: Antonio Scerri, "Centro sperimentale elettroacustico 'Dr. Hermann Scherchen," Gravesano," *Gravesaner Blätter* 1 (1955): 4.

Fig. 232 Luigi Nono and Marino Zuccheri in front of a reel machine in the RAI Studio of Phonology in Milan. Source: Rai Archives, C26PS223/5.

Fig. 233 Polish Radio Experimental Studio, Warsaw, interior view of the studio designed by Team Ten member Oskar Hansen. Source: Photo: Andrzej Zborski. © Museum of Modern Art in Warsaw.

Fig. 234 BBC Radiophonic Workshop, Maida Vale, London, 1958. This suite, comprising a studio and workshop, specialized in the construction of new sounds derived from everyday sources. The first experimental exploitation of these sounds in the BBC was a poem, "Private Dreams and Public Nightmares," specially written by Frederick Bradnum, and produced for the BBC's Third Programme. In this picture, Donald McWhinnie listens to a montage of electronic effects, with Desmond Briscoe operating filter units in association with particular oscillator sounds controlled by electronic music pioneer Daphne Oram. Source: BBC Photo Archives, ID: 43879.

Fig. 235 View of the paper-based RCA synthesizer in the Columbia-Princeton Electronic Music Center in New York City. Source: Columbia University Computer Music Center.

Fig. 236 Engineer Ernst Schreiber and composer Hans Hendrik Wehding in front of the Subharchord synthesizer. Source: Personal archive of Gerhard Steinke.

Fig. 237 Sales brochure for the Subharchord synthesizer developed by Ernst Schreiber at the Labor für akustisch-musikalische Grenzprobleme. Source: Akademie der Künste Berlin (AdK) (Ost), Historisches Archiv, 4546.

Fig. 238 Siemens Studio für elektronische Musik, Munich, interior view. Source: Siemens Historical Institute.

Fig. 239 Block diagram of the Subharchord synthesizer developed by Ernst Schreiber. Source: Akademie der Künste Berlin (AdK) (Ost), Historisches Archiv, 4546.

Fig. 240 Block diagram of a studio for sound and noise generation using the Subharchord synthesizer. Source: Gerhard Steinke, "Experimental Music with the 'Subharchord' Subharmonic Sound Generator," *Journal of the Audio Engineering Society* 14, no. 2 (1966): 142.

Fig. 241 Block diagram showing the Funkhaus Berlin-Nalepastrasse. Source: Bundesarchiv. DM 301/11321.

Fig. 242 Block diagram showing the microphone system for Saal I in the Funkhaus Berlin-Nalepastrasse. Source: Bundesarchiv. DM 301/11250.

Fig. 243 Block diagram of the SM8/H1 transmission station in Köpenick, East Berlin. Source: Deutsche Rundfunkarchiv Babelsberg. F 210-00-00/0007.

PART 3
THE LABORATORY

Ordinary sounds contain the deepest mysteries, and I started to follow some of them when I boarded the M21 tram at Frankfurter Tor in Berlin on January 14, 2017. The tram meandered through the streets of Friedrichshain in former East Berlin before coursing through the district of Köpenick, with its imposing industrial buildings, until it reached the Funkhaus in Nalepastrasse, where I had an appointment to meet Gerhard Steinke, a sound engineer and founding director, in 1956, of the sonorously named Labor für akustisch-musikalische Grenzprobleme (Laboratory for Border Problems in Musical Acoustics). The first of its kind in the Eastern Bloc, the laboratory was set up to bridge the gap between the aesthetic and technological dimensions of the emerging field of electronic music.

I met the eighty-year-old Steinke in the Milch Bar at the Funkhaus. Before long, however, he decided we had to move to another room with a more suitable acoustic environment for our conversation. I followed Steinke as he moved unerringly through the labyrinthine corridors— he had worked in the building since it had opened in the early 1950s—and eventually we arrived at a beautiful room overlooking the River Spree. With my digital recorder running, we chatted at length about his role during the pioneering years of electronic music in the GDR. Amongst other things, Steinke told me how he had persuaded a North American composer to cross the Wall to record at his studio and experiment with a new electronic synthesizer they had just developed. Fascinated by the story, I asked if there was an audio recording of the exchange. He said there was bound to be one somewhere on the tapes he had donated to the Academy of Arts (AdK, Akademie der Künste) after the fall of the Berlin Wall. It was this early conversation that made me realize that what I was really interested in locating were the spatial traces of the interaction of buildings and sounds, as registered by recording devices—traces that are quite different from any written language and also distinct from the symbolic traces of orthographic drawings.

Taking up my quest for the elusive indicators of what is referred to here as the architecture of radio—a form of architecture with unique spatial, technological, and political implications—I arrived several months later at the AdK's archives, just next to the Charité Hospital in former East Berlin. After complying with the proverbial Kafkaesque bureaucracy, I was directed to a "reading room." Except there were no written records to read. Instead, I found myself in front of a Magnetophon 15a tape recorder and a stack of quarter-inch tapes containing the traces of experiments conducted at the acoustic laboratory founded by Steinke. Those traces were potentially the only way I could access the sonic vibrations and reverberations produced by the objects I was investigating. Here, too, I found a means to experience the construction of space via the architecture of radio—a construction that emerges from the entangled relationships between building and electronic media technologies, and is only intelligible through the sense of hearing.

213 → P. 306

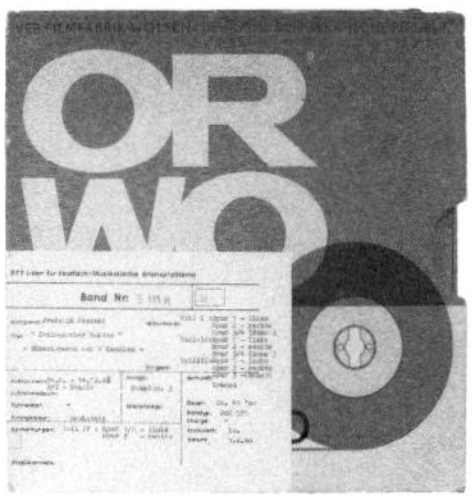

214 → P. 306

Those tapes gave me a sampling of what socialist acoustic space actually sounded like at the time. I listened, for example, to tests of wind instruments in Saal III in the Funkhaus on Nalepastrasse, taking account of the fact that those experiments had been transmitted as a collective lesson in teaching people how to listen.[1] I heard lectures about the history of music and technology broadcast by the national station, Deutschlandsender,[2] as well as programs produced by its sister institution, the Polish Radio Experimental Studio in Warsaw.[3] Equally, I listened to broadcasts from the West, from the RAI studio in Milan, among many others.[4] I heard early compositions by Karlheinz Stockhausen and Herbert Eimert recorded in 1958 in a studio in Cologne, West Germany, and broadcast by RIAS, the radio station established by the United States in Berlin.[5] The architecture of radio was something recorded, and thus experienced, on both sides of the Iron Curtain.

(Fig. 213, 214) In Germany, policies concerning sound archives tend to be stricter than those for

1 Gerhard Steinke and Dieter Boeck, "Die Musikwerkstatt, 1964," in Akademie der Künste, Berlin (AdK), Sammlung Audiovisuelle Medien (AVM), Tonbänder (AVM-31) 0801.1-2, 73'25.

2 Gerhard Steinke, "Die historische Entwicklung der elektronischen Musikinstrumente: vom Thermenvox zur Hammondorgel, 1965," in Akademie der Künste, Berlin (AdK), Sammlung Audiovisuelle Medien (AVM), Tonbänder (AVM-31) 0819, 44'.

3 Gerhard Steinke, "Besuch im Studio für experimentelle Musik in Warschau, 1965," in Akademie der Künste, Berlin (AdK), Sammlung Audiovisuelle Medien (AVM), Tonbänder (AVM-31) 0823, 37'.

4 Gerhard Steinke, "Das elektronische Studio der RAI Mailand, 1965," in Akademie der Künste, Berlin (AdK), Sammlung Audiovisuelle Medien (AVM), Tonbänder (AVM-31) 0837.

5 "Elektronische Musik, 1958," in Akademie der Künste, Berlin (AdK), Sammlung Audiovisuelle Medien (AVM), Tonbänder (AVM-31) 0814.

written material. Since I was not allowed to make any recordings of the archive tapes, all those listening sessions had essentially to be consigned to the storage capacity of my own memory. In addition, as a material medium, the tapes themselves are extremely fragile, making the subtle spatial traces stored in them even harder to pin down. As the tapes spun through the tape recorder, a fine dust of aubergine-colored powder was a telltale sign that the magnetic bands were eroding before my very eyes. After a short time, the tape would break like an untamed snake dancing in the air and would have to be glued back together using an improvised kit of tools.

The tenuous materiality of the tapes storing the effects of the architecture of radio opened onto a realm of the real that is by nature elusive. Essentially, I had accessed a different form of the present—a present that had had its time earlier, but was nevertheless cached on those tapes. More to the point, it was an experiential form of the present different to the notion of the historical "past," since I was confronted with the phenomenon immediately and in real time. Although I had originally aimed to write a historical treatment of the subject matter, there was nothing strictly historical about this particular material encounter, for it yielded a pure presence of that present in the here and now. Sitting in front of a Telefunken Magnetophon 15a tape recorder, listening to the vibrations that had been captured by microphones as they reverberated within the rooms of buildings and stored on ORWO tapes (with Russian labels), I was listening to a moment when the interaction between buildings, the mechanical phenomenon of sound, and electronic technical media was conveyed into the present by, and through, electricity.

In effect, I was listening to the traces of electrified space, a phenomenon made possible several years before the present that had been stored in those devices. In terms of physical distance, however, this phenomenon had originated not so far from where I was sitting at the Academy of Arts. In addition, all of those machines—synthesizers, microphones, tape recorders—as well as the institutions that hosted those exchanges, and even the practices carried out by particular individuals, had all begun several years prior to that sustained present to which I was listening, all in the same city. In this context, the process by which space was electrified is crucial for understanding the cultural, technological, and political dimensions of the architecture of radio. And unpacking those dimensions requires an analysis of the space and practices of the electroacoustic laboratories that originated in Berlin during the Weimar Republic, for it was in those laboratories that space would effectively be electrified.

I

The origins of the architecture of radio can be traced back to 1887, when a young Heinrich Hertz transformed the physics auditorium of the University of Karlsruhe into an experimental construction site. With wood, pipes, walls, sheets of zinc, and a set of drawings, Hertz configured for the first time the necessary conditions for electrifying space. He had begun to work with electromagnetism two years before, in 1885, after taking up a position as professor of physics at the Karlsruhe Technische Hochschule.[6] In part, his interest was motivated by a prize his mentor, Hermann von Helmholtz, had established under the auspices of the Academy of Science in Berlin. Helmholtz, a physicist and philosopher who made major contributions to several fields of science, offered the prize to anyone who could empirically settle the dispute over electromagnetism as hypothesized in theories by Wilhelm Weber and James Clerk Maxwell.[7]

Hertz's aim was to prove that electricity could travel through space without the aid of a transmitting material medium like wires. In parallel, he set out to demonstrate that space itself—that empty substance existing between walls, floors, and ceilings—could function as the *medium* for electricity transmission through the electromagnetic spectrum. To translate

6 Salvo D'Agostino, *A History of the Ideas of Theoretical Physics: Essays on the Nineteenth and Twentieth Century Physics* (Dordrecht: Kluwer, 2000), 140.

7 Wolfgang Hagen, *Das Radio: Zur Geschichte und Theorie des Hörfunks Deutschland/USA* (Munich: Wilhelm Fink, 2005), 20–23.

Weber and Maxwell's purely theoretical phenomenon into "a visible and almost tangible form," he needed to *construct* that conductive space. The procedures he carried out in his improvised laboratory in Karlsruhe would be meticulously described in a paper he published in 1888, "Electromagnetic Waves in Air and Their Reflection."[8] First, he built an improvised transmitting antenna that he called the primary conductor. This apparatus comprised "a cylindrical brass body, 3 cm in diameter and 26 cm long, interrupted midway along its length by a spark gap whose poles on either side are formed by spheres of 2 cm in diameter."[9] When applying electric current to the induction coil, a spark of light would glow in the spark gap. The receiver, or secondary conductor, as Hertz called it, consisted of a copper wire ring that also had a gap. He discovered that whenever the spark glowed in the primary conductor, the same spark (in German *Funk*, the same term used for radio) was registered in the secondary conductor of the ring as well, even if both were placed a few meters apart and separated only by air.

Hertz, however, would have been aware from his work with Helmholtz that there are always reflection planes, thresholds, and limits to free space, and that sounds are subjected to similar material and spatial constraints.[10] He writes, "I had often observed phenomena which seemed to point to a reflection of the induction action from the walls of the building."[11] Elsewhere, he notes, "When I placed the primary conductor in one corner of a large lecture room 14 m long and 12 m broad, the sparks could be perceived in the farthest parts of the room; the whole room seemed filled with the oscillations of the electric force."[12] At the same time, however, "it is true that in the neighbouring rooms the action could not be perceived even at small distances; solid walls exercise a powerful damping effect upon it."[13] Here, Hertz is acknowledging that electromagnetic waves are part of, and subject to, the built environment. In recognizing this phenomenon, he was well on his way to becoming the first architect of radio.

8 Heinrich Hertz, "Electromagnetic Waves in Air and Their Reflection," in *Electric Waves*, trans. D. E. Jones (London: Macmillan, 1893), 124–135, here 124. Originally published in German as "Über Elektrodynamische Wellen im Luftraume und deren Reflexion," *Wiedemann Annalen* 34 (1888): 610.

9 Hertz, *Electric Waves*, 172.

10 On Helmholtz's early experiments with sound, see Hermann von Helmholtz, *The Sensations of Tones: As a Physiological Basis for the Theory of Music*, trans. Alexander J. Ellis (London: Longmans, Green & Co., 1895).

11 Hertz, *Electric Waves*, 124.

12 Hertz, *Electric Waves*, 91.

13 Ibid.

Further detailed observations point to his transition to this newfound role of architect:

> The physics lecture room in which these experiments were carried out is about 15 m long, 14 m broad, and 6 m high. Parallel to the two longer walls there are two rows of iron pillars, each of which behaves much like a solid wall towards the electromagnetic action, so that the parts of the room which lie outside these cannot be taken into consideration. Thus only the central space, 15 m long, 8.5 m broad, and 6 m high, remained for the purpose of experiment. From this space I had the hanging parts of the gas-pipes and the chandeliers removed, so that it contained nothing except wooden tables and benches which could not well be removed. No objectionable effects were to be feared from these, and none were observed. The front wall of the room, from which the reflection was to take place, was a massive sandstone wall in which were two doorways, and a good many gas-pipes extended into it. In order to give the wall more of the nature of a conducting surface a sheet of zinc 4 m high and 2 m broad was fastened on to it; this was connected by wires with the gas-pipes and with a neighbouring water pipe, and especial care was taken that any electricity that might accumulate at the upper and lower ends of the sheet should be able to flow away as freely as possible.[14]

14 Hertz, *Electric Waves*, 125.

After arranging the surfaces of his experimental construction site, the idea was to place the primary conductor of the antenna in a corner of the room and to move the secondary conductor of the ring throughout the modified auditorium, testing the intensity of the sparks in order to identify their various wavelengths and speed of propagation. In this respect, the room became the material scaffolding for making the invisible visible. The electromagnetic space that Hertz defined, however, was not flat. He acknowledges, "The direction of the conducting wire

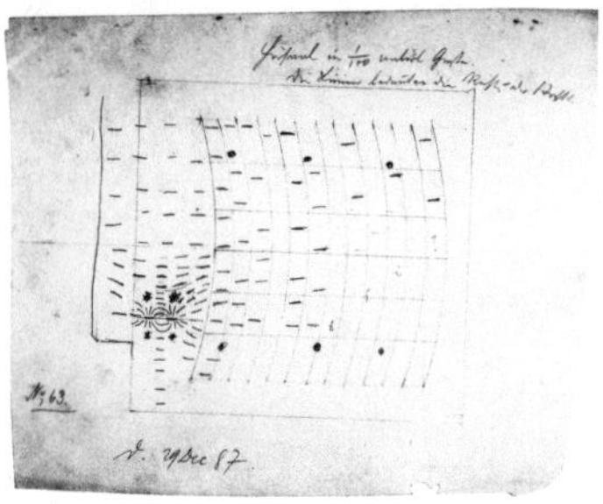

215 → P. 307

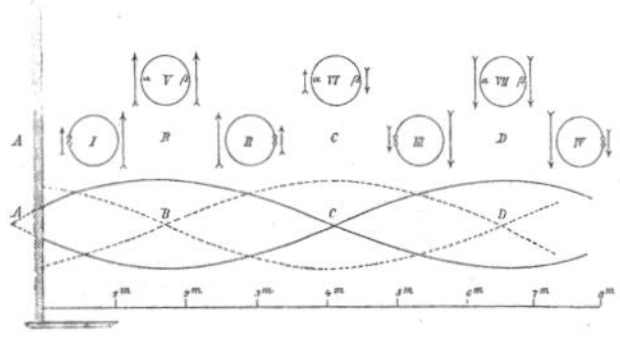

216 → P. 308

was now vertical; hence the forces which have here to be considered oscillate up and down in a vertical direction. The middle point of the primary conductor was 2.5 m above the level floor."[15] This positioning allowed the waves to oscillate up and down without interference from the ground. Consequently, "the observations were also carried out at the same distance above the floor, a gangway for the observer being built up with tables and boards at a suitable height."[16] We can picture in our mind's eye this first architecture of radio, as it took form in an auditorium clad in reflecting zinc sheets, with Heinrich Hertz walking along his improvised gangway, trialing for the first time the traces of the invisible electromagnetic spectrum. To my knowledge, we have no surviving images of Hertz's design; there are, however, two planimetric drawings that serve to confirm Hertz's proficiency as a spatially minded and media-conscious thinker.

(Fig. 215, 216) What we already see in the first drawing, in plan, is a room that is differentiated, demarcated, and defined by new topological relationships that go beyond the limits of the human senses—a space rationalized according to the geometric principles of electromagnetism. The grid of seats in the auditorium appears in stark contrast to another order dictated by the propagation and reflection of the waves in the room. To make electromagnetic waves visible, Hertz had to devise a new medium: in the process, he redefined the very idea of how the room, now transformed into space, could be delineated and, ultimately, occupied. In the second drawing, in section, we see the first description of an electromagnetic wave, its existence demarcated by the floor of the room and the reflecting zinc wall. In both drawings, we see how space becomes the medium of radio, of *Funk*. We see a space that is no longer a room, nor the abstract intellectual construct of theoretical physics—a space constructed as much by the invisible substance of conductive air as it is by conventional floors, walls, and ceilings. In the sense that he radically redefined the way in which a space is understood and represented, Hertz can be designated the first architect of

15 Hertz, *Electric Waves*, 126.

16 Hertz, *Electric Waves*, 126.

radio. His drawings conceive the disposition of material elements in space by seeing them and their agency in relation to electromagnetic oscillations. From now on, waves would engender an entirely novel way of drawing in (and with) space.[17]

In 1897, exactly ten years after Heinrich Hertz conducted his experiments in Karlsruhe, the German physicists Adolf Slaby and Georg von Arco used the material infrastructure of the belltower of the Heilandskirche in Potsdam (designed by Karl Friedrich Schinkel's student, Ludwig Persius) as a station to produce the first successful radio transmission of sounds through free space outside the controlled setting of the laboratory.[18] (Fig. 217) With this experiment, Slaby and Arco provided not only a material and empirical basis for what James Clerk Maxwell had theorized in 1865, but also a territorially scaled-up version of what Heinrich Hertz had first tested in his laboratory in 1887.[19] Hertz's pioneering experiments with electromagnetism had engendered an entirely new field of researching and understanding space. Indeed, the birth of the scientific idea of thinking and operating through space formed the technical precondition for the emergence of a modern idea of space that would have a profound impact on architectural discourse. The modern idea of space—that is, the intellectual effort to theorize what is in between things and objects—took shape in the laboratory even before it did in the lecture hall, beginning with a talk given by the art historian August Schmarsow at the University of the Arts in Leipzig in 1893.[20] From the moment that electromagnetism left the theoretical space of mathematics and intersected the concrete space of the laboratory, building elements would inevitably modulate its performance. In due course, electromagnetism would in turn leave the laboratory, intersecting cities and landscapes, electrifying the entire globe at the speed of light. It was electricity that first made the modern idea of *space* legible and occupiable.

In the second and third decades of the twentieth century, the experimental practices initiated by Hertz

217 → P. 308

17 Hagen, *Das Radio,* 3–39.

18 Georg von Arco would later become the first director of Telefunken, the telecommunications subsidiary of AEG and Siemens & Halske. For more about Arco's role in the history of German radio, see Margot Fuchs, *Georg von Arco (1869–1940)* (Berlin: Diepholz, 2004).

19 Fuchs, *Georg von Arco*, 36–40; 262–289.

20 August Schmarsow, *Das Wesen der architektonischen Schöpfung: Antrittsvorlesung, gehalten in der Aula der K. Universität Leipzig am 8. November 1893* (Stuttgart: Hiersemann, 1984). One of the first examples to echo the impact of electromagnetism in the ideas of space within architectural discourse can be found in Siegfried Ebeling, *Der Raum als Membran* (Dessau: C. Dünnhaupt, 1926). Ebeling's publication is full of references to radio and the electromagnetic spectrum, and such influences are even visible on the cover of the original publication. Ebeling was intellectually indebted to botanist and microbiologist Raoul Heinrich Francé, who established several parallels between technology and plants. See Raoul Heinrich Francé, *Die Pflanze als Erfinder* (Stuttgart: Kosmos, 1920); Raoul Henrich Francé, *Die technischen Leistungen der Pflanzen* (Leipzig: Veit & Comp., 1919). One of the first theories about radio would be published in the same series as Francé's early publications: see Hanns Günther, *Radiotechnik: Das Reich der elektrischen Wellen* (Stuttgart: Kosmos, 1921). For more on the impact of Francé's ideas on the work of Siegfried Ebeling, László Moholy-Nagy, and Ludwig Mies van der Rohe, see Fritz Neumeyer, *The Artless Word: Mies van der Rohe on the Building Art*, trans. Mark Jarzombek (Cambridge, MA: MIT Press, 1994), 170–193. For a broader historical account on the impact of Francé's ideas at the Bauhaus, see Oliver A. I. Botar, "László Moholy-Nagy's New Vision and the Aestheticization

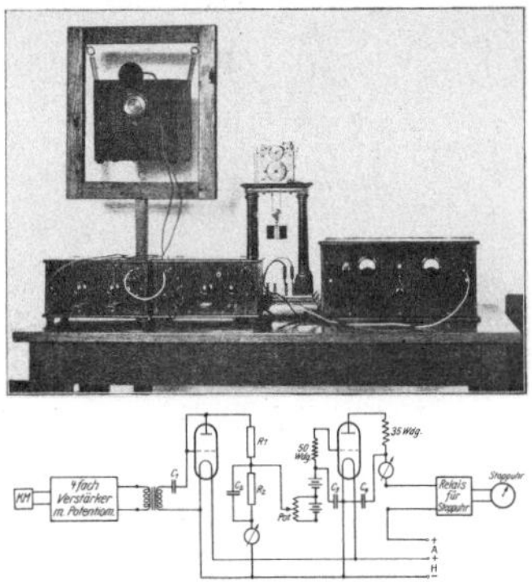

219 → P. 309

in his laboratory would be developed further in civilian and military sectors on both sides of the Atlantic. Radio, which previously had served as the technical principle for point-to-point communications, would undergo a decisive transformation, from *technical medium* to *mass media*. Throughout this political, institutional, and technological transformative process, laboratories remained the trans-scalar construction site for the further electrification of space.

In 1930, Erwin Meyer, head of the Department of Acoustics at the newly created Heinrich-Hertz-Institut für Schwingungsforschung (Heinrich Hertz Institute for Research on Oscillations, or HHI), presented a new measuring device for determining the reverberation time of a given room.[21] Describing his invention in the *Zeitschrift für technische Physik*, he gave equal attention to the machine and the conductive capacities of an ordinary room, as Hertz had in his seminal experiments. Meyer, a physicist, had fought in World War I before studying in Breslau under Erich Waetzmann, defending his dissertation on a study of the forces exerted on membranes by sound waves. Part of a generation trained in the 1920s at various technical universities and research institutions, he had worked at the Reich Telegraph Technical Office prior to joining the Heinrich Hertz Institute.[22]

Meyer's new measuring device was designed to automate the "specifically subjective method" of measuring reverberation defined by Wallace Sabine on the other side of the Atlantic several years earlier, whereby "a stopwatch and hearing measure the time it takes for a certain sound level to drop below the threshold of perceptibility."[23] For Meyer, it was essential for laboratories researching acoustics to use a more objective method. (Fig. 219) In order to enhance understanding of the phenomenon of reverberation time, his machine relied on a sound-emitting source—essentially a loudspeaker—that produced a sonic impulse at a specific frequency. When a button was pressed to emit the sound, a stopwatch was activated, and a condenser microphone inserted in the machine

of Scientific Photography in Weimar Germany," *Science in Context* 17, no. 4 (2004): 525–556, and "The Bauhaus of Nature," in Peder Anker, *From Bauhaus to Ecohouse: A History of Ecological Design* (Baton Rouge: Louisiana State University Press, 2010), 9–23. For a historical account of the relationship between sound and space analyzed in a longer timespan, see Gascia Ouzounian, *Stereophonica: Sound and Space in Science, Technology, and the Arts* (Cambridge, MA: MIT Press, 2020), 1–61.

21 Erwin Meyer, "Ein neues automatisches Verfahren der Nachhallmessung," *Zeitschrift für technische Physik*, no. 7 (1930): 253–259, here 253. Author's translation. It is worth noting that Erwin Meyer was not the only engineer to reach those conclusions around 1929. Indeed, in the same year that Meyer published his invention, Carl Eyring published the description of a similar device and procedure developed in the United States. For more on Eyring's device, see E. C. Wente and E. H. Bedell, "A Chronographic Method of Measuring Reverberation Time," *Journal of the Acoustical Society of America* 1 (April 1930): 422–427. This source is referenced after Emily Thompson, *The Soundscape of Modernity: Architectural Acoustics and the Culture of Listening in America, 1900–1933* (Cambridge, MA: MIT Press, 2002), 286–293, which analyzes Eyring's experiments.

22 Roland Wittje, *The Age of Electroacoustics: Transforming Science and Sound* (Cambridge, MA: MIT Press, 2016), 161.

23 Meyer, "Ein neues automatisches Verfahren," 253–259, here 254. Author's translation.

registered the reverberation of the mechanical impulse of the sound within the room. Mediated by the condenser microphone, the mechanical phenomenon of sound was turned into an electric signal. When the sound dropped below a specified intensity, as measured in decibels, a tube relay serving as a threshold then stopped the timer. (Fig. 220) Looking at the room in which the device was used, we can see that what was actually being tested was the performance of a particular square-shaped material on the floor, since different materials have different coefficients of absorption.

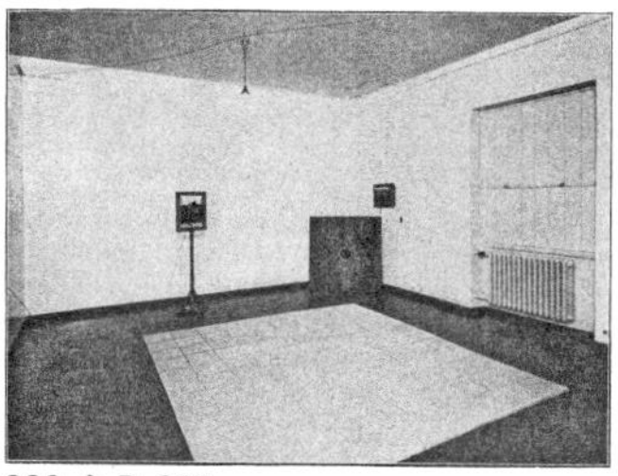

220 → P. 311

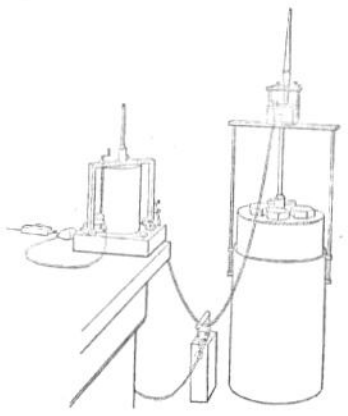

221 → P. 312

It is important to recall that Meyer's experiments were devised to perfect the influential method of testing the reverberation time of different rooms developed by Sabine at Harvard University in the late 1890s. Sabine invented an instrument consisting of an organ pipe connected to a switch.[24] (Fig. 221) When the switch was pressed, the pipe would emit a sound and Sabine would activate a stopwatch. He would listen carefully until he could no longer hear the reverberation of the sound within the room, at which point he manually stopped the watch. Repeating this technique in several rooms, he paid close attention not only to the size and geometry of the space, but to the materiality of the surfaces. After testing dozens of rooms, Sabine arrived at the eponymous formula for determining reverberation time.[25] Meyer's innovation was to replace the observer's ear with the capsule of a condenser microphone, and the observer's finger with a tube relay that stopped the watch. This same substitution of electronic technical media for human sensory capacities—of an electroacoustic process for a purely mechanical one—would also be accomplished by radio, as both a technical medium and mass media. In a broadcasting house, the mechanical phenomenon of sound is converted into an electrical current by microphones and then channeled through wires and other machines to the transmission stations, which in turn amplify that current and translate it into electromagnetic waves that travel through space at the speed of light until they reach the receiver. Again,

24 For more on Sabine's experimental procedures pertaining to acoustics, see Wallace C. Sabine, *Collected Papers on Acoustics* (Cambridge, MA: Harvard University Press, 1922). For more on the impact of his method in North America, see Thompson, *The Soundscape of Modernity*. An analysis of alternatives to Sabine's model for measuring reverberation time can be found in Viktoria Tkaczyk, "The Shot Is Fired Unheard: Sigmund Exner and the Physiology of Reverberation," *Grey Room* 60 (Summer 2015): 66–81.

25 Reverberation time, usually measured in seconds, is the duration between the emission of a sound and the decrease in its intensity below the level of human perception in a given room. Sabine's formula demonstrated that reverberation time is equal to the volume of the room divided by the sum of absorbing surfaces. For further reading on sound testing machines see Viktoria Tkaczyk, Mara Mills, and Alexandra Hui, eds., *Testing Hearing: The Making of Modern Aurality* (Oxford: Oxford University Press, 2020). To read more about laboratory practices as they relate to sound and room acoustics see Sabine von Fischer, "A Visual Imprint of Moving Air: Methods, Models, and Media in Architectural Sound Photography, ca. 1930," *Journal of the Society of Architectural Historians* 76, no. 3 (September 2017): 326–348, and Sabine von Fischer, *Das akustische Argument: Wissenschaft und Hörerfahrung in der Architektur des 20. Jahrhunderts* (Zurich: gta Verlag, 2019).

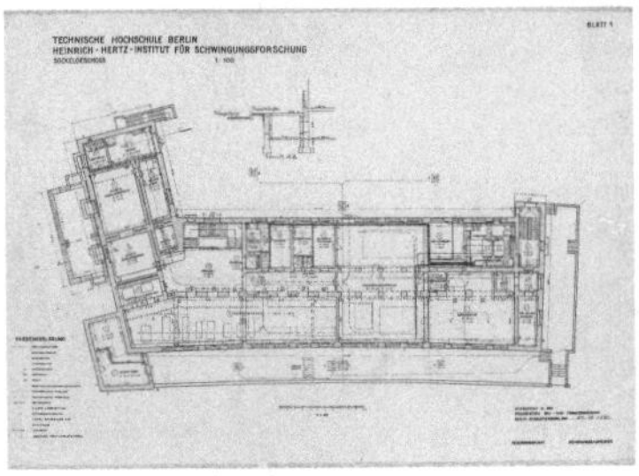

222 → P. 312

223 → P. 313

by following the experiments initiated by Hertz and pursued further in other laboratories, we can begin to understand the conditions of possibility for the electrification of sound and space.

The notion of reverberation time serves as a suitable entry point for examining the redefinition of space following the advent of electroacoustics. Recall that reverberation time registers the interaction of sound and space, being determined by the volume of a room and the materiality of its interior surfaces. In simple terms, it was sound, not vision, that turned the relatively abstract notion of space that had entered architectural discourse in the late nineteenth century into a measurable concept with distinct parameters.[26] In this respect, the automated process defined by Meyer spoke to how newly parameterized space could be understood, manipulated, and ultimately designed so that it could be widely disseminated through radio broadcasts. And within the laboratory, the electrified space envisioned by Hertz was being primed as the quintessential substance of broadcasting.[27]

(Fig. 222, 223) It was no coincidence that the shift in the approach to measuring reverberation time, from Sabine's mechanical method to an automated and electrified process, was defined, both in theory and practice, at the Heinrich Hertz Institute, one of a new breed of German institutions that aimed to coordinate and consolidate scientific research into electroacoustic technologies while at the same time responding to the interests of industry.[28] The institute had been founded in 1930 by the engineer Karl Willy Wagner, with the support of the Reich Broadcasting Corporation (Reichs-Rundfunk-Gesellschaft, RRG), producers of electrical equipment such as AEG and Telefunken, and the Technische Hochschule Berlin-Charlottenburg.[29] It was Wagner who invited Erwin Meyer to lead the Department of Acoustics. Meyer would soon become one of Germany's most prominent acousticians, gaining international recognition for his work on adapting music recording for radio broadcasts.[30]

26 For more on the introduction of the idea of space into architectural discourse, see the entry on "Space" in Adrian Forty, *Words and Buildings: A Vocabulary of Modern Architecture* (London: Thames & Hudson, 2004), 256–275.

27 For more on the early experimental practices conducted in Heinrich Hertz's laboratory in the late nineteenth century, see Hagen, *Das Radio*, 3–39.

28 For more on the origins of the Heinrich-Hertz-Institut, see Karl Willy Wagner, "Das Heinrich-Hertz-Institut für Schwingungsforschung," *Elektrische Nachrichten-Technik* 7, no. 5 (1930): 174–191, and Karl Willy Wagner, "Das Heinrich-Hertz-Institut für Schwingungsforschung," *Funk* 11 (1930): 43–44. For more on the creation of institutions related to electroacoustics, see Wittje, *The Age of Electroacoustics*.

29 Wagner published an influential treatise about the scientific basis of radio transmission. See Karl Willy Wagner, ed., *Die wissenschaftliche Grundlagen des Rundfunkempfangs* (Berlin: Julius Springer, 1927).

30 Meyer's most influential publication was translated into English following a series of lectures he gave at the University of London in 1938; see Erwin Meyer, *Electro-Acoustics* (London: G. Bell and Sons, 1939).

A recognition that "technology reshapes, in the most profound manner, the way people live" was one of the driving forces behind Wagner's founding of the Heinrich Hertz Institute.[31] Putting science at the center of its activities, the institute sought to reconcile the different strands of technological research: electrical engineering, telephony and telegraphy, radio frequency technology, acoustics, and mechanics. There was another research body, however, that aimed to bridge the more fundamental gap between technology and culture, an ambition particularly germane to radio broadcasting. Founded by Georg Schünemann in 1929, and affiliated with the Academy of Music, the purpose of the Rundfunkversuchsstelle (Radio Research Laboratory) was to research the cultural implications of emergent sound recording technologies and, in doing so, to tap into the new acoustic space latent in radio.[32] "Just as the sound of a symphony needs a specific space and a large audience, so radio needs a new music and a new experience," was how Schünemann explained the rationale of the new laboratory.[33] For him, "voice and instrument must be studied as they behave in front of the microphone in both musical and technical terms, to define a basis for music on the radio [*Funkspiel*]."[34] (Fig. 224, 225) By considering the microphone as the decisive agent that redefines the relationship between sound and space, the Rundfunkversuchsstelle took the experiments initiated by Erwin Meyer further, exploring their specific relation to music. Just as Meyer had automated and electrified the process of measuring reverberation invented by Sabine, so the Rundfunkversuchsstelle proposed an understanding of music as a potentially electrical phenomenon. One of its most influential figures, Friedrich Trautwein, would bypass the mechanical realm of instruments altogether in 1930, inventing an electronic synthesizer that would be pivotal in the later development of electronic music.[35] Through these laboratory practices, sound and space were electrified: in the process, sound was liberated from its historical ties to the symbolic order of music notation, and space from its exclusive dependence on orthographic projection.

224 → P. 314

225 → P. 314

31 Wagner, "Das Heinrich-Hertz-Institut," 174–191, here 174. Author's translation.

32 For more on the origins of the Rundfunkversuchsstelle, see Dietmar Schenk, *Die Hochschule für Musik zu Berlin: Preussens Konservatorium zwischen romantischem Klassizismus und neuer Musik, 1869–1932/33* (Stuttgart: Franz Steiner, 2004), 238–272. For more on Schünemann's use of gramophone recordings within the Rundfunkversuchsstelle, see Viktoria Tkaczyk, "Archival Traces of Applied Research: Language Planning and Psychotechnics in Interwar Germany," *Technology and Culture* 60, no. 2 (April 2019): 64–95.

33 Georg Schünemann, "Die Funkversuchsstelle bei der Staatlich Akademischen Hochschule für Musik in Berlin," *Rundfunk Jahrbuch* (1929): 277–293, here 281. Author's translation.

34 Schünemann, "Die Funkversuchsstelle," 279.

35 For more on Friedrich Trautwein's role in the Rundfunkversuchsstelle, see the meeting protocols in Archiv Universität der Künste Berlin (UdK-Archiv), Bestand 1b (Rundfunkversuchsstelle), 9, 12, and 13. Trautwein's connections with the National Socialist Party are widely studied. For more, see Jennifer Iverson, *Electronic Inspirations: Technologies of the Cold War Musical Avant-Garde* (New York: Oxford University Press, 2019), 11, and Benedikt Brilmayer, "Das Trautonium: Prozesse des Technologietransfers im Musikinstrumentenbau" (PhD diss., University of Augsburg, 2014).

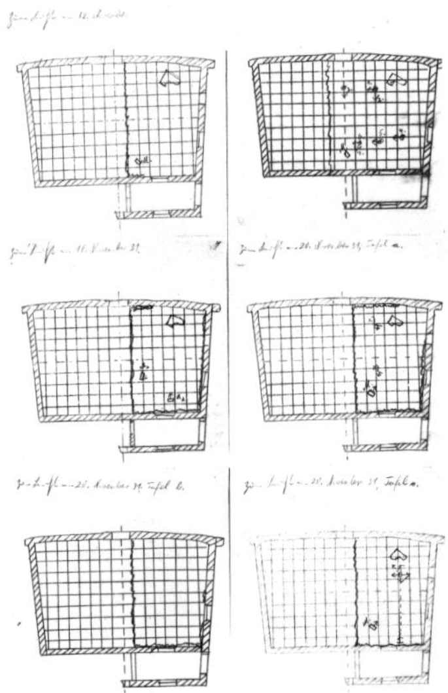

226 → P. 315

The trajectories of electrified sound and space and modern architecture would intersect in 1929, when Hans Poelzig won the competition to design the new broadcasting house for the RRG, which actively supported the Heinrich Hertz Institute, as noted above, and promoted the research efforts of the Rundfunkversuchsstelle. The building designed by Poelzig created a new typological framework for materially resolving the interaction between electroacoustics and mechanical architectural elements.[36] It would be portrayed in the press as the logical outcome of the studies developed by the acousticians and engineers who were advancing a new electroacoustic episteme.[37] In fact, it was the other way around: the Haus des Rundfunks was not the outcome of existing scientific and engineering research, but rather was conceived as a laboratory in its own right. More specifically, it was not scientific *knowledge* produced in a laboratory that influenced the conception of the new building, but rather the *practices* and *procedures* of the laboratory that provided the necessary framework for rethinking the relationship between electroacoustic and mechanical elements. The Haus des Rundfunks was not the direct product of positivistic parameters derived from science—such references are nowhere to be found in the drawings produced by the architect. Rather, it was modeled as an open form in which real spatial and sonic relationships could be studied and tested in situ—its spaces subjected to experimental procedures that expanded on trials carried out in controlled laboratory settings. Supporting evidence for this claim can be found in a series of experimental studies developed by the influential composer Hermann Scherchen to investigate the capacities of radio as "an instrument whose technical features are absolutely measurable."[38] Throughout the month of October 1931, Scherchen conducted a series of experiments with musical instruments, microphones, and a curtain in Saal IIa of the recently inaugurated Haus des Rundfunks.

(Fig. 226) Though the graphic records of these particular experiments might appear simple—traces

36 For more on Poelzig's design of the Haus des Rundfunks, see chapter 2.

37 For more on the contemporary reception of the Haus des Rundfunks, see Hans-Joachim von Braunmühl, "Uber die Intensitätsverhaltnisse von natürlichen Klangbildern mit besonderer Berücksichtigung der Rundfunksendung," *Zeitschrift für technische Physik*, no. 14 (1933): 68–70; Ludwig Kapeller, "Die neue Senderäume der Berliner Funk Stunde," *Funk* 38 (1928): 269–270; Walter Schäffer, "Raumakustik für Rundfunkdarbietungen," *Funktechnische Monatshefte*, no. 2 (1932): 51–54.

38 Hermann Scherchen, "Rechtlinien für Tonmeister, 1931," in Akademie der Künste, Berlin (AdK), Hermann-Scherchen-Archiv, Scherchen 327. This and subsequent quotes from the Scherchen Archive are translated by the author.

not unlike those sketches coming from Heinrich Hertz's laboratory—the whole architecture of radio was mobilized here in order to assess the spatial qualities afforded by Poelzig's building. With the help of the broadcasting house's sound engineers, Scherchen set out to test different positions for the instruments in relation to the microphone in the room.[39] As he described it, his method was initially "subjective (only with the ear), determining insufficient, mediocre, and optimal locations in the room itself for the positioning of sound sources (piano, violin, voice)."[40] He later repeated the experiment "with the microphone as an objective ear." Comparing the results, he concluded that "the most important requirement that needed to be met was: a satisfactory spatial effect [*Raumwirkung*]."[41] The material characteristics of the space were modified by the position and extension of a curtain that served as a damping material. That curtain, used as a mechanical architectural element, and the microphone as an electroacoustic one, were both understood as finishings, that is, as the final layer modulating the spatial effect afforded by the Haus des Rundfunks.

Within Saal IIa, the grid was the guiding organizational principle—appearing with the same intensity in Scherchen's sketches as it did in Poelzig's drawings of the building, or indeed in the geometricized laboratory space constructed by Hertz, from where it seemed to have migrated. As an organizing principle, the grid served as a trusted gauge for understanding the unpredictable behavior of sound oscillations—whether mechanical or electromagnetic—in space. The acoustic space tests developed by Scherchen were recorded by gramophone recorders in order to evaluate objectively the "satisfactory spatial effect" perceived by the microphone.[42] In the first scenario, Scherchen "found [the sound] to be very dry and too short in the reverberation, pointed and uncharacteristic in sound, without any substance." Intrigued, he asked himself, "[Was this perhaps] caused by blowing directly onto the damping curtain?"[43]

39 From the early 1930s, Hermann Scherchen was interested in the potential of electroacoustic technologies as they related to music, and therefore to space. Evidence of his engagement can be found in the notes he made at the conference on "Rundfunkmusik" in Munich in 1931, which address, among other things, the presentations of Erwin Meyer on "sound analysis," of Max Dessoir on "the psychological basis," of Georg Schünemann on the experiments undertaken at the Rundfunkversuchsstelle, of Friedrich Trautwein on "Electronic Music," of Karl Willy Wagner on the scientific research conducted at the Heinrich Hertz Institute, and of Kurt von Boeckmann on the "sociology of radio." See Hermann Scherchen, "Tagung für Rundfunkmusik München 1931," in Akademie der Künste, Berlin (AdK), Hermann-Scherchen-Archiv, Scherchen 358.

40 Hermann Scherchen, "Versuchsberichte über Saal 2a, Oktober 1931," in Akademie der Künste, Berlin (AdK), Hermann-Scherchen-Archiv, Scherchen 372.

41 Scherchen, "Versuchsberichte über Saal 2a."

42 Scherchen, "Versuchsberichte über Saal 2a."

43 Hermann Scherchen, "Akustische Raumprobe Saal 2a, Oktober 1931," in Akademie der Künste, Berlin (AdK), Hermann-Scherchen-Archiv, Scherchen 371.

44 Scherchen, "Akustische Raumprobe Saal 2a."

45 Scherchen, "Akustische Raumprobe Saal 2a."

46 Scherchen, "Versuchsberichte über Saal 2a."

Forthcoming about the tentative character of the overall experiment, Scherchen then tried something altogether different:

> Unfortunately, we only had half an hour (delayed because the amplifier was used for a recorded concert). We could not rehearse. We just came to familiarize ourselves with the adjusted room. Since the microphone used to record the previous samples was under repair and had been replaced by another, the sound on the sample 11 (at which I was not present) had such absolutely new and worse sound effects that we were forced to make do and rehearse again just with the grand piano. In order to achieve a better spatial effect, I moved the free-hanging curtain forward, so that a smaller portion of the entire room was enclosed. The grand piano remained in its best position so far (B). We placed the microphone in one of its former best analog positions. Except for the dividing curtain, all walls were undamped except for about 0.5 m damping behind the microphone.[44]

Later, he reported "the grand piano is now positioned 1.5 m further into the corner. Sound impression even better, but the whole thing a bit distant and blurry. Transmission improved by shifting the microphone 1 m in the direction of the diagonal to the grand piano, getting closer to an optimal condition."[45] He concluded, "optimal grand piano and microphone positions resulted on the diagonal of the room when the grand piano took one corner and the microphone the other. The corner behind the grand piano must be free, all other walls must be dampened by felt curtains," with one exception: "Mozart's piano music and its characteristic closeness require about 1 m wall at the microphone to be cleared of damping."[46]

The lessons and consequences of Scherchen's experiments are extremely relevant in a twofold sense. First, they attest to the nature of the building proposed by Poelzig, helping us to read anew the relationship between the procedures initiated in the laboratory and their impact on architecture. Rather than understanding the scientific systematization of the relationship between sound and space—initiated by Wallace Sabine and electrified by Erwin Meyer—as an unwavering, measured outcome, we might understand them as a series of experimental practices that prescribed a wholly different set of parameters for architectural design. Thus, when looking at the Haus des Rundfunks, we see that Poelzig's scheme is indeed an open form, in which the practices and procedures dictated by electroacoustics are largely defined in the untried, ever itinerant margins of experimentation. To reiterate, rather than being a precise, objective, and positivistic outcome of the laboratory, Poelzig's typological scheme should be read as a laboratory itself, as a type open and full of imperfections, yet still exact enough for coordinated experimentation to take place.

The experiments conducted by Scherchen also offer evidence of his early interest in electroacoustics and its relation to sound and space, though he would not have the opportunity to experiment for much longer at the Haus des Rundfunks. In 1933, when the Nazis came to power, he emigrated to Switzerland. After the war, he would set up the UNESCO-funded Centro sperimentale elettroacustico in Gravesano, one of the most influential institutions devoted to electroacoustics and part of a series of built artifacts that would become active agents within a complex process of *technology transfer* spanning from the Weimar Republic to the postwar years.[47] Specifically in this context, the type created by Hans Poelzig, the practices developed in the Heinrich Hertz Institute, and the institutional framework defined by Rundfunkversuchsstelle would come to serve as the model for constructing acoustic space after the end of World War II.

47 For more on the notion of technology transfer in the context of this study, see Friedrich Kittler, "Unconditional Surrender," in *The Truth of the Technological World*, ed. Hans Ulrich Gumbrecht (Stanford: Stanford University Press, 2013), 195–208. See also chapter 1 of this volume.

II

(Fig. 213, 214) It was a few years after meeting Gerhard Steinke, founder of the Labor für akustisch-musikalische

Grenzprobleme, that I came across recordings of a piece of music in the lab's archives at the Academy of Arts, and realized I had finally located the material traces of the East–West encounter he had described to me.[48] "Zoologischer Garten," by the US composer Frederic Rzewski, speaks directly to the presence of the Berlin Wall and to the ways in which the electrified space of radio was occupied and politically contested in the postwar years. In an account of the genesis of the piece published in the Swiss magazine *Du atlantis*, Rzewski describes how in August 1965 he began commuting across a divided Berlin almost every day, going to work on a piece commissioned by the GDR station, Berliner Rundfunk. The composer's unsanctioned commute between East and West testifies, on the one hand, to the huge presence of the American cultural intelligentsia in West Berlin. But it is also a clear indicator that individual politics, and particularly the politics of artists, tend to be far more unpredictable than institutions would like to pretend.

Cultural diplomacy was a weapon deployed on both the left and the right. The Russian-born composer Nicolas Nabokov—cousin of the writer Vladimir—had begun his tenure as secretary general of the CIA-sponsored Congress of Cultural Freedom (CCF) in 1951.[49] Ten years later, with the building of the Berlin Wall, he thought "the time was ripe to play Berlin's cultural card, but play it ably and firmly on a broad, international scale."[50] Among the "cards" envisioned by Nabokov—and financed by a generous grant from the Ford Foundation—was the presence in Berlin of Rzewski, still in his twenties, but already with an international reputation as a pianist and avant-garde composer. Supported by the US establishment, but committed, from those years on, to the political ideas of the left, Rzewski described his daily commute as a "hardly-believable transition between two different worlds coexisting side by side."[51] His composition was conceived precisely to transgress the contentious divide constructed and mediated by the Wall—to be transmitted, via the electromagnetic field of radio, across all physical and political borders. He wrote:

48 For the original composition, listen to Frederic Rzewski, "Zoologischer Garten: Einzelspuren auf zwei Kanälen, 1965," in Akademie der Künste, Berlin (AdK), Sammlung Audiovisuelle Medien (AVM), Tonbänder (AVM-31) 0825.1-2, 23'05". For the noise samples developed for the composition, listen to Frederic Rzewski, "Zoologischer Garten: Beispiele zum Geräuschkatalog, 1965," in Akademie der Künste, Berlin (AdK), Sammlung Audiovisuelle Medien (AVM), Tonbänder (AVM-31) 0826.1-3. To listen to Frederic Rzewski's "Zoologischer Garten" and other programs broadcast by DDR Radio in relation to electronic music and the work performed at the Labor für akustisch-musikalische Grenzprobleme go to the following link https://soundandscience.net/contributor-essays/labor-fur-akustisch-musikalische-grenzprobleme-a-critical-selection/ or use the QR Code placed below.

49 For a comprehensive historical, visual, and intellectual account of the role of the Congress of Cultural Freedom in Berlin see Anselm Franke, Nida Ghouse, Paz Guevara, and Antonia Majaca, eds., *Para Politics: Cultural Freedom and the Cold War* (Berlin: Sternberg Press, 2021).

50 Nicolas Nabokov, *Bagázh: Memoirs of a Russian Cosmopolitan* (New York: Atheneum, 1975), 257.

51 Frederic Rzewski, "Notizen zu Zoologischer Garten," *Du atlantis* 26 (1966): 942. Author's translation.

"Crossing the border at Friedrichstrasse could take ... minutes or it could take an hour. The time I spent in the S-Bahn, and getting on and getting off the train at the Zoo station, must have imprinted on my consciousness a certain sense of duration and change, which, I see now, has become the actual material of my composition." In East Berlin, Rzewski had access to a new medium of composition, producing sounds that would have been an absolute novelty for listeners on both sides of the Iron Curtain.[52] The Subharchord synthesizer, which he described as "one of the major advances in the field of electronic instruments" and an "extension of the human body,"[53] had been developed at the research lab by East German engineer Ernst Schreiber in 1961.[54]

The synthesizer was assigned a specific (and pivotal) function: that of producing a new kind of sound through electronic synthesis and, with it, a new kind of acoustic space.[55] The resulting musical composition is twenty-three minutes long and consists of six seemingly unrelated fragments bound together by the medium of magnetic tape. Based on the experience of confronting two radically different realities in East and West, it provides a prescient commentary on the early effects of the Berlin Wall at the scale of the city. Rzewski recorded how the Wall divided the city and simultaneously created contrasting proximities. At the same time, the technological, aesthetic, and political conditions under which the music was conceived made it possible for the work to transgress the Wall, going straight through it.[56] Rzewski's composition thus bears critically on other ideologically laden walls, which for too long have been overlooked. By carefully tracing the connections between the various institutional, mechanical, electronic, human and non-human actors related to those sounds, the piece reveals the intricate relationship between buildings, synthetic sounds, and electromagnetic waves that open onto a media archaeology of Berlin and its walls (including *the* Wall) in the age of radio.[57]

When Rzewski crossed from West to East Berlin, he stepped straight into the electrified space of the East

52 See Gerhard Steinke, "Experimental Music with the 'Subharchord' Subharmonic Sound Generator," *Journal of the Audio Engineering Society* 14, no. 2 (1966): 140–144.

53 Rzewski, "Notizen zu Zoologischer Garten," 942.

54 For more on the origins of the Labor für akustisch-musikalische Grenzprobleme, see Tatjana Böhme-Mehner, "Berlin Was Home to the First Electronic Studio in the Eastern Bloc: The Forgotten Years of the Research Lab for Inter-disciplinary Problems in Musical Acoustics," *Contemporary Music Review* 30, no. 1 (2011): 33–47.

55 The notion of acoustic space emerged in the context of the Culture and Communications seminar organized by Marshall McLuhan and Edmund Carpenter in Toronto. Architectural historian Sigfried Giedion initiated a debate about the concept with McLuhan, and the historical traces of that debate can be found in Reto Geiser, *Giedion and America: Repositioning the History of Modern Architecture* (Zurich: gta Verlag, 2018), 372–404. Architect and urban planner Jaqueline Tyrwhitt also contributed significantly to that debate: for more on her role, see Michael Darroch, "Bridging Urban and Media Studies: Jaqueline Tyrwhitt and the Explorations Group, 1951–1957," *Canadian Journal of Communication* 33, no. 2 (2008): 147–169.

56 For more on the reception of Rzewski's piece in the context of academic classical music in the former GDR, see Korrespondenz, Protokolle der Sitzungen am 18.3. Vorschläge für Zuwahlen und die Verleihung des Nationalpreises, Information über die Auslandsarbeit der Sektion, Konzertprogramm; Meisterschüler-Angelegenheiten, 1966, in Akademie der Künste Berlin (AdK) (Ost), Historisches Archiv, 0636.

57 See Wolfgang Ernst, *Digital Memory and the Archive* (Minneapolis: University of Minnesota Press, 2013), 55–73.

Berlin acoustic laboratory, passing through the Wall and beneath the skein of electromagnetic waves that traveled unimpeded between East and West. Listening to his piece, we hear a number of simultaneous sonic situations that blend the electronically synthesized sounds produced by the Subharchord with samples recorded and manipulated through magnetic tape. Simply stated, we hear the sub-harmonic tones produced by the synthesizer as well as a female voice that fades in and out of the acoustic space. The sub-harmonic tones situate the listener within an ambiguous space that resonates with a certain sci-fi imaginary, as well as with a much darker and uncanny atmosphere. The sub-harmonic tones also enhance the lower frequency of the generated sounds and could be compared with the soundtrack composed by Oskar Sala for Alfred Hitchcock's film *The Birds* (1963), which used a version of the Trautonium, the synthesizer invented in the 1930s by Friedrich Trautwein at the Rundfunkversuchsstelle.

The oscillation between a sci-fi and a horror sensation is less an ambiguous analogy and more a precise technological effect: most of the scores for the GDR's sci-fi films were produced with the Subharchord synthesizer. In this sense, Rzewski's piece appears—however inadvertently—as a most apt soundtrack for a divided Berlin: A US composer supported by the Ford Foundation and the Congress of Cultural Freedom crosses the Wall into East Berlin. The composer opts for an instrument that is the technological offspring of one that originated during the Weimar Republic and was later promoted by the Nazis. The composition comments—abstractly and figuratively—on the experience of inhabiting a divided city. In the last three minutes, the figurative references disappear: we no longer hear the female voice as the composition acquires a deeper and darker tone. Throughout, however, what we are essentially listening to is the space of the laboratory, a new site and locus of spatial production originating in Berlin in the 1930s. The composition was meant to be broadcast in February 1966, using the electronic and electromagnetic apparatus of FM radio to modify the acoustic space of the entire city, but it was never aired as planned.

As mentioned earlier in this chapter, the music piece composed by Rzewski opens onto an environmental history of information of Berlin, but equally registers the trans-scalar agency of the laboratory, reinforcing Bruno Latour's claim, "I hope to convince the reader that the very difference between the 'inside' and the 'outside' and the difference of scale between 'micro' and 'macro' levels, is precisely what laboratories are built to destabilize or undo."[58] In this sense, the trans-scalar laboratory practices initiated by Heinrich Hertz were a fundamental precondition for both the material and immaterial constitution of the architecture of radio. Without them, space arguably would have never been electrified. In 1887, Heinrich Hertz redefined the relationship between matter and energy inside the laboratory. During the 1930s, the experiments conducted by Erwin Meyer in his newly established laboratory served as a model for reformulating the relationship between buildings and electronic recording devices. In the postwar period, laboratory practices related to the production of electronic sound served as the vehicle for occupying space, destabilizing, in Latour's terms, the difference between the micro and the macro scales, while annulling the distinction between inside and outside effected by those borders. Thus, when we listen to the piece composed by Rzewski, we are not only listening to a site of production, in the general sense of a laboratory, but experiencing a highly specific and politically informed site of construction of space. If the electrification of space constructed by Hertz is understood by historians as an *invention*, the electrification of space during the postwar period was preoccupied with *differentiation*.[59]

Both Rzewski's composition and the Subharchord synthesizer were constructed in the Labor für akustisch-musikalische Grenzprobleme. Like most of the equivalent laboratories in the West, it was dependent on radio institutions and thus functioned

58 Bruno Latour, "Give Me a Laboratory and I Will Move the World," in *Science Observed: Perspectives on the Social Study of Science*, ed. K. Knorr and M. Mulkay (London: Sage, 1983), 141–170, here 143.

59 For more on the laboratory as a site of production related to sound and other forms of creative practices, see Alexandra Hui, Julia Kursell, and Myles W. Jackson, "Music, Sound, and the Laboratory from 1750 to 1980," *Osiris* 28 (2013): 1–11. See also Sabine von Fischer, "A Visual Imprint of Moving Air: Methods, Models, and Media in Architectural Sound Photography, ca. 1930," *Journal of the Society of Architectural Historians* 76, no. 3 (September 2017): 326–348; and Peter Galison and Caroline Jones, "Factory, Laboratory, Studio: Dispersing Sites of Production," in *The Architecture of Science*, ed. Peter Galison, Shimon Edelman, and Emily Thompson (Cambridge, MA: MIT Press, 1999), 497–534. See also Bruno Latour, *Laboratory Life: The Construction of Scientific Facts* (Princeton: Princeton University Press, 1979).

227 → P. 317

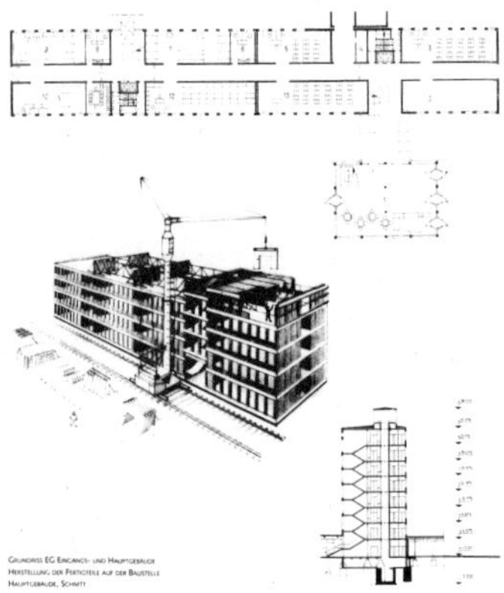

228 → P. 318

within a much larger network of local and global practices. Within the institutional hierarchy of the GDR, the Labor was part of the Rundfunk- und fernsehtechnisches Zentralamt (RFZ), which was originally set up as the eastern section of the Heinrich Hertz Institute in 1950, under the name of Betriebslaboratorium für Rundfunk und Fernsehen (BRF).[60] The RFZ in turn reported to the Ministry of Post and Telecommunications. (Fig. 227)

Construction of the new RFZ headquarters was completed in 1963. Designed by architect Peter Flierl, who worked for the state-owned VEB Industrieprojektierung, the building was located in Adlershof, 7 km south of the Funkhaus, in an area populated by facilities built for the GDR's Academy of Science.[61] (Fig. 228) The architect described the program as having three main components: the development of new projects and technologies (main building), the construction of new components (workshop), and the construction and maintenance of broadcasting vans (shed).[62] These three elements—with some seven hundred workspaces—were integrated within an outwardly generic eight-story building made of pre-cast concrete panels (*Plattenbau*) clad with a repetitive pattern of standardized aluminum windows developed by a Dresden company specializing in aeronautical engineering.[63]

In a protocol signed by the director of the RFZ, ten points clearly describe the central functions of the institution, with emphasis placed on achieving a "unity between politics, technology, and the economy."[64] The document states that, above all, the RFZ was "responsible for implementing the decisions made by the Socialist Unity Party of Germany, implementing the Constitution, the laws and other legislations of the GDR."[65] Legal and political duties aside, "the main task of the RFZ [was] to be a fitting scientific leader for mastering the technical revolution in radio communications."[66] A key mandate was the scientific advancement of radio communications and information, but equally important for the institution

60 The institute founded by Karl Willy Wagner in 1930 underwent a National Socialist "cleansing" in 1936 that erased the name of Heinrich Hertz. Its headquarters was destroyed at the end of the war but the institute was refounded in 1945 before being split by the political division of Berlin in 1949.

61 Peter Flier designed several industrial projects and later worked on the Marx-Engels-Forum and on Museum Island. For more on his career, see Angelika Petruschat, *Peter Flier Architekt* (Berlin: form+zweck, 2010).

62 Peter Flier, "Rundfunk- und Fernsehtechnisches Zentralamt Berlin-Adlershof," *Deutsche Architektur* 11 (November 1964): 662–664.

63 Petruschat, *Peter Flier Architekt,* 49.

64 See "Ordnung über die Aufgaben und Arbeitsweise des RFZ, August 1968," 9, Personal Archive Siegfried Köppen, Berlin, Germany. I would like to thank Siegfried Köppen, a development engineer at RFZ for over two decades, for providingd access to his personal archive for the purposes of this research. This and subsequent quotes are the author's translation.

65 "Ordnung über die Aufgaben und Arbeitsweise des RFZ," 4.

66 "Ordnung über die Aufgaben und Arbeitsweise des RFZ," 5.

was its duty "to increase its contribution to the national income by developing economically oriented research that contributes to the progress of socialist life."[67] Furthermore, to make radio and communication technologies functional and profitable, the RFZ was to enable "the rapid and smooth transfer of research-and-development results into practice, as well as the application of new technologies for the Ministry of Post, techniques, and work processes, in particular by issuing guidelines, procedures, operating, maintenance, and care instructions, and to secure changes in technical and technological innovations."[68]

Put more directly, the RFZ was expected to develop the "techno-scientific and economic analysis for future planning in parallel with the introduction of scientific management methods and the organization of science and data processing for the radio system."[69] At a national and international level, the RFZ would "exercise the function of a control agent for standardization, making the means of production, technologies, and working methods of the Ministry of Post's radio system more compatible." But more relevant for the architecture of radio, it was also supposed "to perfect and expand the material and technical basis of radio communications, as well as the application of new technical equipment, devices, technologies, and to further develop working methodologies," while also assuming "the function of a control agent, acting as a specialized project manager for the Ministry of Post's radio communication system during the preparation of its investments." Therefore, the RFZ was "responsible for preparing and implementing the centralized investments of the Ministry of Post's radio system as well as developing and approving new facilities for the Ministry's radio stations."[70]

In the context of the GDR program for improving industrial productivity (*Neuererbewegung*),[71] the RFZ was likewise in charge of "the production of sound recording devices, as well as the construction of equipment and facilities and their proper installation

67 "Ordnung über die Aufgaben und Arbeitsweise des RFZ," 5.

68 "Ordnung über die Aufgaben und Arbeitsweise des RFZ," 5.

69 "Ordnung über die Aufgaben und Arbeitsweise des RFZ," 6.

70 "Ordnung über die Aufgaben und Arbeitsweise des RFZ," 6.

71 For more on the economic policies in the GDR, see Hartmut Berghoff, *The East German Economy, 1945–2010: Falling Behind or Catching Up* (Cambridge: Cambridge University Press, 2013).

based on research-and-development evidence." Along with the responsibilities for developing material infrastructure and the drafting of patents, it assumed the role of a "leader agency in the field of communications" and exercised "the function of a control agency for radio information and documentation."[72] In other words, the RFZ was both a political institution and a scientific test center: on the one hand, it was committed to developing the knowledge and technologies necessary for improving the radio communication system of the GDR; on the other, it produced new components and technologies for sale to other countries, whether in the Eastern or Western Blocs.

It was under the auspices of this multipurpose institution that the Labor für akustisch-musikalische Grenzprobleme was founded in 1956, using parts of the facilities of the former BRF. In view of the important political and aesthetic role that radio played in Germany at the time, the remit of the lab was "to research new aesthetic laws through unexplored sound possibilities, to study information-theory laws related to music, and to enrich the sonic means of expression in radio and television through developing a specifically radiophonic sound art."[73] In other words, this laboratory was committed to researching and fabricating the technological means of a new art, electronic music, devised specifically for dissemination through radio. Concurrently, it was the place where the aesthetic characteristics of this new art were to be engendered.

According to its founding director, Gerhard Steinke, the Labor was modeled after the Rundfunkversuchsstelle created in Berlin in the late 1920s, which had direct ties to the RRG and the Heinrich Hertz Institute.[74] But regardless of this connection with its interwar predecessor, it was established amidst a global proliferation of such laboratories, most of them related to radio institutions.[75] The pioneer of this new type of institution—which relied on people, as well as on the

72 "Ordnung über die Aufgaben und Arbeitsweise des RFZ," 6.

73 Transcript of a report sent by Gerhard Steinke, March 1, 1973, in Akademie der Künste Berlin (AdK) (Ost), Historisches Archiv, 4546. Author's translation.

74 Gerhard Steinke held different positions within the RFZ. Throughout, he was not only an avid promoter of the aesthetic and technological exchanges between music and electroacoustic technology, but also a committed bridge-maker between East and West. Steinke pushed for bridging the gap between classical and electronic music with many actors and authorities of the GDR's Academy of Fine Arts, the department of music at DDR Rundfunk (GDR Radio), and the department of music at the Staatlichen Rundfunkkomitee (State Radio Committee), while also seeking to secure a place for this new art within the context of technological experimentation at the RFZ. For more on Steinke's promotion of this laboratory and of various institutions, see the meeting protocols and correspondences in Akademie der Künste Berlin (AdK) (Ost), Historisches Archiv, 4546.

75 Gerhard Steinke, "Akustisch-musikalische Grenzprobleme in der Rundfunk- und Fernsehtechnik," *Technische Mitteilungen des RFZ* 11 (1967): 149–159, here 149.

76 For more on the origins of electronic music in Germany in the Cologne region, see Lowell Cross, "Electronic Music, 1948–1953," *Perspectives of New Music* 7, no. 1 (Autumn–Winter 1968): 32–65. For more on the characteristics of the Studio für elektronische Musik itself, see *Sonderheft über elektronische Musik* (Cologne: Hausmitteilungen des Nordwestdeutschen Rundfunks, 1954). To my knowledge, the most comprehensive study on the history and politics of the experimental studio in Cologne can be found in Jennifer Iverson, *Electronic Inspirations: Technologies of the Cold War Musical Avant-Garde* (New York: Oxford University Press, 2019). For more on the relationship between the NWDR and the Haus des Rundfunks, see Maral Herbst, *Demokratie und Maulkorb: Der Deutsche Rundfunk in Berlin zwischen Staatsgründung und Mauerbau* (Berlin: Vistas, 2001), 41–55.

77 For more on the history and the architecture of the studio in Gravesano, see Antonio Scerri, "Centro sperimentale elettroacustico 'Dr. Hermann Scherchen,' Gravesano," *Gravesaner Blätter* 1 (July 1955): 5–15.

78 David Crowley, "Spatial Music: Design and the Polish Radio Experimental Studio," https://post.at.moma.org/content_items/335-spatial-music-design-and-the-polish-radio-experimental-studio.

79 F. C. Brooker, "The BBC Radiophonic

model envisioned by the Rundfunkversuchsstelle—was the Studio für elektronische Musik, founded in Cologne in 1951 as an offshoot of NWDR (Nordwestdeutscher Rundfunk), the same body that would take over Hans Poelzig's Haus des Rundfunks in 1953.[76] (Fig. 229–236) It was followed, within less than a decade, by the creation of the Groupe de Recherches Musicales in Paris (1951), founded by Pierre Schaeffer under the auspices of Radiodiffusion-Télévision Français (RTF); the Centro sperimentale elettroacustico in Gravesano (1954);[77] the RAI Studio of Phonology in Milan (1955); the Polish Radio Experimental Studio led by Jozef Patkowski and designed by Polish architect and Team Ten member Oskar Hansen in Warsaw (1957);[78] the BBC Radiophonic Workshop in London (1958);[79] the Columbia-Princeton Electronic Music Center in New York (1959);[80] and the Siemens Studio für elektronische Musik in Munich (1959).[81]

Of this roll-call of institutions, it was the Centro sperimentale elettroacustico in Gravesano that served as the most influential prototype for the Berlin laboratory. As described earlier in the chapter, its founder, Hermann Scherchen, had worked in close collaboration with the RRG during the early 1930s, conducting experiments at the Haus des Rundfunks before going into exile in Switzerland in 1933. Supported financially by UNESCO, the studio in Gravesano became both an intellectual and experimental hub, developing further Scherchen's early interest in electroacoustics as it related to sound and space. Scherchen not only led the studio—where he was described by Sigfried Giedion as a "great experimenter"[82]— but also oversaw the *Gravesaner Blätter*, a journal that published works and articles addressing the *Grenzprobleme* (border problems) between music, space, and electroacoustics. The journal brought together a disparate cast of artists, scientists, and intellectuals, among them Le Corbusier,[83] Iannis Xenakis,[84] Theodor W. Adorno,[85] Lothar Cremer,[86] Hans-Joachim von Braunmühl,[87] and Pierre Schaeffer.[88] It was this very idea of *Grenzprobleme* that inspired the name of the studio in East Berlin, tracing an arc

Workshop," *Electronics and Power* 11 (January 1965): 24–29. See also Teresa Winter, "Thoughts on Autonomy, Sound and Artistic Compromise: Three Experimental Radio Broadcasts of Poetry with Radiophonic Sound by Delia Derbyshire (1965–1966)," *Sound Studies* 3, no. 1 (2017): 3–16.

80 Nick Patterson, "The Archives of the Columbia-Princeton Electronic Music Center," *Notes* 67, no. 3 (March 2011): 483–502.

81 Stefan Schenk, *Das Siemens-Studio für elektronische Musik Geschichte, Technik und kompositorische Avantgarde um 1960* (Tutzing: Hans Schneider, 2016).

82 See correspondence between Sigfried Giedion and Hermann Scherchen, October 6, 1965, in the gta Archive, 43-K-1965-10-06(G):1, 1 Bl./S.

83 Le Corbusier, "The 'Modulor,'" *Gravesaner Blätter* 9, no. 3 (1957): 2–5.

84 Iannis Xenakis, "La crise de la musique sérielle," *Gravesaner Blätter* 1, no. 1 (1955): 2–4; Iannis Xenakis, "Corbusier's 'Electronic Poem,'" *Gravesaner Blätter* 9, no. 3 (1957): 43–54.

85 For Adorno's contribution to the journal, see Theodor W. Adorno, "Zum Verhältnis von Musik und Technik heute," *Gravesaner Blätter* 11/12, no. 4 (1958): 36–61.

86 Lothar Cremer was a sound engineer associated with the Heinrich Hertz Institute in the 1920s and 1930s. He assisted Hermann Scherchen in making the improvements to the acoustics of the Haus des Rundfunks. For more, see Meyer Erwin and Lothar Cremer, "Über die Hörsamkeit Holtzausgekleideter Raüme," *Zeitschrift für technische Physik* 14 (1933): 500–507. Cremer was a frequent contributor to the *Gravesaner Blätter*: the publication of his work as the sound engineer on Hans Scharoun's competition-winning project for the Berlin Philharmonie was particularly relevant. See Lothar Cremer, "Drei Entwürfe zur Berliner Philharmonie," *Gravesaner Blätter* 3, no. 9 (1957): 55–74, and "Die Grenzen planmäßiger raumakustischer Gestaltung," *Gravesaner Blätter*, no. 2/3 (1956): 10–32. The influence of Lothar Cremer extended even to the Disney Concert Hall designed by Frank Gehry in Los Angeles in the early 2000s. For more, see Carol McMichael Reese, "Gehry in Conversation," in *Late Thoughts: Reflections on Artists and Composers at Work*, ed. Karen Painter and Thomas Crow (Oxford: Oxford University Press, 2008), 97–113.

87 Hans-Joachim von Braunmühl was a sound engineer at the RRG: he assisted in the acoustic design of the Haus des Rundfunks and was one of the inventors of

237 → P. 323

238 → P. 324

of influences between the Weimar period and postwar Berlin. The studio in Gravesano, which was no doubt influenced by the model of the Rundfunkversuchsstelle, in turn influenced the laboratory founded by Steinke back in East Berlin. In the Cold War context, the entangled relationships among radio, electroacoustics, and sound became a complex and trans-scalar spatial and political problem, while also becoming, especially in Berlin, an archetypal border problem as well.[89]

III

(Fig. 237) The questions raised by these newly discovered *Grenzprobleme* called for a new architecture. A photograph of a room in the East Berlin research lab shows a combination of traditional, and almost banal, architectural elements such as two doors, a linoleum floor, hanging lamps; there are no windows, but we do see a generic dropped ceiling. In addition to these elements, there are a number of other electroacoustic elements, and the image intentionally grants more prominence to them. In the foreground we see tape recorders; on the wall on the right, compressors and filters, and on the left, just next to the door, a mixing board. But at the center of the diagonal composition of the image is the Subharchord, the analog synthesizer developed within this very lab. (Fig. 238) Another photo shows the inventor of that instrument, the engineer Ernst Schreiber, dressed in signature white lab coat. He is sitting next to the composer Hans Hendrik Wehding, who is wearing a suit. Although seemingly generic in its qualities, what this room was articulating was actually a complex form of architecture. The lab coat and the suit, science and fine art, were the sort of *Grenzprobleme* that the walls of this room were supposed to address. Looking back at the previous image, the Subharchord synthesizer was not only placed at the center of this picture, but also figured as a central agent of what is referred to here as the architecture of radio.

Developed by Schreiber in 1961, the Subharchord could be seen as the prodigal offspring of the

the Magnetophone tape recorder under the Third Reich. His "Neuere raumakustische Erkenntnisse und Gestaltungen" appeared in *Gravesaner Blätter* no. 2/3 (1956): 10–33.

88 The French composer Pierre Schaeffer founded the Groupe de Recherches Musicales in Paris. His "Zur Situation der Versuche elektroakustischer Klanggestaltung" appeared in *Gravesaner Blätter* 6, no. 2 (1956): 110–116.

89 The word *Grenzprobleme* was used in the subtitle of the first issues of the *Gravesaner Blätter* to describe the field in which the studio's research was positioned. Jack Bornoff, representing UNESCO, an institution that financially supported both the studio and the publication, referred to the relevance of this new cultural and technological problem. See Jack Bornoff, "Eine Vierteljahresschrift für musikalische, elektroakustische und schallwissenschaftliche Grenzprobleme," *Gravesaner Blätter* no. 1 (1955): 1. In an interview, Gerhard Steinke recalled that Hermann Scherchen invited him on several occasions to visit the studio in Gravesano. For more, see Tatjana Böhme-Mehner, "Interview with Gerhard Steinke, October 12, 2010, Steinke's Apartment, Berlin," *Contemporary Music Review* 30, no. 1 (2011): 15–23.

laboratory because it not only gave material expression to the scientific knowledge produced there, but provided the laboratory with its own sonic signature, one that no other test site of its kind had acquired.[90] As a "central sound and noise generator," the synthesizer was unique in that it was "simple to handle, allowing *any* kind of sound structure to be repeated and recorded very quickly and without complications."[91] Specially developed "for use in studios of broadcasting, film, and television organizations, in theaters and opera houses, the apparatus [was] able to operate with *any* kind of sound effect."[92] These relatively straightforward claims, included in a paper published in an audio engineering journal as well as in the promotional brochure for the instrument, signaled an architectural revolution in the making.

By introducing a new paradigm of electroacoustics, the lab, like others of its kind, reshaped the historical bond between sound and space, bypassing the mechanical phenomenon of room acoustics that for millennia had framed the relationship between architecture and music. By developing its own electronic instruments, it created a medium through which sound acquires a spatial signature (*Raumeindruck*). Combined with cultural and technological activities, this spatial signature provided the foundation for the definition of an acoustic space. When Frederic Rzewski crossed the border in 1965 to compose and record "Zoologischer Garten"; when Siegfried Matthus recast Bertolt Brecht's *Life of Galileo* as an electronic sonic experience featuring overlapping tape-recorded voices combined with marimba sounds and noises from the Subharchord;[93] when the DEFA sci-fi production *Signale* opened with the characteristic sub-harmonic tones of the Subharchord, creating a unique soundscape to express what outer space, technology, and the future meant for the GDR,[94] they were contributing to the creation of a distinctive acoustic space with the capacity to extend well beyond the borders of the GDR, reshaping the established boundaries between buildings and sound.

90 The Subharchord represents another specific example of technology transfer that had direct links to the Trautonium developed by Friedrich Trautwein in the Rundfunkversuchsstelle in the early 1930s. For more on the invention of the Subharchord and how it relates to Trautwein's invention, see Ernst Schreiber, "Ein neuartiger elektronischer Klang- und Geräuscherzeugung," *OIRT-Zeitschrift für Rundfunk und Fernsehen* 2 (1965): 35–39; Steinke, "Experimental Music with the 'Subharchord,'" 140–144.

91 Steinke, "Experimental Music with the 'Subharchord,'" 140–141, here 141.

92 "Druckbelege, betr. Vorstellung des Gerätes 'Subharchord II,' 1967," in Akademie der Künste Berlin (AdK) (Ost), Historisches Archiv, 4546. Author's translation.

93 Siegfried Matthus was a composer of classical music in the GDR. In an interview, he recalls, "Even with the Berlin Wall, the radio waves were always able to cross the border. Back then, I sat in front of the radio every evening and listened to all the broadcasts of contemporary music. Boulez, Nono, or Stockhausen—I knew all of them, but only through listening." For the entire interview, see Tatjana Böhme-Mehner, "Interview with Siegfried Matthus, 20 February 2009, Stolzenhagen," *Contemporary Music Review* 30, no. 1 (2011): 53–60. The original composition can be heard at Siegfried Matthus, "Galilei: für Singstimme, Instrumente und elektronische Klänge über Worte von Bertolt Brecht," April 25, 1966, in Akademie der Künste, Berlin (AdK), Sammlung Audiovisuelle Medien (AVM), Tonbänder (AVM-31) 0789, 12'10". For more on the reception of the composition within the environment of the Akademie der Künste in Berlin, see "Probleme der elektronischen Klangerzeugung, 1966," in Akademie der Künste Berlin (AdK) (Ost), Historisches Archiv, 0636.

94 The film *Signale* was screened for the first time in the GDR in 1970. Produced by DEFA, most of the soundtrack and sound design of the film were composed and produced using the Subharchord synthesizer. For a general history of DEFA, see Sean Allan and John Sandford, eds., *DEFA: East German Cinema, 1946–1992* (New York; Oxford: Berghahn Books, 1999).

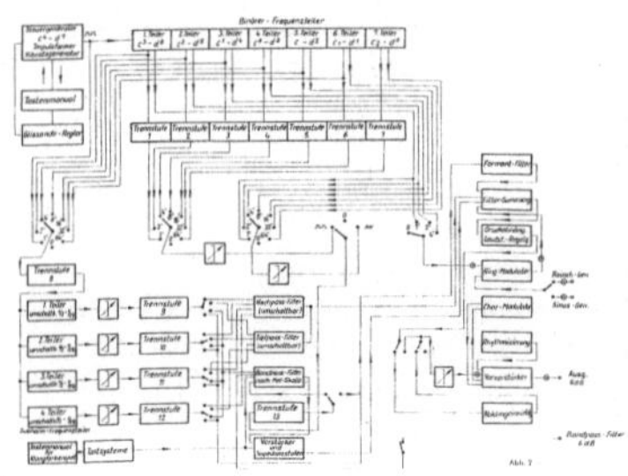

239 → P. 325

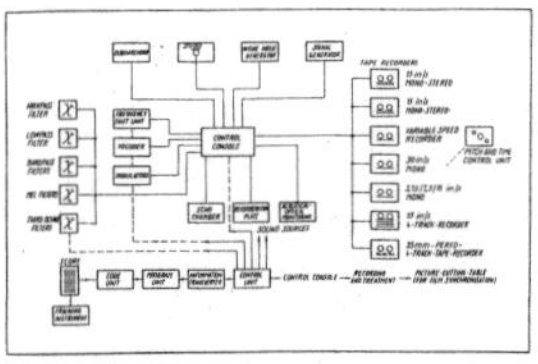

240 → P. 326

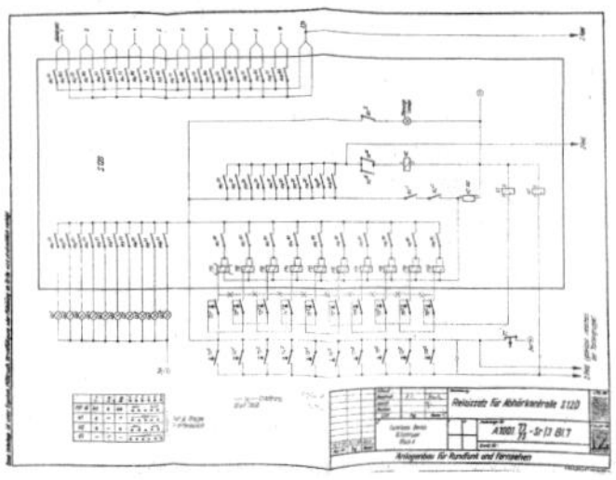

241 → P. 327

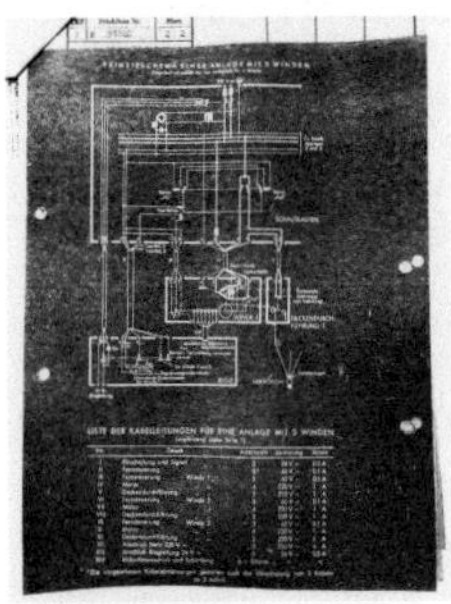

242 → P. 328

As I have demonstrated throughout this book, the historical transformation that recalibrated the role of architecture in the Cold War years did not occur overnight, but was a prolonged process in which the production of space was intersected by scientific knowledge and experimentation, mixed with institutional policies, artistic interests, and political motivations. The production of space was modified, in essence, by the material transformations affecting the production, transmission, and reception of sounds and information. Not *despite*, but precisely *because* of its material essence, the resulting epistemological shift did not lead to the wholesale replacement of buildings in this process of producing, transmitting, and receiving information via sounds. Rather, this shift established the conditions of possibility for the emergence of a new form of architecture that related buildings and sounds in a new way. To illustrate the point, the image just analyzed shows how the basic elements of architecture—ceilings, doors, floors, and walls—work together to house and shelter an electronic machine, a synthesizer. (Fig. 239, 240) Yet, from the circuit diagrams we see that the conventional reading of this image could just as well be reversed, with the machine itself determining the nature of the architectural components. The first diagram explains the circuit of the Subharchord, showing how the electric impulse is amplified and modulated through various filters, "producing other interesting sound and noise structures."[95] By adding a sine tone through the ring modulator, and by potentially giving a vibrato effect, "the complex signal produces an ensemble effect."[96] In short, what the diagram illustrates is how each of the knobs of the machine can convert spatial-sonic effects into a particular electrical signal. What we see in the second diagram is the topological disposition of elements, which confirms that the laboratory in which the Subharchord is situated is but one part of a more complex system. In this case, a standard orthographic floor plan would not be sufficient to explain the laboratory. Adopting the language of the circuit diagram, pieces of buildings like the "echo chamber" and the "studio" are represented as if

95 Steinke, "Experimental Music with the 'Subharchord,'" 140–141, here 142.

96 Steinke, "Experimental Music with the 'Subharchord,'" 140–141, here 142.

they were electronic components sitting adjacent to reverberation plates, control consoles, filters, and tape recorders, and instead of being connected through corridors or other like spaces, these pieces are topologically linked via a single line, which in most cases signifies a wire circuit.

The adoption of the graphic language of electroacoustics to explain spatial effects and dispositions—whether electrical, mechanical, or otherwise—suggests that electroacoustics instigated an entirely different way of thinking that tied sound and space together. In his book *The Age of Electroacoustics*, historian of science Roland Wittje describes how electroacoustics evolved from a laboratory practice into a new scientific and aesthetic practice that created a new kind of sound.[97] Taking this line of thought further, my central argument here is that this unprecedented coupling of buildings and circuits, of electromagnetic waves and music, of politics and science, created not only a new kind of sound but also a novel kind of space, which I refer to as the *architecture of radio*.

(Fig. 241, 242) In a divided Cold War Berlin, different artifacts at different scales—the circuit diagrams of the Subharchord and massive buildings like the Funkhaus on Nalepastrasse—were all regarded in the same way. (Fig. 212, 243) Transmission stations like the one at Köpenick were also conceived in the same language. Indeed, an entire city and nation were apprehended and experienced through one and the same medium—the architecture of radio, a specific system of relations allowing space and sound to be conceptualized at various scales, extending beyond the boundaries of building, city, and nation. (Fig. 69, 78) So, the first impression given by that photograph of the laboratory—of a ceiling and a floor sheltering an electronic instrument in a defined place—is actually deceptive, for this was but one stage of an electroacoustic transformation that would eventually facilitate an architecture of radio "conceptually stretched to blanket the earth."[98]

212 → P. 253

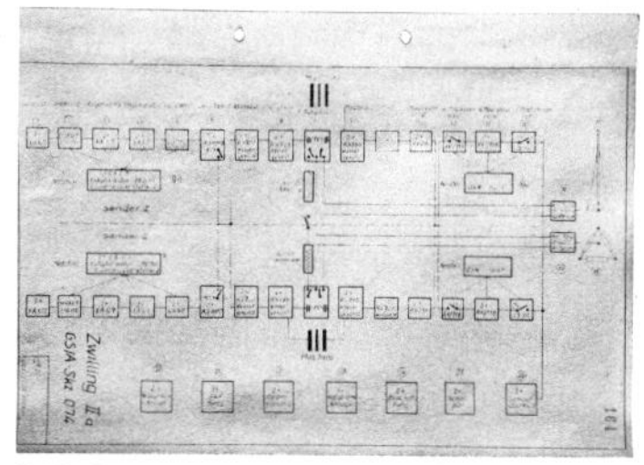
243 → P. 329

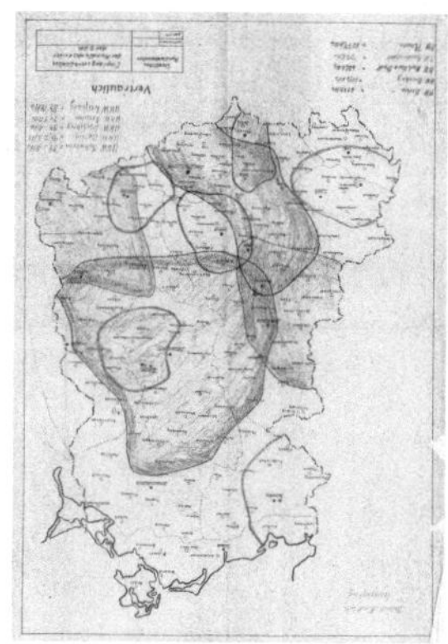
69 → P. 104

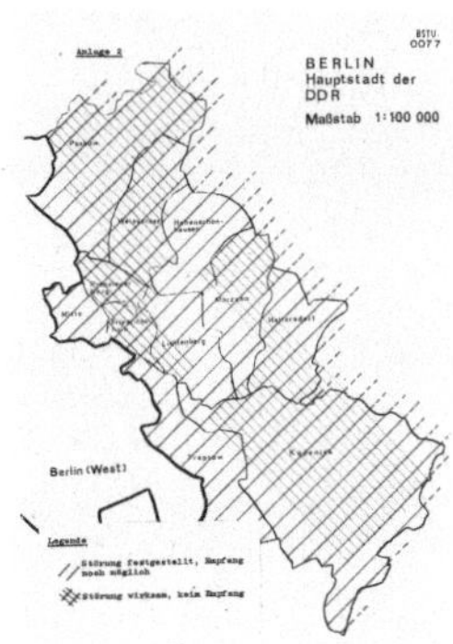

78 → P. 113

97 Wittje, *The Age of Electroacoustics*. See also Roland Wittje, "The Electrical Imagination: Sound Analogies, Equivalent Circuits, and the Rise of Electroacoustics, 1863–1939," *Osiris* 28 (2013): 40–64.

98 Peter Galison, "The Ontology of the Enemy: Norbert Wiener and the Cybernetic Vision," *Critical Inquiry* 21, no. 1 (Autumn 1994): 228–266, here 235.

99 Lothar Kiebs, "Perspektiven für eine raumbezogene Rundfunkübertragung," *Technische Mitteilungen aus dem BRF* 1 (March 1960): 2–20.

100 Gerhard Steinke and Dieter Boeck, *Die Musikwerkstatt*, radio program recorded in Saal III of the Funkhaus, November 6, 1964, in Akademie der Künste, Berlin (AdK), Sammlung Audiovisuelle Medien (AVM), Tonbänder (AVM-31) 0801.1-2. Author's translation.

Returning to the Funkhaus on Nalepastrasse designed by Franz Ehrlich in 1952—and discussed in detail in the previous chapter—there is a close relationship between the effects of walls in the building and the effects produced by machines like the Subharchord. The twofold performance of the walls within the building—partitioning to create distinct spaces and heterogeneity on the one hand, while giving a sonic signature to the mechanical signals on the other—could be understood to constitute a spatial symbiosis. Through their geometric disposition and material constitution, the walls in the Funkhaus mechanically fulfill a function equivalent to the one attributed to the Subharchord synthesizer used by Rzewski to compose his piece "Zoologischer Garten": namely, they fabricate sounds that otherwise would not exist. Although walls operate mechanically and the synthesizer uses an electrical sonic impulse, both act acoustically as "machines" by imparting spatial qualities to sounds, creating what acousticians refer to as *Raumeindruck* (space impression).[99] In a radio program broadcast in 1964, Gerhard Steinke explained the main thrust of the research carried out by the Labor für akustisch-musikalische Grenzprobleme in relation to the recently completed Funkhaus: "What we want to do is to broadcast the space. And it's not that you will be in the concert room, but that you will feel that the room is there, behind that wall, that loudspeaker wall."[100] Through the medium of its walls, the Funkhaus created spaces that were designed to be recorded and transmitted across Berlin—to enter the domestic realms and workspaces of its inhabitants and, in doing so, to modulate the very spatiality of the city.

Buildings like the Funkhaus on Nalepastrasse or the Labor für akustisch-musikalische Grenzprobleme in East Berlin, or the Haus des Rundfunks in West Berlin, and the numerous transmission stations scattered across the city—all functioned as the material armature of immaterial radio waves that freely traversed a divided Berlin. Analyzing this transgressive power of radio, we can begin to understand not only the important role that architects actually played in the design of those facilities, but also the intrinsic agency of an architecture of radio, that is, radio *as* architecture in its own right. This inimitable architecture is understood as a system of distended relations, a non-local, tectonic arrangement woven across different institutions, scales, and technologies. Rethinking architecture in this way requires us to look again at elements such as windows, doors, walls, and columns in relation to electronic technical media such as tape recorders, mixing boards, vacuum tubes, transistors, and microphones. We need to analyze the internal organization of building types and how they relate to these elements, and understand their disposition and location within the structure of the city. Furthermore, the architecture of radio is hardly monolithic in its constitution. It is not, in other words, a top-down project designed by a specific architect, planner, or political agent alone. Rather, it constitutes a trans-scalar technological ensemble comprising electronic and mechanical elements, building types, and, ultimately, urban structures that make the tangled relationships among these different scales of operation binding in their effect.

The political agency of this form of architecture hinges on how it transgresses, redefines, and redistributes limits and thresholds that have traditionally been understood as stable and impermeable. Rather than manifesting a compliant and monumental edifice—onto which supposedly secure political constructs are applied—the architecture of radio is a medium through which assertions of power and ideology were redefined, negotiated, and calibrated. It redistributed and destabilized the order of the *polis* and thus exerted a political power that neither East nor West could fully monopolize. Architecture, to be sure, is always political, but the politics of architecture do not necessarily coincide with the political positions of those who conceive it. As contended throughout this book, the walls of Berlin once operated in a dual way: as the sensory medium upon which the architecture of radio left its traces, and as the very agents enabling this form of architecture to perform and proliferate. Thus, Berlin and its walls seem to be the perfect ruin from which to understand and learn how to deal with the strange way in which human beings, through architecture, have rendered their world inhabitable by circumscribing themselves and forgetting about all those parts of it that offend them.

APPENDIX

ACKNOWLEDGMENTS

One writes to see what one thinks, and the writing of this book has shown me how transformative this journey can be, thanks to the generous support of mentors, colleagues, and friends along the way. Philip Ursprung has been an inspiring guide ever since I met him under unusual circumstances in Princeton, New Jersey. Nothing that I have achieved in this process would have been possible without Philip's knowledge, patience, and support; he has always been available when I needed him. Jean-Louis Cohen saw the very first seeds of this project when I was his student at Princeton. From the outset, he was endlessly generous in sharing his vast wisdom. Jean-Louis departed from this world too soon to see the research materialized in printed form in this volume, which is in part dedicated to him.

From my experience in the United States, I would also like to thank Thomas Y. Levin for teaching me a new form of listening during his beautiful seminar; Anthony Vidler, who read my very first essay on the subject matter, written while I was taking his class; Beatriz Colomina for enhancing both curiosity and fun in her seminars; Elizabeth Diller for helping me to leave unresolved the distinction between thinking and building; and Michael Meredith for sharing our common fascination for (non) composition and repetition. Mark Lee has been a friend and an inspiration since I met him, and some of the ideas present in this book were sharpened and refined through the conversations I had with students at the Graduate School of Design at Harvard University. I am grateful to Mark for inviting me to teach at the GSD and to the "Acoustic Space" seminar students for willfully and courageously diving into these ideas. That this book exists at all is ultimately due to the wit, openness, and imagination of Thomas Weaver, who saw potential in a clumsily arranged dissertation. With his and Pamela Johnston's precise revisions, this manuscript slowly took its current shape.

Developing from loose ideas until it reached the form of a book, this project grew while I was part of different stimulating and inspiring communities. The Institute for the History and Theory of Architecture (gta) at ETH Zurich is a fantastic community that both welcomed me and fostered my intellectual development. The culture built around the seminars taught by Laurent Stalder was fundamental for this project, and the courses led by Ita Heinze-Greenberg brought the Institute together with an expanded idea of what the history of architecture could and should mean today. Without a doubt, the most valuable resource at the Institute was my colleagues, and I would like to thank Anne Kockelkorn for offering bits of her rigorous brain even before I landed in Zurich, as well as Davide Spina, Laila Seewang, André Tavares, and Sarah Nichols for engaging in lengthy and passionate conversations. Wulf Böer, who revised parts of this manuscript, shared with me his intellectual curiosity and his passion for music, and offered his couch to sleep on multiple occasions.

Sebastian Bonhoeffer provided me with a year of peace and a beautiful room in the Semper-designed Sternwarte to polish ideas that otherwise would not be in this book. I am grateful to him and the entire community of the Collegium Helveticum, who supported my work uniquely and generously. This project also grew in Italy, where I spent a year as a fellow at Villa Massimo, the German Academy in Rome. Julia Draganović provides an ideal setting for friendship, ideas, sounds, insects, and plants to coexist. This book was completed inside the Laboratory for History and Theories of Architecture, Technology, and Media at EPFL in Lausanne. Pedro Correa, Ella Neumaier, and Xavier Nueno criticized and supported this manuscript until the last minute, making our laboratory the best environment to finish projects and start new ones.

To understand what I wanted to do, I moved to live on the site: Berlin. Among all the people I met here, I would like to thank Viktoria Tkaczyk for sharing her tuned interest in the sonic environment and for having me as a fellow at the Max Planck Institute. Wolfgang Ernst was generous when he invited me to present still very rough ideas of this project in his colloquium at the Humboldt University, precisely where Hegel used to live in Berlin. Carolyn Birdsall, Michael Osman, and Niklas Maak, each with their sensibilities, read early versions of this manuscript. Combining a patience with ignorance with a passion for ideas, their comments simply made this book better. Leonard Streich offered his delicate vision to help with the design of this book, and Justus Menten offered his friendship from the time of my arrival in Berlin.

Before I left my native Chile, Rodrigo Pérez de Arce seeded in me an intense curiosity for these strange objects called buildings. Beatriz Bories instilled an early interest in the mysteries that vibrations secretly hold. Working with Ariel Bustamante is always a new lesson about the materiality of sound, and this project would not have been possible without those exchanges. Isabel Margarita Riesco and Francisco Thiermann have always been by my side, and everything I do has been a collaborative effort ever since we orbit around the sun, holding hands as a family, with Michelle Gravel, Pedro Tristán, and Juan Nataniel. The origins of my project lie in my first visit to the city of Berlin in 2015 with my sister and brother, Beatriz and Pablo Thiermann. The day before I was due to leave, Pablo insisted that I visit a strange building he had come across on the eastern edge of the city, by the Spree. "I'm sure you'll like it," he said, and I surely did (it was the Funkhaus on Nalepastrasse). The intervening years have been a humble attempt to make sense of that encounter.

ARCHIVAL SOURCES

Archives Akademie der Künste Berlin, Robert-Koch-Platz 10, 10115, Berlin, Germany. With gratitude to Uta Simmons and Rene Engelmann.

Archives of the Bauhaus Dessau Foundation, Mies-van-der-Rohe-Platz 1, 06844, Dessau-Rosslau, Germany. With gratitude to Claudia Parren and Lutz Schube.

Archives of the Karlsruhe Institute of Technology, Kaiserstrasse 12, 76131 Karlsruhe, Germany. With gratitude to Elke Leinenweber and Klaus Nippert.

Archives of the Museum of Architecture, Technical University Berlin, Strasse des 17. Juni 150, 10623, Berlin, Germany. With gratitude to Hans-Dieter Nägelke.

Archives of the University of the Arts Berlin, Einsteinufer 43–53, 10587, Berlin, Germany. With gratitude to Dietmar Schenk.

Archivo de Fondos y Colecciones Museo de la Memoria y los Derechos Humanos, Matucana 501, 8500000 Santiago, Región Metropolitana, Chile.

Canadian Centre for Architecture Archival Collection, 1920 rue Baile, QC H3H 2S6, Montreal, Canada. With gratitude to Giovanna Borasi.

Federal Commissioner for the Records of the State Security Service of the former German Democratic Republic, Karl-Liebknecht-Str. 31/33, 10178, Berlin, Germany.

German Broadcasting Archive Babelsberg, Marlene-Dietrich-Allee 20, 14482, Potsdam, Germany. With gratitude to Jörg-Uwe Fischer.

German Broadcasting Archive Frankfurt, Bertramstrasse 8, 60320, Frankfurt am Main, Germany. With gratitude to Karin Langer.

German Federal Archives Dienstelle Berlin, Finckensteinallee 63, 12205, Berlin, Germany.

gta Archives, ETH Zurich, Stefano-Francini-Platz 5, 8093, Zurich, Switzerland. With thanks to Bruno Maurer and Muriel Perez.

Houghton Library and Frances Loeb Library, Harvard University, Quincy Street & Harvard St, Cambridge, MA 02138, United States. With thanks to Ines Zalduendo.

BIBLIOGRAPHY

Adorno, Theodor W. "Culture Industry Reconsidered." Translated by Anson G. Rabinbach. *New German Critique*, no. 6 (Autumn 1975): 12–19.

Adorno, Theodor W. *Current of Music: Elements of a Radio Theory*. Edited by Robert Hullot-Kentor. Frankfurt am Main: Suhrkamp, 2006.

Adorno, Theodor W. *Obra Completa en 20 tomos*. Vol. 7, *Teoría Estética*. Madrid: Akal, 2004.

Adorno, Theodor W. *Obra Completa en 20 tomos*. Vol. 10/2, *Crítica de la Cultura y Sociedad II*. Madrid: Akal, 2009.

Adorno, Theodor W. "Technique, Technology, and Music To-day." *Gravesaner Blätter.* 11/12, no. 4 (1958): 51–61.

Adorno, Theodor W. "Zum Verhältnis von Musik und Technik heute." *Gravesaner Blätter* 11/12, no. 4 (1958): 36–61.

Adorno, Theodor W, and Max Horkheimer. *Dialectic of Enlightenment*. Stanford: Stanford University Press, 2020.

AEG, ed. *25 Jahre AEG-Dampfturbinen*. Berlin: VDI-Verlag, 1928.

AEG, ed. *Zehn Jahre Forschungs-Institut der AEG*. Berlin: AEG-Forschungs-Institut, 1938.

Agamben, Giorgio. *What is an Apparatus? and Other Essays*. Stanford: Stanford University Press, 2009.

Aggregate (Group), ed. *Governing by Design: Architecture, Economy, and Politics in the Twentieth Century*. Pittsburgh: University of Pittsburgh Press, 2012.

Allais, Lucia. *Design of Destruction: The Making of Monuments in the Twentieth Century*. Chicago; London: University of Chicago Press, 2018.

Alexander, Zeynep Çelik. *Kinaesthetic Knowing: Aesthetics, Epistemology, Modern Design*. Chicago; London: University of Chicago Press, 2017.

Alexander, Zeynep Çelik. "The Larkin's Architectural Technologies of Trust." *Journal of the Society of Architectural Historians* 77, no. 3 (September 2018): 300–318.

Alexander, Zeynep Çelik, and John May, eds. *Design Technics: Archaeologies of Architectural Practice*. Minneapolis: University of Minnesota Press, 2020.

Allan, Sean, and John Sandford, eds. *DEFA: East German Cinema, 1946–1992*. Oxford: Berghahn Books, 1999.

Åman, Anders. *Architecture and Ideology in Eastern Europe during the Stalin Era: An Aspect of Cold War History*. Cambridge, MA: MIT Press, 1992.

Amhoff, Tilo, Katie Lloyd Thomas, and Nick Beech. *Industries of Architecture*. New York: Routledge, 2016.

Anderson, Benedict. *Buried City, Unearthing Teufelsberg: Berlin and Its Geography of Forgetting*. New York: Routledge, 2017.

Anderson, Stanford. *Peter Behrens and a New Architecture for the Twentieth Century*. Cambridge, MA: MIT Press, 2000.

Anker, Peder. *From Bauhaus to Ecohouse: A History of Ecological Design*. Baton Rouge: Louisiana State University Press, 2010.

Architekten- und Ingenieur-Verein zu Berlin, ed. *Berlin und seine Bauten*. Pt. 10, Vol. B: *Anlagen und Bauten für den Verkehr (4): Post und Fernmeldewesen*. Berlin: Ernst & Sohn, 1987.

Arendt, Hannah. *The Human Condition*. Chicago: University of Chicago Press, 1958.

Arnheim, Rudolf. *Radio: An Art of Sound*. London: Faber & Faber, 1936.

Ashbridge, Noel. "The Acoustical Problems of Broadcasting Studios." *Engineering* 132 (1931): 505–539.

Attali, Jacques. *Noise: The Political Economy of Music*. Minneapolis: University of Minnesota Press, 1977.

Aureli, Pier Vittorio. *The Project of Autonomy: Politics and Architecture within and against Capitalism*. New York: Princeton Architectural Press, 2008.

Aureli, Pier Vittorio. *Architecture and Abstraction*. Cambridge, MA: MIT Press, 2023.

Baacke, Rolf-Peter.*Lichtspielhausarchitektur in Deutschland: von der Schaubude bis zum Kinopalast*. Berlin: Frölich und Kaufmann, 1982.

Badenoch, Alexander. *Voices in Ruins: West German Radio across the 1945 Divide*. New York: Palgrave Macmillan, 2008.

Badenoch, Alexander, Andreas Fickers, and Christian Henrich-Franke, eds. *Airy Curtains in the European Ether: Broadcasting and the Cold War*. Baden-Baden: Nomos, 2013.

Bajpai, Anadita. *Cordial Cold War: Cultural Actors in India and the German Democratic Republic.* New Delhi, Sage, 2021.

Balbi, Gabriele, and Maria Rikitianskaia. "The Mediatization of the Air: Wireless Telegraphy and the Origins of a Transnational Space of Communication, 1900–1910s." *Journal of Communication* 68, no. 4 (2018): 758–779.

Banham, Reyner. *The Architecture of the Well-Tempered Environment.* Chicago: University of Chicago Press, 1969.

Banham, Reyner. *A Concrete Atlantis: US Industrial Building and European Modern Architecture 1900–1925.* Cambridge, MA: MIT Press, 1989.

Banham, Reyner. *Los Angeles: The Architecture of Four Ecologies.* Berkeley, CA: University of California Press, 1971.

Banham, Reyner. *Theory and Design in the First Machine Age.* New York: Praeger, 1967.

Bauernfeind, Wolfgang. *Tonspuren: Das Haus des Rundfunks in Berlin.* Berlin: Ch. Links, 2010.

Bausch, Hans, ed. *Rundfunk in Deutschland.* Munich: Deutscher Taschenbuch Verlag, 1980.

Begoll, Günter. '"Eine Nacht, die wir nicht vergessen werden' 1952 Blockade des Funkhauses in der Westberliner Masurenallee—ein Interview." *RFZ-Frequenz* 11(1952): 2.

Begun, S. J. *Magnetic Recording.* London: Murray Hill Books, 1949.

Bergmeier, Horst J. P., and Rainer E. Lotz. *Hitler's Airwaves: The Inside Story of Nazi Radio Broadcasting and Propaganda Swing: Jazz, Swing, and Nazi Radio Propaganda.* New Haven: Yale University Press, 1997.

Behling, Klaus, and Andreas Jüttemann. *Der Berliner Teufelsberg: Trümmer, Truppen und Touristen.* Berlin: Berlin Story, 2014.

Behne, Adolf. *Der Moderne Zweckbau.* Berlin: Drei Masken, 1926.

Behne, Adolf. *The Modern Functional Building.* Translated by Michael Robinson. Santa Monica: Getty Research Institute for the History of Art and the Humanities, 1996.

Benjamin, Walter. *Gesammelte Schriften.* Frankfurt am Main: Suhrkamp, 1980.

Benjamin, Walter. *Radio Benjamin.* Edited by Jonathan Lutes, Lisa Harries Schumann, and Diana Reese. London: Verso, 2014.

Benjamin, Walter. *The Work of Art in the Age of Its Technological Reproducibility and Other Writings on Media.* Edited by Michael W. Jennings, Brigid Doherty, and Thomas Levin. Cambridge, MA: Harvard University Press, 2008.

Bergdoll, Barry. *Karl Friedrich Schinkel: An Architecture for Prussia.* New York: Rizzoli, 1994.

Berghahn, Volker R. *America and the Intellectual Cold Wars in Europe.* Princeton: Princeton University Press, 2001.

Berghoff, Hartmut, and Uta Andrea Balbier, eds. *The East German Economy, 1945–2010. Falling Behind or Catching Up.* Cambridge: Cambridge University Press, 2013.

Berlage, Hendrik Petrus. *4 Vorträge gehalten im Kunstgewerbemuseum zu Zürich.* Berlin: Bard, 1908.

Berman, Marshall. *All That is Solid Melts into Air: The Experience of Modernity.* New York: Penguin Books, 1982.

Bessire, Lucas, and Daniel Fisher, eds. *Radio Fields: Anthropology and Wireless Sound in the 21st Century.* New York: New York University Press, 2012.

Bevans, Charles I., ed. *Treaties and Other International Agreements of the United States of America, 1776–1949.* Vol. 3. Washington: Department of State, 1969.

Biraghi, Marco. "Poelzig and the Golem." *AA files* 75 (2017): 190–201.

Birdsall, Carolyn. *Nazi Soundscapes: Sound, Technology and Urban Space in Germany, 1933–1945.* Amsterdam: Amsterdam University Press, 2012.

Blunck, Erich. "Das Haus der deutschen Funkindustrie in Berlin-Charlottenburg." *Deutsche Bauzeitung* 59, no. 14 (February 1925):105–112.

Böer, Wulf. "Air-Conditioning, Architecture, and Modernism: On the History of the Controlled Environment, 1911–1952." Dr. Sc. diss., ETH Zurich, 2019.

Böhme-Mehner, Tatjana. "Berlin was Home to the First Electronic Studio in the Eastern Bloc: The Forgotten Years of the Research Lab for Inter-disciplinary Problems in Musical Acoustics." *Contemporary Music Review* 30, no. 1 (2011): 33–47.

Böhme-Mehner, Tatjana. "Interview with Gerhard Steinke, 12 October 2010, Steinke's Apartment, Berlin." *Contemporary Music Review* 30, no. 1 (2011): 15–23.

Böhme-Mehner, Tatjana. "Interview with Siegfried Matthus, 20 February 2009, Stolzenhagen." *Contemporary Music Review* 30, no. 1 (2011): 53–60.

Bollé, Michael. *Die Grossfunkstation Nauen und ihre Bauten von Hermann Muthesius.* Berlin: Willmuth Arenhövel, 1996.

Bornoff, Jack. "Eine Vierteljahresschrift für musikalische, elektroakustische und schallwissenschaftliche Grenzprobleme." *Gravesaner Blätter* 1, no. 1 (1955): 1.

von Borries, Friedrich, and Jens-Uwe Fischer. *Gefangen in der Titotalitätsmaschine: Der Bauhäusler Franz Ehrlich.* Berlin: Suhrkamp, 2022.

Botar, Oliver A. I. "László Moholy-Nagy's New Vision and the Aestheticization of Scientific Photography in Weimar Germany." *Science in Context* 17, no. 4 (2004): 525–556.

Brady, Erika. *A Spiral Way: How the Phonograph Changed Ethnography.* Jackson: University of Mississippi Press, 1999.

Brandelhuber, Arno, Florian Hertweck, and Thomas Mayfried, eds. *The Dialogic City: Berlin wird Berlin.* Cologne: Walther König, 2015.

Bratton, Benjamin H. *The Stack.* Cambridge, MA: MIT Press, 2016.

Braunmühl, Hans-Joachim von. "Neuere raumakustische Erkenntnisse und Gestaltungen." *Gravesaner Blätter* 6, no. 2 (1956): 7–14.

Braunmühl, Hans-Joachim von. "Uber die Intensitätsverhaltnisse von natürlichen Klangbildern mit besonderer Berücksichtigung der Rundfunksendung." *Zeitschrift für Technische Physik* 14 (1933): 68–70.

Brech, Martha. *Der hörbare Raum: Entdeckung, Erforschung und musikalische Gestaltung mit analoger Technologie.* Bielefeld: [transcript] Verlag, 2015.

Brecht, Bertolt. "Der Rundfunk als Kommunikationsapparat." *Blätter des Hessischen Landestheaters*, no. 16 (July 1932).

Bressler, Eva Susanne. *Von der Experimentierbühne zum Propagandainstrument: Die Geschichte der Funkausstellung von 1924 bis 1939.* Cologne: Böhlau, 2009.

Brilmayer, Benedikt. *Das Trautonium: Prozesse des Technologietransfers im Musikinstrumentenbau.* PhD diss., University of Augsburg, 2014.

British Broadcasting Corporation. *The Broadcasting House*. Plaistow: Curwen Press, 1932.

British Broadcasting Corporation. *A Technical Description of Broadcasting House*. Bungay: Chaucer Press, 1932.

Broadbent, Philip and Sabine Hake. *Berlin Divided City, 1945–1989*. Oxford: Berghahn Books, 2010.

Brooker, F. C. "The BBC Radiophonic Workshop." *Electronics and Power* 11 (January 1965): 24–29.

Browne, Don R. "RIAS Berlin: A Case Study of a Cold War Broadcast Operation." *Journal of Broadcasting* 10, no. 2 (1966): 119–135.

Bucciarelli, Piergiacomo. *Fritz Höger: Hanseatischer Baumeister 1877–1949*. Berlin: Vice Versa, 1992.

Bundeszentrale für Politische Bildung, ed. "DDR-Rundfunk- und Fernsehkomitee: Aufgaben und Geschichte." In *Tele-Visionen: Fernsehgeschichte Deutschlands in Ost und West*, 1–15. Bonn: BpB; Grimme-Inst, 2011.

Butter, Andreas, and Ulrich Hartung. *Ostmoderne: Architektur in Berlin 1945–1965*. Berlin: Jovis, 2004.

Büttner, Fritz L. *Das Haus des Rundfunks in Berlin*. Berlin: Haude & Spenersche Verlagsbuchhandlung, 1965.

Canguilhem, Georges. *Knowledge of Life*. Translated by Stefanos Geroulanos and Daniela Ginsburg. New York: Fordham University Press, 2008.

Carpenter, Edmund, and Marshall McLuhan, eds. *Explorations in Communications: An Anthology*. London: Beacon Press, 1960.

Cary, Jonathan. *Techniques of the Observer: On Vision and Modernity in the Nineteenth Century*. Cambridge, MA: MIT Press, 1992.

Castro, J. Justin. *Radio in Revolution: Wireless Technology and State Power in Mexico, 1897–1938*. Lincoln: University of Nebraska Press, 2016.

Cavallo, Ascanio, and Margarita Serrano. "Las 24 horas que estremecieron a Chile." *La Tercera Reportajes*, August 24, 2023, 8–23.

Cavallo, Ascanio, and Margarita Serrano. *Golpe 11 de Septiembre 1973: Las 24 horas más dramáticas del siglo 20*. Santiago: uqbar, 2013.

Certeau, Michel de. *The Practice of Everyday Life*. Berkeley: University of California Press, 1984.

Cheng, Irene, Charles L. Davis II, and Mabel O. Wilson. *Race and Modern Architecture: A Critical History from the Enlightenment to the Present*. Baltimore: Project Muse, 2020.

Ciucci, G., Francesco Dal Co, Mario Manieri-Elia, and Manfredo Tafuri. *The American City: From the Civil War to the New Deal*. Cambridge, MA: MIT Press, 1979.

Clarke, Joseph L. *Echo's Chambers: Architecture and the Idea of Acoustic Space*. Pittsburgh: University of Pittsburgh Press, 2021.

Classen, Christopher. "Captive audience? GDR Radio in the Mirror of Listener's Mail." *Cold War History* 13, no. 2 (2013): 239–254.

Classen, Christoph, and Klaus Arnold, eds. *Zwischen Pop und Propaganda: Radio in der DDR*. Berlin: Christoph Links, 2004.

Cohen, Jean-Louis. *Architecture in Uniform: Designing and Building for the Second World War*. Paris: Hazan Editions, 2011.

Cohen, Jean-Louis. *Scenes of the World to Come: European Architecture and the American Challenge 1893–1960*. Paris: Flammarion, 1995.

Colomina, Beatriz. *Privacy and Publicity: Modern Architecture as Mass Media*. Cambridge, MA: MIT Press, 1996.

Combes, Muriel. *Gilbert Simondon and the Philosophy of the Transindividual*. Translated by Thomas LaMarre. Cambridge, MA: MIT Press, 2013.

Conley, Patrick. *Der parteiliche Journalist: Die Geschichte des Radio-Features in der DDR*. Berlin: Metropol Verlag, 2012.

Conrads, Ulrich. *Programs and Manifestoes on 20th Century Architecture*. Cambridge MA: MIT Press, 1970.

Constantine, Helen, ed. *Berlin Tales*. Oxford: Oxford University Press, 2009.

Cremer, Lothar. "Drei Entwürfe zur Berliner Philharmonie." *Gravesaner Blätter* 9, no. 3 (1957): 55–74.

Cremer, Lothar. "Die Grenzen planmässiger raumakustischer Gestaltung." *Gravesaner Blätter* 2/3 (1956): 10–32.

Cross, Lowell. "Electronic Music, 1948–1953." *Perspectives of New Music* 7, no. 1 (1968): 32–65.

Crowley, David. "Spatial Music: Design and the Polish Radio Experimental Studio." Museum of Modern Art, January 14, 2014. https://post.moma.org/spatial-music-design-and-the-polish-radio-experimental-studio/.

Crowley, David, and Susan E. Reid, eds. *Pleasures in Socialism: Leisure and Luxury in the Eastern Bloc*. Evanston: Northwestern University Press, 2010.

Crowley, David, and Susan E. Reid, eds. *Socialist Spaces: Sites of Everyday Life in the Eastern Bloc*. Oxford: Berg Publishers, 2002.

D'Agostino, Salvo. *A History of the Ideas of Theoretical Physics: Essays on the Nineteenth and Twentieth Century Physics*. Dordrecht: Kluwer, 2000.

Darroch, Michael. "Bridging Urban and Media Studies: Jaqueline Tyrwhitt and the Explorations Group, 1951–1957." *Canadian Journal of Communication* 33, no. 2, (2008): 147–169.

Daston, Lorraine, et al., eds. *How Reason Almost Lost Its Mind: The Strange Career of Cold War Rationality*. Chicago: University of Chicago Press, 2013.

Davis II, Charles L. *Building Character: The Racial Politics of Modern Architectural Style*. Pittsburgh: University of Pittsburgh Press, 2019.

Debry, Régis. *Lob der Grenze*. Hamburg: LAIKA, 2016.

Debry, Régis. *Media Manifestos: On the Technological Transmission of Cultural Forms*. London: Verso, 1996.

Deleuze, Gilles. *Difference and Repetition*. Translated by Paul Patton. New York: Columbia University Press, 1994.

Diacon, Todd A. *Stringing Together a Nation: Cândido Mariano Da Silva Rondon and the Construction of a Modern Brazil, 1906–1930*. Durham: Duke University Press, 2004.

Döblin, Alfred. *Berlin Alexanderplatz*. Translated by Michael Hofmann. London: Penguin, 2019.

Douglas, Susan J. *Listening In: Radio and the American Imagination*. Minneapolis: University of Minnesota Press, 1999.

Dussel, Konrad. *Deutsche Rundfunkgeschichte*. Konstanz: UVK Medien, 1999.

Ebeling, Siegfried. *Der Raum als Membran*. Dessau: C. Dünnhaupt Verlag, 1936.

Editorial. "English and Foreign Precedents." *Architectural Review*, no. 429 (August 1932): 44.

Editorial. "The New Tower of London." *Architectural Review*, no. 429 (August 1932): 43.

Edwards, Paul N. *A Vast Machine: Computer Models, Climate Data, and the Politics of Global Warming*. Cambridge, MA: MIT Press, 2010.

Ehrlich, Franz. "Aufnahme- und Studiogebäude des Staatlichen Rundfunkkomitees." *Deutsche Architektur* 9 (1956): 399–409.

Ehrlich, Franz. "Bauhaus und Totaltheater." *Wissenschaftliche Zeitschrift der Hochschule für Architektur und Bauwesen* 29, nos. 5/6 (1983): 424.

Eik, Jan. *Besondere Vorkommnisse: Politische Affären und Attentate in der DDR*. Berlin: Das Neue Berlin, 2006.

Elcott, Noam M. *Artificial Darkness: An Obscure History of Modern Art and Media*. Chicago: University of Chicago Press, 2016.

Enzensberger, Hans Magnus. "Constituents of a Theory of Media." *New Left Review* 64 (1970): 13–36.

Epstein, Margot. "Das Neue Haus des Rundfunks in Berlin." *Deutsche Tonkünstler Zeitung* 4 (1931): 52–53.

Ericson, Steffan, and Kristina Regent, eds. *Media Houses: Architecture, Media, and the Production of Centrality*. New York: Peter Lang, 2010.

Ernst, Wolfgang. *Chronopoetics: The Temporal Being and Operativity of Technological Media*. Translated by Anthony Enns. London: Rowman and Littlefield International, 2016.

Ernst, Wolfgang. *Digital Memory and the Archive*. Minneapolis: University of Minnesota Press, 2013.

Ernst, Wolfgang. *Sonic Time Machines: Explicit Sound, Sirenic Voices, and Implicit Sonicity*. Amsterdam: Amsterdam University Press, 2016.

Evans, Robin. *The Fabrication of Virtue: English Prison Architecture, 1750–1840*. Cambridge: Cambridge University Press, 1982.

Evans, Robin. *The Projective Cast: Architecture and Its Three Geometries*. Cambridge, MA: MIT Press, 1995.

Evans, Robin. "The Rights of Retreat and the Rites of Exclusion: Notes towards the Definition of Wall." *Architectural Design* 41, no. 6 (1971): 335–339.

Evans, Robin. *Translations from Drawing to Building and Other Essays*. London: Architectural Association Publications, 1997.

Falser, Michael. *Deutsche-koloniale Baukulturen: Eine globale Architekturgeschichte in 100 visuellen Primärquellen*. Passau: Dietmar Klinger, 2023.

Fanon, Frantz. *A Dying Colonialism*. Translated by Haakon Chevalier. New York: Grove Press, 1967.

Fischer, Sabine von. *Das akustische Argument: Wissenschaft und Hörerfahrung in der Architektur des 20. Jahrhunderts*. Zurich: gta Verlag, 2019.

Fischer, Sabine von. "A Visual Imprint of Moving Air: Methods, Models, and Media in Architectural Sound Photography, ca. 1930." *Journal of the Society of Architectural Historians* 76, no. 3 (2017), 326–348.

Flier, Peter. "Rundfunk- und Fernsehtechnisches Zentralamt Berlin-Adlershof." *Deutsche Architektur* 11 (1964), 662–664.

Forensic Architecture. *Forensis: The Architecture of Public Truth*. Berlin: Sternberg Press, 2014.

Forster, Kurt W. *Schinkel: A Meander through His Life and Work*. Basel: Birkhäuser, 2018.

Forty, Adrian. *Objects of Desire: Design and Society from Wedgwood to IBM*. New York: Pantheon Books, 1986.

Forty, Adrian. "Wireless Style: Symbolic Design and the English Radio Cabinet, 1928–1933." *Architectural Association Quarterly* 4 (1972): 22–31.

Forty, Adrian. *Words and Buildings: A Vocabulary of Modern Architecture*. New York: Thames and Hudson, 2000.

Foster, Hal. "Neo-Futurism: Architecture and Technology." *AA files* 14 (1987): 28–35.

Foucault, Michel. *The Archaeology of Knowledge*. New York: Pantheon Books, 1972.

Foucault, Michel. *The Birth of the Clinic: An Archaeology of Medical Perception*. Translated by A. M. Sheridan Smith. New York: Vintage Books, 1994.

Foucault, Michel. *Discipline and Punish: The Birth of the Prison*. Translated by Alan Sheridan. New York: Vintage Books, 1995.

Foucault, Michel. *Historia de la Locura en la Época Clásica*. Bogotá: Fondo de la Cultura Económica, 1994.

Foucault, Michel. *Power/Knowledge: Selected Interviews and Other Writings, 1972–1977*. Edited by C. Gordon. New York: Pantheon Books, 1980.

Francé, Raoul Henrich. *Die Pflanze als Erfinder*. Stuttgart: Kosmos-Verlag, 1920.

Francé, Raoul Henrich. *Die technischen Leistungen der Pflanzen*. Leipzig: Verlag von Veit & Comp., 1919.

Franke, Anselm, Nida Ghouse, Paz Guevara, and Antonia Majaca, eds. *Para Politics: Cultural Freedom and the Cold War*. Berlin: Sternberg Press, 2021.

Frankl, Paul. *Die Entwicklungsphasen der neuern Baukunst*. Stuttgart: B. G. Teubner, 1914.

Frankl, Paul. *Principles of Architectural History: The Four Faces of Architectural Style 1420–1900*. Translated by James F. O'Gorman. Cambridge, MA: MIT Press, 1968.

Frick, Heinz. *Mein Gloria Palast: Das Kino vom Kurfürsterdamm*. Munich: Universitas Verlag, 1986.

Fuchs, Margot. *Georg von Arco (1869–1940)*. Berlin: Diepholz, 2004.

Fuller, Matthew. *Media Ecologies: Materialist Energies in Art and Technoculture*. Cambridge, MA: MIT Press, 2005.

Fuller, R. Buckminster. *Nine Chains to the Moon*. New York: Anchor Books, 1971.

Fürst, Artur. *Ätherwellen: Die Eroberung der Erde durch die drahtlose Telegraphie*. Berlin: Comeniu, 1925.

Fürst, Artur. *Im Bannkreis von Nauen. Die Eroberung der Erde Durch Drahtlose Telegraphie*. Stuttgart; Berlin: Deutsche Verlags-Anstalt, 1923.

Galison, Peter. "The Ontology of the Enemy: Norbert Wiener and the Cybernetic Vision." *Critical Inquiry* 21, no. 1 (Autumn 1994), 228–266.

Galison, Peter, Shimon Edelman, and Emily Thompson, eds. *The Architecture of Science*. Cambridge, MA: MIT Press, 1999.

Galison, Peter, and Caroline Jones. "Factory, Laboratory, Studio: Dispersing Sites of Production." In *The Architecture of Science*, edited by Peter Galison, Shimon Edelman and Emily Thompson, 497–534. Cambridge, MA: MIT Press, 1999.

Gay, Peter. *Weimar Culture: The Outsider as Insider*. New York: W. W. Norton, 1968.

Geiser, Reto. *Giedion and America: Repositioning the History of Modern Architecture*. Zurich: gta Verlag, 2018.

Gethmann, Daniel, and Florian Sprenger. *Die Enden des Kabels: Kleine Mediengeschichte der Übertragung*. Berlin: Kadmos 2015.

Gethmann, Daniel, and Markus Stauff. *Politiken der Medien*. Zurich: Diaphanes, 2005.

Giedion, Sigfried. *Building in France Building in Iron Building in Ferroconcrete*. Translated by J. Duncan Berry. Santa Monica: Getty Center Publications, 1995.

Giedion, Sigfried. *The Eternal Present: The Beginnings of Art*. New York: Pantheon Books, 1962.

Giedion, Sigfried. *Mechanization Takes Command: A Contribution to Anonymous History*. Minneapolis: University of Minnesota Press, 2013.

Giedion, Sigfried. *Space, Time and Architecture: The Growth of a New Tradition*. Cambridge, MA: Harvard University Press, 1962.

Gilliam, Bryan. *Music and Performance during the Weimar Republic*. Cambridge, Cambridge University Press, 1994.

Ginzburg, Moisei. *Dwelling: Five Years' Work on the Problem of the Habitation*. London: Fontanka, 2017.

Gleich, Moritz. *Inhabited Machines: Genealogy of an Architectural Concept*. Basel: Birkhäuser, 2023.

Godel, Addison. "When Windows Were Wires: The Projection of Network Invulnerability and the Architecture of the AT&T Long Lines." *Grey Room* 61 (Fall 2015): 34–65.

Goldhagen, Erich. "The Glorious Future—Realities and Chimeras." *Problems of Communism* 9, no. 6 (November 1960): 10–18.

Gombrich, E. H. *Myth and Reality in German War-Time Broadcasts*. London: Athlone Press, 1970.

Goodman, Steve. *Sonic Warfare: Sound, Affect, and the Ecology of Fear*. Cambridge, MA: MIT Press, 2010.

Götze, H., and F. Täumer. "Problem der Planung von Lang- und Mittelwellenrundfunksendern für die afrikanische Rundfunkzone." *Technische Mitteilungen des RFZ* 9, no. 1 (1965): 49–53.

Göx, Andreas, and Hannes Wanderer. *Die Rote Burg: Das Rundfunkgelände an der Nalepastrasse*. Berlin: Peperoni Books, 2007.

Graudenz, Karlheinz, and Hanns-Michael Schnidler. *Die deutschen Kolonien: 100 Jahre Geschichte in Wort, Bild und Karte*. Augsburg: Weltbild Verlag, 1989.

Gumbrecht, Hans Ulrich, and Ludwig K. Pfeiffer, eds. *Materialities of Communication*. Stanford, CA: Stanford University Press, 1994.

Günther, Hanns. *Radiotechnik. Das Reich der elektrischen Wellen*. Suttgart: Kosmos-Verlag, 1921.

Hadamovsky, Eugen. *Dein Rundfunk; das Rundfunkbuch für alle Volksgenossen*. Munich: Zentralverlag der NSDA, 1934.

Hagen, Wolfgang. *Das Radio: Zur Geschichte und Theorie des Hörfunks—Deutschland/USA*. Munich: Wilhelm Fink, 2005.

Halpern, Orit. *Beautiful Data: A History of Vision and Reason since 1945*. Durham: Duke University Press, 2014.

Hambrock, Heike. *Hans und Marlene Poelzig: Bauen im Geist des Barock— Architekturphantasien, Theaterprojekte und moderner Festbau (1916–1926)*. Delmenhorst: Aschenbeck & Holstein, 2005.

Hammar, Peter, and Don Ososke, "The Birth of the German Magnetophon Tape Recorder 1928–1945." *db Magazine* 16, no. 3 (1982): 47–52.

Hanson, O. B. "The Plan and Construction of the National Broadcasting Company Studios." *Architectural Forum* (August 1932): 153–160.

Hanson, O. B. "Planning the NBC Studios for Radio City." *Proceedings of the Institute of Radio Engineers* 20, no. 8 (August 1932): 1296–1309.

Hanson, O. B., and R. M. Morris. "The Design and Construction of Broadcast Studios." *Proceedings of the Institute of Radio Engineers* 19 (January 1931): 17–34.

Harwood, John. *The Interface: IBM and the Transformation of Corporate Design 1945–1976*. Minneapolis: University of Minnesota Press, 2011.

Harwood, John. "Wires." In *Architecture / Machine: Programs, Processes, and Performances*, edited by Moritz Gleich and Laurent Stalder. Zurich: gta Verlag, 2018.

Harwood, John. "Wires, Walls and Wireless: Notes toward an Investigation of Radio Architecture." *Journal of the New Media Caucus* 10, no. 1 (Spring 2014).

Hays, K. Michael, ed. *Architecture Theory since 1968*. Cambridge, MA: MIT Press, 1998.

Hegemann, Werner. "Der Reichskanzlerplatz." *Wasmuths Monatshefte für Baukunst* 14, no. 4 (1930): 436–439.

Helmholtz, Hermann von. *The Sensations of Tones: As a Physiological Basis for the Theory of Music*. Translated by Alexander J. Ellis. London; New York: Longmans, Green, & Co., 1895.

Herbst, Maral. *Demokratie und Maulkorb: Der Deutsche Rundfunk in Berlin zwischen Staatsgründung und Mauerbau*. Berlin: Vistas, 2001.

Hertz, Heinrich. *Electric Waves*. Translated by D. E. Jones. London: Macmillan & Co., 1893.

Hertz, Heinrich. "Über Elektrodynamische Wellen im Luftraume und deren Reflexion." *Wiedemann Annalen* 34 (1888): 610.

Herzog, Gisela. "Der Studioneubau des Funkhauses Berlin." *Technische Mitteilungen des RFZ* 10, no. 4 (1966): 141–148.

Heuss, Theodor. *Hans Poelzig, Bauten und Entwürfe: Das Lebensbild eines deutschen Baumeisters*. Berlin: E. Wasmuth, 1948.

Hill, Alexandra Merley, ed. *Germany in the Loud Twentieth Century: An Introduction*. Oxford: Oxford University Press, 2012.

Hoffmann-Axthelm, Dieter, and Ludovica Scarpa. *Berliner Mauern und Durchbrüche*. Berlin: Museumspädagogischer Dienst, 1987.

Hofmannstahl, Hugo von. *Selected Tales: The Tale of the 672nd Night, A Cavalry Tale, Marshal de Bassompierre's Adventure, Letter from Lord Chandos, and Other Narratives*. Translated by J. M. Q. Davies. London: Angel Books, 2007.

Hood, Raymond. "The Design of Rockefeller City." *Architectural Forum* 56, no. 1 (January 1932): 2–9.

Hui, Alexandra, Julia Kursell, and Myles W. Jackson. "Music, Sound, and the Laboratory from 1750 to 1980." *Osiris* 28 (2013): 1–11.

Huysmans, Joris-Karl. *Against Nature (À rebours)*. New York: Penguin, 2004.

Huyssen, Andreas. *After the Great Divide: Modernism, Mass Culture, Postmodernism*. Bloomington: Indiana University Press, 1986.

Huyssen, Andreas. *Present Past: Urban Palimpsests and the Politics of Memory*. Stanford: Stanford University Press, 2003.

Innis, Harold Adams. *Empire and Communications*. Oxford: Oxford University Press,1950.

Isherwood, Christopher. *Mr. Norris Changes Trains and Goodbye to Berlin.* London: Chatto & Windus, 1985.

Iverson, Jennifer. *Electronic Inspirations: Technologies of the Cold War Musical Avant-Garde.* Oxford: Oxford University Press, 2019.

Jackson, Myles W. *Harmonious Triads: Physicists, Musicians, and Instrument Makers in Nineteenth-Century Germany.* Cambridge, MA: MIT Press, 2008.

Jacobs, Bart. "Maximator: European Signals Intelligence Cooperation, from a Dutch Perspective." *Intelligence and National Security* 35, no. 5 (2020): 659–668.

Jaeggi, Rahel. *On the Critique of Forms of Life.* Cambridge, MA: Harvard University Press, 2018.

Jasper, Sandra. "Sonic Refugia: Nature, Noise Abatement and Landscape Design in West Berlin." *Journal of Architecture* 23, no. 6 (2018): 936–60.

Jebb, Joshua. *Modern Prisons: Their Construction and Ventilation.* London: John Weale, 1857.

Jebb, Joshua. *Observations on the Separate System of Discipline Submitted to the Congress Assembled at Brussels, on the Subject of Prison Reform.* London: William Clowes & Sons, 1847.

Jebb, Joshua. *Report to the Surveyor-General of Prisons on the Construction, Ventilation, and Details of Pentonville Prison.* London: William Clowes & Sons, 1844.

Jelavich, Peter. *Berlin Alexanderplatz: Radio, Film, and the Death of Weimar Culture.* Berkeley: University of California Press, 2006.

John, Richard R. *Network Nation: Inventing American Telecommunications.* Cambridge, MA: Harvard University Press, 2010.

Johnson, James H. *Listening in Paris: A Cultural History.* Berkeley: University of California Press, 1995.

Kaes, Anton, Martin Jay, and Edward Dimendberg, eds. *The Weimar Republic Sourcebook.* Berkeley: University of California Press, 1994.

Kahn, Douglas. *Earth Sound Earth Signal: Energies and Earth Magnitude in the Arts.* Berkeley: University of California Press, 2013.

Kahn, Douglas. *Noise, Water, Meat: A History of Sound in the Arts.* Cambridge, MA: MIT Press, 2001.

Kahn, Douglas, and Gregory Whitehead, eds. *Wireless Imagination: Sound, Radio, and the Avant-Garde.* Cambridge, MA: MIT Press, 1992.

Kapeller, Ludwig. "Die neue Senderäume der Berliner Funk Stunde." *Funk* 38 (1928): 269–270.

Kater, Michael H. *Weimar: From Enlightenment to the Present.* New Haven: Yale University Press, 2014.

Katz, Mark. *Capturing Sound: How Technology Has Changed Music.* Berkeley: University of California Press, 2010.

Kaufmann, Emil. *Von Ledoux bis Le Corbusier: Ursprung Und Entwicklung der Autonomen Architektur.* Stuttgart: G. Hatje, 1985.

Kennedy, Hollyamber. '"Infrastructures of "Legitimate Violence': The Prussian Settlement Commission, Internal Colonization, and the Migrant Remainder." *Grey Room* 76 (2019): 58–97.

Kestenberg, Leo. *Kunst und Technik.* Berlin: Volksverband der Bücherfreunde, Wegweiser-Verlag, 1930.

Ketterer, Ralf. *Funken, Wellen, Radio. Zur Einführung eines technischen Konsumartikels duch die deutsche Rundfunkindustrie 1923–1939.* Berlin: Vistas, 2003.

Kiebs, Lothar. "Perspektiven für eine raumbezogene Rundfunkübertragung." *Technische Mitteilungen aus dem BRF* 1 (March 1960): 2–20.

Kittler, Friedrich A. "The City is a Medium." *New Literary History* 27 (1996): 717–729.

Kittler, Friedrich A. *Gramophone, Film, Typewriter.* Translated by Geoffrey Winthrop-Young. Stanford: Stanford University Press, 1999.

Kittler, Friedrich A. *Optical Media: Berlin Lectures 1999.* Cambridge: Polity Press, 2010.

Kittler, Friedrich A. "Stuttgarter Rede über Architektur." *Zeitschrift für Medien-und Kulturforschung* 1 (2012): 97–104.

Kittler, Friedrich A. *The Truth of the Technological World.* Edited by Hans Ulrich Gumbrecht. Stanford: Stanford University Press, 2013.

Klawitter, Gerd. *100 Jahre Funktechnik in Deutschland: Funksendestellen rund um Berlin.* Berlin: Funk Verlag Bernhard Hein, 2004.

Klein-Arendt, Reinhard. *"Kamina ruft Nauen!": Die Funkstellen in den deutschen Kolonien 1904–1918.* Dessau: Herbst Verlag, 2010.

Knigge, Volkhard, et al., eds. *Franz Ehrlich: Ein Bauhäusler in Widerstand und Konzentrationslager.* Weimar: Stiftung Gedenkstätten Buchenwald und Mittelbau-Dora, 2009.

Knuppe, Helmut. *Ein Riese unter Riesen: Aufstieg und Fall des Deutschlandsenders III in Herzberg (Elster).* Leipzig: Leipziger Verlagsgesellschaft, 2003.

Koch, Hans Jürgen, and Hermann Glaser. *Ganz Ohr: Eine Kulturgeschichte des Radios in Deutschland.* Cologne: Böhlau Verlag, 2005.

Kohlenbach, Bernhard. "Franz Ehrlich: Ein Architekt zwischen Bauhaustradition und DDR-Baudoktrin." *ICOMOS—Hefte des Deutschen Nationalkomitees* 20 (2015): 44–47.

Koolhaas, Rem. *Delirious New York: A Retroactive Manifesto for Manhattan.* New York: Monacelli Press, 1994.

Koolhaas, Rem. *S,M,L,XL: Office for Metropolitan Architecture.* New York: Monacelli Press, 1995.

Kopperschmidt, Josef. *Hitler der Redner.* Munich: Brill I Fink, 2003.

Kracauer, Siegfried. *Escritos sobre Arquitectura.* Edited by Daniel Pisani. Barcelona: Sd ediciones, 2011.

Kracauer, Siegfried. *The Mass Ornament: Weimar Essays.* Edited by Thomas Y. Levin. Cambridge, MA: Harvard University Press, 1995.

Kracauer, Siegfried. "Sendestation: Das Haus." *Die Frankfurter Zeitung,* January 23, 1931.

Kracauer, Siegfried. *Strassen in Berlin und anderswo.* Frankfurt am Main: Suhrkamp 2009.

Ladd, Brian. *The Ghosts of Berlin: Confronting German History in the Urban Landscape.* Chicago: University of Chicago Press, 1997.

Larkey, Edward. "GDR Rock Goes West: Finding a Voice in the West German Market." *German Politics & Society* 23, no. 4 (Winter 2005): 45–68.

Larkin, Brian. *Signal and Noise: Media, Infrastructure, and Urban Culture in Nigeria.* Durham: Duke University Press, 2008.

Latour, Bruno. "Give Me a Laboratory and I Will Move the World." In *Science Observed: Perspectives on the Social Study of Science*, edited by K. Knorr and M. Mulkay, 141–170. London: Sage, 1983.

Latour, Bruno. *We Have Never Been Modern*. Translated by Catherine Porter. Cambridge, MA: Harvard University press, 1993.

Latour, Bruno, and Steve Woolgar. *Laboratory Life: The Construction of Scientific Facts*. Princeton: Princeton University Press, 1986.

Leach, Andrew. *What is Architectural History?* Cambridge: Polity Press, 2010.

Le Corbusier. *Het Electronische Gedicht*. Eindhoven: Philips, 1958.

Le Corbusier. "The 'Modulor.'" *Gravesaner Blätter* 9, no. 3 (1957): 2–5.

Lefebvre, Henri. *Critique of Everyday Life*. London: Verso, 2014.

Lefebvre, Henri. *The Production of Space*. Cambridge: Blackwell, 1991.

Lefebvre, Henri. *Rhythmanalysis: Space, Time and Everyday Life*. London: Bloomsbury, 2004.

Lerner, Ben. *10:04*. New York: Picador Farrar, Straus & Giroux, 2014.

Levin, Thomas Y. "Elements of a Radio Theory: Adorno and the Princeton Radio Research Project." *Musical Quarterly* 78, no. 2 (1994): 316–324.

Lewis, Thomas S. W. *Empire of the Air: The Men Who Made Radio*. New York: Harper Collins, 1991.

Lommers, Suzanne. *Europe on Air. Interwar Projects for Radio Broadcasting*. Amsterdam: Amsterdam University Press, 2012.

Lovell, Stephen. *Russia in the Microphone Age*. Oxford: Oxford University Press, 2015.

Lyotard, Jean-François. "After Six Months of Work... (1984)." In *30 Years after Les Immatériaux: Art, Science, and Theory*, edited by Yuk Hui and Andreas Broeckmann, 27–66. Lüneburg: Meson Press, 2015.

Mallgrave, Harry Francis. "Gustav Klemm and Gottfried Semper: The Meeting of Ethnological and Architectural Theory." *Anthropology and Aesthetics* 9 (Spring, 1985): 68–79.

Marchessault, Janine. *Ecstatic Worlds: Media, Utopias, Ecologies*. Cambridge, MA: MIT Press, 2017.

Martin, Reinhold. "Atrocities. Or, Curtain Wall as Mass Medium." *Perspecta* 32 (2001): 66–75.

Martin, Reinhold. "The Dialectics of the University: His Master's Voice." *Grey Room* 60 (Summer 2015): 82—109.

Martin, Reinhold. *Mediators: Aesthetics, Politics, and the City*. Minneapolis: University of Minnesota Press, 2015.

Martin, Reinhold. *The Organizational Complex: Architecture, Media, and Corporate Space*. Cambridge, MA: MIT Press, 2003.

Martin, Reinhold. *The Urban Apparatus: Media Politics and the City*. Minneapolis: University of Minnesota Press, 2016.

Mattern, Shannon. *Deep Mapping the Media City*. Minneapolis: University of Minnesota Press, 2015.

Mattern, Shannon. "Fluid Text, Total Design: The Woodberry Poetry Room as Idea, Collection, and Place." *Space and Culture* 14, no. 1 (2011): 27–50.

Mauss, Marcel. *Techniques, Technology and Civilisation*. Oxford: Berghahn Books, 2006.

Mayer, Birgit. *Studien zu Hans Poelzig: Bauten und Projekte der 20er Jahre*. Munich: n.p., 1991.

McLeod, Kembrew, and Peter Dicola. *Creative License: The Law and Culture of Digital Sampling*. Durham: Duke University Press, 2011.

McLuhan, Marshall. "Classroom Without Walls." *Explorations: Studies in Culture and Communication* 7 (March 1957): 22–29.

McLuhan, Marshall. *The Mechanical Bride: Folklore of Industrial Man*. London: Routledge, 1951.

McLuhan, Marshall. "Media Fit the Battle of Jericho." *Explorations: Studies in Culture and Communication* 6 (July1956): 15–19.

McLuhan, Marshall. *Understanding Media: The Extensions of Man*. Toronto: McGraw-Hill, 1964.

Medina, Eden. *Cybernetic Revolutionaries: Technology and Politics in Allende's Chile*. Cambridge, MA: MIT Press, 2014.

Mehlman, Jeffrey. *Walter Benjamin for Children: An Essay on His Radio Years*. Chicago: University of Chicago Press, 1993.

Menzel, Werner. "Blick in Deutschlands Rundfunk-Zukunft." *Funk*, no. 18 (1929): 75–76.

Meuser, Philipp. *Schlossplatz Berlin: Vom Staatsratsgebäude zur European School of Management and Technology*. Berlin: DOM, 2016.

Meyer, Erwin. *Electro-Acoustics*. London: G. Bell & Sons, 1939.

Meyer, Erwin. "Ein Neues Automatisches Verfahren der Nachhallmessung." *Zeitschrift für Technische Physik*, no. 7 (1930): 253–259.

Meyer, Erwin. "Technische Grundlagen und Bedingungen in der mechanischen Musik." In *Kunst und Technik*, edited by Leo Kestenberg, 97–116. Berlin: Volksverband der Bücherfreunde, Wegweiser-Verlag, 1930.

Meyer, Erwin, and Lothar Cremer, "Über die Hörsamkeit Holtzausgekleideter Raüme." *Zeitschrift für Technische Physik* 14 (1933): 500–507.

Meyrowitz, Joshua. *No Sense of Place: The Impact of Electronic Media on Social Behavior*. New York: Oxford University Press, 1985.

Moholy-Nagy, László. *Von Material zu Architektur*. Mainz: Florian Kupferberg, 1929.

Moneo, Rafael. "On Typology." *Oppositions*, no. 13 (1978): 22–45.

Moorman, Marissa J. *Powerful Frequencies: Radio, State Power, and Cold War in Angola, 1931–2002*. Athens: Ohio University Press, 2019.

Moreno, Joaquim. *The University is Now on Air: Broadcasting Modern Architecture*. Heijiningen: Jap Sam Books, 2018.

Moten, Fred. *Black and Blur*. Durham: Duke University Press, 2017.

Muller, K. E. "Eine Neuartige Kurzwellenrichtantenne." *Technische Mitteilungen RFZ* 9, no. 1 (1965): 1–6.

Müller, Roswitha. *Bertolt Brecht and the Theory of Media*. Lincoln: University of Nebraska Press, 1989.

Mullin, John T. "Creating the Craft of Tape Recording." *High Fidelity Magazine* 26, no. 4 (1976): 62–67.

Mumford, Lewis. *Technics and Civilization*. Chicago: University of Chicago Press, 1934.

Musil, Robert. *Five Women*. Translated by Eithne Wilkins and Ernst Kaiser. Boston: Nonpareil Books, 1966.

Musil, Robert. *Posthumous Papers of a Living Author*. Translated by Peter Wortsman. London: Archipelago Books, 2006.

Muthesius, Hermann, and Henry van de Velde. "Werkbund Theses and Antitheses" [1914]. In *Programs and Manifestoes on Twentieth-Century Architecture*, ed. Ulrich Conrads. Cambridge, MA: MIT Press, 1971, 28–31.

Nabokov, Vladimir Vladimirovich. *Speak, Memory: An Autobiography*. New York: G. P. Putnam, 1966.

Nägelke, Hans-Dieter. "Poelzig vor Poelzig." In *Hans Poelzig—Architekt Lehrer Künstler*, edited by Wolfgang Pehnt and Matthias Schirren, 84–93. Stuttgart: DVA, 2007.

Neumeyer, Fritz. *The Artless Word: Mies van der Rohe on the Building Art*. Translated by Mark Jarzombek. Cambridge, MA: MIT Press, 1994.

Nietzsche, Friedrich. *Der Wille zur Macht. Eine Auslegung alles Geschehens*. Leipzig: Alfred Kröner Verlag, 1917.

Nordwestdeutschen Rundfunk. *Sonderheft über elektronische Musik*. Cologne: Hausmitteilungen des Nordwestdeutschen Rundfunks, 1954.

Nörig, Hermann, et al. *2000 Jahre Nachrichten aus dem Krieg: Technik—Medien—Kunst*. Göttingen: Vandenhoeck und Ruprecht, 2009.

Ochoa Gautier, Ana María. *Aurality: Listening and Knowledge in Nineteenth-Century Colombia*. Durham: Duke University Press, 2014

Odermann, Heinz. *Wellen mit tausend Klängen: Geschichte rund um den Erdball in Sendungen des Auslandsrundfunks der DDR Radio Berlin International*. Berlin: Vistas Verlag, 2003.

Osayimwese, Itohan, ed. *German Colonialism in Africa and its Legacies: Architecture, Art, Urbanism, and Visual Culture*. London; New York; Dublin: Bloomsbury Visual Arts, 2023.

Osman, Michael. *Modernism's Visible Hand: Architecture and Regulation in America*. Minneapolis: University of Minnesota Press, 2018.

Oswalt, Philipp, ed. *Hannes Meyer neue Bauhauslehre: Von Dessau bis Mexiko*. Basel: Birkhäuser, 2019.

Ouzounian, Gascia. *Stereophonica: Sound and Space in Science, Technology, and the Arts*. Cambridge, MA: MIT Press, 2020.

Painter, Karen, and Thomas Crow, eds. *Reflections on Artists and Composers at Work*. Oxford: Oxford University Press, 2008.

Papapetros, Spyros. *On the Animation of the Inorganic: Art, Architecture, and the Extension of Life*. Chicago: University of Chicago Press, 2012.

Parikka, Jussi. *What is Media Archaeology?* Cambridge: Polity Press, 2012.

Parikka, Jussi, and Erkki Huhtamo, eds. *Media Archaeology: Approaches, Applications, and Implications*. Berkeley: University of California Press, 2011.

Partheil, Gustav. *Die drahtlose Telegraphie und Telephonie: Nach Geschichte, Wesen und Bedeutung für Militär und Marine, Verkehr und Schule*. Berlin: Gerdes & Hödel, 1907.

Patterson, Nick. "The Archives of the Columbia-Princeton Electronic Music Centre." *Notes* 67, no. 3 (March 2011): 483–502.

Pehnt, Wolfgang, and Matthias Schirren, eds. *Hans Poelzig—Architekt Lehrer Künstler*. Stuttgart: DVA, 2007.

Peters, John Durham. *The Marvelous Clouds: Toward a Philosophy of Elemental Media*. Chicago: University of Chicago Press, 2015.

Peters, John Durham. *Speaking into the Air: A History of the Idea of Communication*. Chicago: University of Chicago Press, 1999.

Petruschat, Angelika. *Peter Flier Architekt*. Berlin: form+zweck, 2010.

Pevsner, Nikolaus. *A History of Building Types*. London: Thames & Hudson, 1976.

Picon, Antoine. "Between Technological and Cultural History." *Construction History*, no. 21 (2005–2006): 5–19.

Poelzig, Hans. *Der dramatische Raum: Hans Poelzig, Malerei, Theater, Film*. Krefeld: Museum Haus Lange, 1986.

Poelzig, Hans. *Hans Poelzig 1869–1936: Bauten Entwürfe Gemälde*. Frankfurt: F. Guhl, 1951.

Poelzig, Hans. *Hans Poelzig: Ein grosses Theater und ein kleines Haus*. Berlin: Aedes, 1986.

Poelzig, Hans. *Hans Poelzig: Die Pläne und Zeichnungen aus dem ehemaligen Verkehrs- und Baumuseum in Berlin*. Berlin: W. Ernst & Sohn, 1989.

Poelzig, Hans. *Hans Poelzig: Gesammelte Schriften und Werke*. Edited by Julius Posener. Berlin: Gebr. Mann, 1970.

Poelzig, Hans. "Haus des Rundfunks." *Bauwelt* 22, no. 16 (1931): 1–8.

Pohl, R. W. *Einführung in die Physik*. Vol. 1, *Mechanik und Akustik*. Berlin: Julius Springer, 1930.

Posener, Julius. *Berlin auf dem Wege zu einer neuen Architektur: Das Zeitalter Wilhelms II*. Munich: Prestel, 1979.

Posener, Julius. "Hans Poelzig 'Haus des Rundfunks,' Berlino 1929–1931." *Domus*, no. 702 (1989): 50–55.

Posener, Julius. *Hans Poelzig: Reflections on his Life and Work*. Cambridge, MA: MIT Press, 1992.

Posener, Julius. *Vorlesungen zur Geschichte der Neuen Architektur*. Berlin: ARCH+, 2013.

Potter, Simon J. *Wireless Internationalism & Distant Listening: Britain, Propaganda and the Invention of Global Radio, 1920–1939*. Oxford: Oxford University Press, 2020.

Potter, Simon J., et al., eds. *The Wireless World: Global Histories of International Radio Broadcasting*. Oxford: Oxford University Press, 2022.

Preciado, Paul (Beatriz). "Pornotopia." In *Cold War Hothouses: Inventing Postwar Culture from Cockpit to Playboy*, edited by Beatriz Colomina, Annmarie Brennan, and Jeannie Kim, 216–253. New York: Princeton Architectural Press, 2004.

Prüwer, Tobias. *Welt aus Mauern. Eine Kulturgeschichte*. Berlin: Klaus Wagenbach, 2018.

Pynchon, Thomas. *Gravity's Rainbow*. London: Vintage Books, 2013.

Rancière, Jacques. *The Emancipated Spectator*. Translated by Gregory Elliott. London: Verso, 2011.

Rancière, Jacques. *The Politics of Aesthetics*. Translated by Gabriel Rockhill. London: Bloomsbury, 2004.

Rankin, William. *After the Map: Cartography, Navigation, and the Transformation of Territory in the Twentieth Century*. Chicago: University of Chicago Press, 2016.

Raschkowitsch, Alexander. "Neue Musik- und Hörspielstudios." *Radio und Fernsehen* 5 (1955): 204–207.

Reese, Carol McMichael. "Gehry in Conversation." In *Late Thoughts: Reflections on Artists and Composers at Work*, edited by Karen Painter and Thomas Crow, 97–113. Oxford: Oxford University Press, 2008.

Regener, Sven. *Berlin Blues*. Translated by John Brownjohn. London: Vintage Books, 2004.

Reichmann, Hans-Peter. *Hans Poelzig: Bauten für den Film*. Frankfurt am Main: Deutsches Filmmuseum, 1997.

RIAS Berlin. *RIAS: A Free Voice of the Free World*. Berlin: Brüder Hartmann, 1962.

Ribbe, Wolfgang. *Geschichte Berlins*. Munich: C. H. Beck, 1987.

Richie, Alexandra. *Faust's Metropolis: A History of Berlin*. New York: Carroll & Graf, 1999.

Riezler, Walter. "Die Atonale Welt." *Die Form* 4, no. 2 (1929): 25–27.

Rossi, Aldo. *The Architecture of the City*. Cambridge, MA: MIT Press, 1984.

Rzewski, Frederic. "Notizen zu Zoologischer Garten." *Du atlantis* 26 (1966): 942.

Sabine, Wallace C. *Collected Papers on Acoustics*. Cambridge, MA: Harvard University Press, 1922.

Saunders, Frances Stonor. *The Cultural Cold War: The CIA and the World of Arts and Letters*. New York: New Press, 2000.

Scerri, Antonio. "Centro sperimentale elettroacustico 'Dr. Hermann Scherchen' Gravesano." *Gravesaner Blätter* 1 (July 1955): 5–15.

Sconce, Jeffrey. *Haunted Media: Presence from Telegraphy to Television*. Durham: Duke University Press, 2009.

Semper, Gottfried. *The Four Elements of Architecture and Other Writings*. Translated by Harry F. Mallgrave and Wolfgang Herrmann. Cambridge: Cambridge University Press, 1989.

Semper, Gottfried. *Der Stil in den technischen und tektonischen Künsten, oder Praktische Aesthetik: Ein Handbuch für Techniker, Künstler und Kunstfreunde*. Frankfurt am Main, Verlag für Kunst und Wissenschaft, 1860.

Semper, Gottfried. *Die vier Elemente der Baukunst: Ein Beitrag zur vergleichenden Baukunde*. Braunschweig: Friedrich Vieweg & Sohn, 1851.

Sender Freies Berlin and Dirk-Jens Rennefeld, eds. *Hans Poelzig Haus des Rundfunks*. Berlin: Ars Nicolai, 1994.

Schäche, Wolfgang. Daniel Ralf Schmitz, and David Pessier. *Building Berlin: Developers Who Shaped the Emerging Metropolis*. Berlin: Jovis, 2019.

Schaeffer, Pierre. "Zur Situation der Versuche elektroakustischer Klanggestaltung." *Gravesaner Blätter* 6, no. 2 (1956): 110–116.

Schafer, Murray R. *The Tuning of the World*. New York: Alfred A. Knopf, 1977.

Schäffer, Walter. "Raumakustik für Rundfunkdarbietungen." *Funktechnische Monatshefte* 2 (1932): 51–54.

Schäffner, Wolfgang. "Architecture of the Openings. Windows, Doors, and Switches." In *Architecture of the Medial Spaces*, edited by Joachim Krausse. Dessau: Stiftung Bauhaus, 2007.

Schäffner, Wolfgang. "Elemente architektonischer Medien." *Zeitschrift für Medien-und Kulturforschung* 1 (2010): 137–49.

Schenk, Dietmar. *Die Hochschule für Musik zu Berlin: Preussens Konservatorium zwischen romantischem Klassizismus und neuer Musik, 1869–1932/33*. Stuttgart: Franz Steiner, 2004.

Schenk, Stefan. *Das Siemens-Studio für elektronische Musik Geschichte, Technik und kompositorische Avantgarde um 1960*. Tutzing: Hans Schneider, 2016.

Schlosser, Nicholas J. "The Berlin Radio War: Broadcasting in Cold War Berlin and the Shaping of Political Culture in Divided Germany, 1945–1961." PhD diss., University of Maryland, 2008.

Schlosser, Nicholas J. *Cold War on the Airwaves: The Radio Propaganda War against East Germany*. Urbana: University of Illinois Press, 2015.

Schlosser, Nicholas J. "Creating an 'Atmosphere of Objectivity': Radio in the American Sector, Objectivity and the United States' Propaganda Campaign against the German Democratic Republic, 1945–1961." *German History* 29, no. 4 (2011): 610–627.

Schmarsow, August. *Das Wesen der architektonischen Schöpfung: Antrittsvorlesung, gehalten in der Aula der K. Universität Leipzig am 8. November 1893*. Stuttgart: Hiersemann, 1984.

Schmidt, Thomas. *Werner March: Der Architekt des Olympia-Stadions 1894–1976*. Basel: Birkhäuser, 1992.

Schöbe, Lutz. "Franz Ehrlich. Beitrag zu einer Monographie." Master's thesis, Humboldt University, 1983.

Schöbe, Lutz. "Von der Fläche zum Raum—Franz Ehrlich." *Form+Zweck*, no. 3 (1986): 27–36.

Schreiber, Ernst. "Ein neuartiger elektronischer Klang- und Geräuscherzeugung." *OIRT-Zeitschrift für Rundfunk und Fernsehen*, no. 2 (1965): 35–39.

Schünemann, Georg. "Die Funkversuchsstelle bei der Staatlich Akademischen Hochschule für Musik in Berlin." *Rundfunk Jahrbuch* (1929): 277–293.

Schwartz, Frederic J. "Commodity Signs: Peter Behrens, the AEG, and the Trademark." *Journal of Design History* 9, no. 3 (1996): 153–185.

Schwartz, Frederic J. *The Werkbund: Design Theory and Mass Culture before the First World War*. New Haven: Yale University Press, 1996.

Seewang, Laila. "Ernst Litfass and the Trickle-Down Effect." *AA files*, no. 75 (2017): 45–58.

Siegert, Bernhard. *Cultural Techniques: Grids, Filters, Doors, and Other Articulations of the Real*. Translated by Geoffrey Winthrop-Young. New York: Fordham University Press, 2014.

Siegert, Bernhard, and Geoffrey Winthrop-Young. "Cacography or Communication? Cultural Techniques in German Media Studies." *Grey Room* 29 (Fall 2007): 26–47.

Sievert, Thomas. *Zwischenstadt: Zwischen Ort und Welt, Raum und Zeit, Stadt und Land*. Braunschweig: Vieweg, 1999.
Simondon, Gilbert. *Comunicación e información (Cursos y conferencias)*. Buenos Aires: Editorial Cactus, 2010.

Silverberg, Laura. "East German Music and the Problem of National Identity." In *The Communist Quest for National Legitimacy in Europe, 1918–1989*, edited by Martin Mevius, 123–145. New York: Routledge, 2011.

Simondon, Gilbert. *On the Mode of Existence of Technical Objects*. Translated by Cecile Malaspina and John Rogove. Minneapolis: Univocal, 2017.

Slaby, Adolf. *Entdeckungsfahrten in den elektrischen Ozean*. Berlin: L. Simion, 1911.

Sloterdijk, Peter. *Extrañamiento del mundo*. Translated by Eduardo Gil Bera. Madrid: Pre-Textos, 2008.

Sloterdijk, Peter. *Temblores de aire*. Translated by Nicolás Sánchez Durá. Madrid: Pre-Textos, 2003.

Speer, Albert. *Inside the Third Reich*. New York: Simon & Schuster, 1997.

Spigel, Lynn. *Make Room for TV: Television and the Family Ideal in Postwar America*. Chicago: University of Chicago Press, 1992.

Stalder, Laurent. "Air, Light, and Air-Conditioning." *Grey Room*, no. 40 (Summer 2010): 84–99.

Stalder, Laurent, and Moritz Gleich, eds. *Architecture/Machine: Programs, Processes, and Performances*. Zurich: gta Verlag, 2018.

Steinberg, Marc, and Alexander Zahlten, eds. *Media Theory in Japan*. Durham: Duke University Press, 2017.

Steinke, Gerhard. "Akustisch-musikalische Grenzprobleme in der Rundfunk- und Fernsehtechnik." *Technische Mitteilungen des RFZ* 11 (1967): 149–159.

Steinke, Gerhard. "Experimental Music with the 'Subharchord' Subharmonic Sound Generator." *Journal of the Audio Engineering Society* 14, no. 2 (1966): 140–144.

Stern, Jonathan. *The Audible Past: Cultural Origins of Sound Reproduction*. Durham: Duke University Press, 2003.

Stern, Jonathan, ed. *The Sound Studies Reader*. London: Routledge, 2012.

Stern, Jonathan. "Space within Space: Artificial Reverb and the Detachable Echo." *Grey Room*, no. 60 (Summer 2015): 111–127.

Stern, Robert A. M., Gregory Gilmartin, and Thomas Mellins. *New York 1930: Architecture and Urbanism between the Two World Wars*. New York: Rizzoli, 1987.

Stone, Rob. *Auditions: Architecture and Aurality*. Cambridge, MA: MIT Press, 2015.

Szendy, Peter. *Listen: A History of Our Ears*. New York: Fordham University Press, 2008.

Tafuri, Manfredo. *The Sphere and the Labyrinth: Avant-Garde and Architecture from Piranesi to the 1970s*. Translated by Pellegrino d'Acierno and Robert Connolly. Cambridge, MA: MIT Press, 1990.

Tafuri, Manfredo. *Theories and History of Architecture*. New York: Harper & Row, 1980.

Taylor, Frederick. *The Berlin Wall: A World Divided, 1961–1989*. New York: Harper Perennial, 2006.

Teyssot, Georges. *A Topology of Everyday Constellations*. Cambridge, MA: MIT Press, 2013.

Thompson, Emily. *The Soundscape of Modernity: Architectural Acoustics and the Culture of Listening in America, 1900–1933*. Cambridge, MA: MIT Press, 2002.

Thurn, Hermann. *Die Funkentelegraphie*. Leipzig: Teubner, 1918.

Tkaczyk, Viktoria. "Archival Traces of Applied Research: Language Planning and Psychotechnics in Interwar Germany." *Technology and Culture* 60, no. 2 (April 2019): 64–95.

Tkaczyk, Viktoria. "The Shot Is Fired Unheard: Sigmund Exner and the Physiology of Reverberation." *Grey Room* 60 (Summer 2015): 66–81.

Tkaczyk, Viktoria, Mara Mills, and Alexandra Hui, eds. *Testing Hearing: The Making of Modern Aurality*. Oxford: Oxford University Press, 2020.

Touloumi, Olga. "Sound in Silence: Design and Listening Cultures in the Woodberry Poetry Room." *Journal of Architecture* 23, no. 6 (2018): 1003–1029.

Trüby, Stephan, ed. *Hertzianismus: Elektromagnetismus in Architektur, Design und Kunst*. Paderborn: Wilhelm Fink, 2009.

Turnovsky, Jan. *The Poetics of a Wall Projection*. Translated by Kent Kleinman. London: AA Publications, 2011.

Uexküll, Jakob von. *A Foray into the Worlds of Animals and Humans*. Translated by Joseph D. O'Neil. Minneapolis: University of Minnesota Press, 2010.

Ursprung, Philip. *Historiografía Performativa*. Santiago: Ediciones ARQ, 2018.

Varas, José Miguel. *Escucha Chile: Radio Moscú*. Santiago: LOM Ediciones, 2012.

Venturi, Robert. *Complexity and Contradiction in Architecture*. New York: Museum of Modern Art, 1966.

Venturi, Robert. *Iconography and Electronics upon a Generic Architecture: A View from the Drafting Room*. Cambridge, MA: MIT Press, 1998.

"Verordnung über die Bildung des Staatlichen Rundfunkkomitees." *Gesetzblatt der Deutschen Demokratischen Republik*, no. 112 (August 1952): 733–734.

Vidler, Anthony. *The Architectural Uncanny: Essays in the Modern Unhomely*. Cambridge, MA: MIT Press, 1992.

Vidler, Anthony. *Histories of the Immediate Present: Inventing Architectural Modernism*. Cambridge, MA: MIT Press, 2008.

Vidler, Anthony. "The Idea of Type: The Transformation of the Academic Ideal 1750–1830." *Oppositions* 8 (Spring 1977): 95–115.

Vidler, Anthony. "The Third Typology." *Oppositions* 7 (Winter 1977): 1–4.

Vidler, Anthony. *Warped Space. Art, Architecture, and Anxiety in Modern Culture*. Cambridge, MA: MIT Press, 2000.

Viollet-le-Duc, Eugène-Emmanuel. *The Habitations of Man in All Ages*. Boston: James R. Osgood and Company, 1876.

Virilio, Paul. *Speed and Politics*. Los Angeles: Semiotext(e), 2006.

Vismann, Cornelia. "Cultural Techniques and Sovereignty." *Theory, Culture & Society* 30, no. 6 (2013): 83–93.

Vismann, Cornelia. *Files: Law and Media Technology*. Stanford: Stanford University Press, 2008.

Vitruvius. *The Ten Books on Architecture*. Translated by Morris Hicky Morgan. Cambridge, MA: Harvard University Press, 1914.

Vogl, Joseph. *On Tarrying*. Translated by Helmut Müller-Sievers. London: Seagull Books, 2011.

Vrachliotis, Georg. *The New Technological Condition: Architecture and Design in the Age of Cybernetics*. Basel: Birkhäuser, 2022.

Vronskaya, Alla. *Architecture of Life: Soviet Modernism and the Human Sciences*. Minneapolis: University of Minnesota Press, 2022.

Wagner, Karl Willy. "Das Heinrich-Hertz-Institut für Schwingungsforschung." *Elektrische Nachrichten-Technik* 7, no. 5 (1930): 174–191.

Wagner, Karl Willy. "Das Heinrich-Hertz-Institut für Schwingungsforschung." *Funk* 11 (1930): 43–44.

Wagner, Karl Willy, ed. *Die Wissenschaftliche Grundlagen des Rundfunkempfangs*. Berlin: Julius Springer, 1927.

Wagner, Klaus, and Wolfgang Hoeg. *Stereofonie-Aufnahmetechnik*. Berlin: VEB Technik, 1970.

Wagner, Martin. "Die Reichsausstellungsstadt Berlin." *Das Neue Berlin* 1 (1929): 7–17.

Walker, Enrique, ed. *The Ordinary: Recordings*. New York: Columbia University Press, 2018.

Warburg, Aby. "A Lecture on Serpent Ritual." *Journal of the Warburg Institute* 2, no. 4 (April 1939): 277–292.

Warburg, Aby. *Schlangenritual. Ein Reisebericht. Bilder aus dem Gebiet der Pueblo-Indianer in Nord-Amerika* [1923]. Berlin: Verlag Klaus Wagenbach, 1988.

Warner, Daniel. *Live Wires: A History of Electronic Music.* London: Reaktion Books, 2017.

Warnke, Stephanie. *Stein gegen Stein: Architektur und Medien im geteilten Berlin 1950–1970.* Frankfurt am Main: Campus Verlag, 2009.

Weiss, Allen S., ed. *Experimental Sound and Radio.* Cambridge, MA: MIT Press, 2001.

Weizman, Eyal. *Hollow Land: Israel's Architecture of Occupation.* London: Verso, 2007.

Weizman, Eyal, and Thomas Keenan. *Mengele's Skull: The Advent of a Forensic Aesthetic.* Berlin: Sternberg Press, 2011.

Weizman, Eyal, and Fazal Sheikh. *The Conflict Shoreline: Colonization as Climate Change in the Negev Desert.* New York: Cabinet Books, 2015.

Wente, E. C., and E. H. Bedell. "A Chronographic Method of Measuring Reverberation Time." *JASA* 1 (April 1930): 422–427.

Wigley, Mark. *Buckminster Fuller Inc.: Architecture in the Age of Radio.* Zurich: Lars Müller, 2015.

Williams, D. Carl. "Acoustic Space." *Explorations: Studies in Culture and Communication*, no. 4 (February 1955): 15–20.

Winner, Langdon. "Do Artifacts Have Politics?" *Daedalus* 109, no. 1 (Winter 1980): 121–36.

Winseck, Dwayne R., and Robert M. Pike. *Communication and Empire: Media, Markets, and Globalization, 1860–1930.* Durham: Duke University Press, 2007.

Winter, Teresa. "Thoughts on Autonomy, Sound and Artistic Compromise: Three Experimental Radio Broadcasts of Poetry with Radiophonic Sound by Delia Derbyshire (1965–1966)." *Sound Studies* 3, no. 1 (2017): 3–16.

Wittje, Roland. *The Age of Electroacoustics: Transforming Science and Sound.* Cambridge, MA: MIT Press, 2016.

Wittje, Roland. "The Electrical Imagination: Sound Analogies, Equivalent Circuits, and the Rise of Electroacoustics, 1863–1939." *Osiris* 28 (2013): 40–64.

Wood, James. *History of International Broadcasting.* London: Institution of Electrical Engineers, 2000.

Wyden, Peter. *Wall: The Inside Story of Divided Berlin.* New York: Simon & Schuster, 1989.

Xenakis, Iannis. "Corbusier's 'Electronic Poem'." *Gravesaner Blätter* 9, no. 3 (1957): 43–54.

Xenakis, Iannis. "La crise de la musique sérielle." *Gravesaner Blätter* 1, no. 1 (1955): 2–4.

Yusaf, Shundana. *Broadcasting Buildings: Architecture on the Wireless, 1927–1945.* Cambridge, MA: MIT Press, 2014.

Yusaf, Shundana. "The Porous Shells of Radiophony, or Towards a Theory of Radio Stations." *Journal of Architecture* 23, no. 6 (2018): 1075–1092.

Zaragoza, Luis. *Voces en las Sombras: Una Historia de las Radios Clandestinas.* Madrid: Ediciones Cátedra: 2016.

INDEX

This index is alphabetized in word-by-word order. Italicized page numbers indicate illustrations and photographs. The parenthetical "(cap.)" indicates the location of captions relating to the images, also cited.

COLOPHON

Radio-activities: Architecture and broadcasting in Cold War Berlin
Alfredo Thiermann Riesco

This book is published with the support of the Swiss National Science Foundation for Open Access Publications and the Laboratory for History and Theories of Architecture, Technology, and Media at the École Polytechnique Fédérale in Lausanne.

Designed by Something Fantastic A.D.

This book was set in Sabon and Arial and printed on 70 gsm Holmen, 70 gsm Speedgloss, and 220g Peydur Feinleinen paper

Printed and bound by DZA Druckerei zu Altenburg in Germany

Library of Congress Cataloging-in-Publication Data
Names: Thiermann Riesco, Alfredo, author.
Title: Radio-activities : architecture and broadcasting in Cold War Berlin / Alfredo Thiermann Riesco.
Description: Cambridge, Massachusetts : The MIT Press, [2024] | Includes bibliographical references and index.
Identifiers: LCCN 2023018116 | ISBN 9780262048705 (hardcover)
Subjects: LCSH: Architecture and technology--Germany--Berlin--History--20th century. | Architecture and society--Germany--Berlin--History--20th century. | Architecture--Political aspects--Germany--Berlin--History--20th century. | Radio broadcasting--Germany--Berlin--History--20th century. | Haus des Rundfunks (Berlin, Germany) | Funkhaus Berlin. | Berlin (Germany)--Buildings, structures, etc. | Berlin (Germany)--History--1945-1990.
Classification: LCC NA2543.T43 T49 2024 | DDC 725/.23--dc23/eng/20231114
LC record available at https://lccn.loc.gov/2023018116

10 9 8 7 6 5 4 3 2 1

A

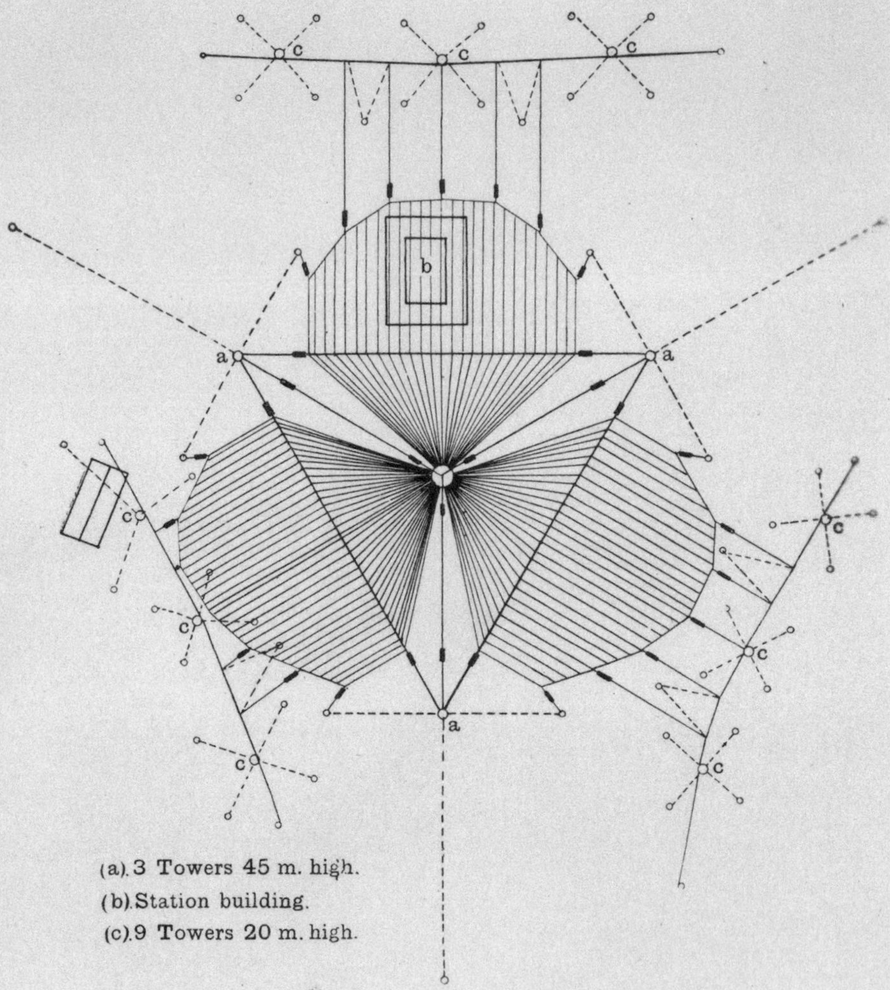

—ANTENNA OF THE WIRELESS STATION MASISEA AND PT. BERMUDEZ, PERU.
Scale 1 : 2,000.

C

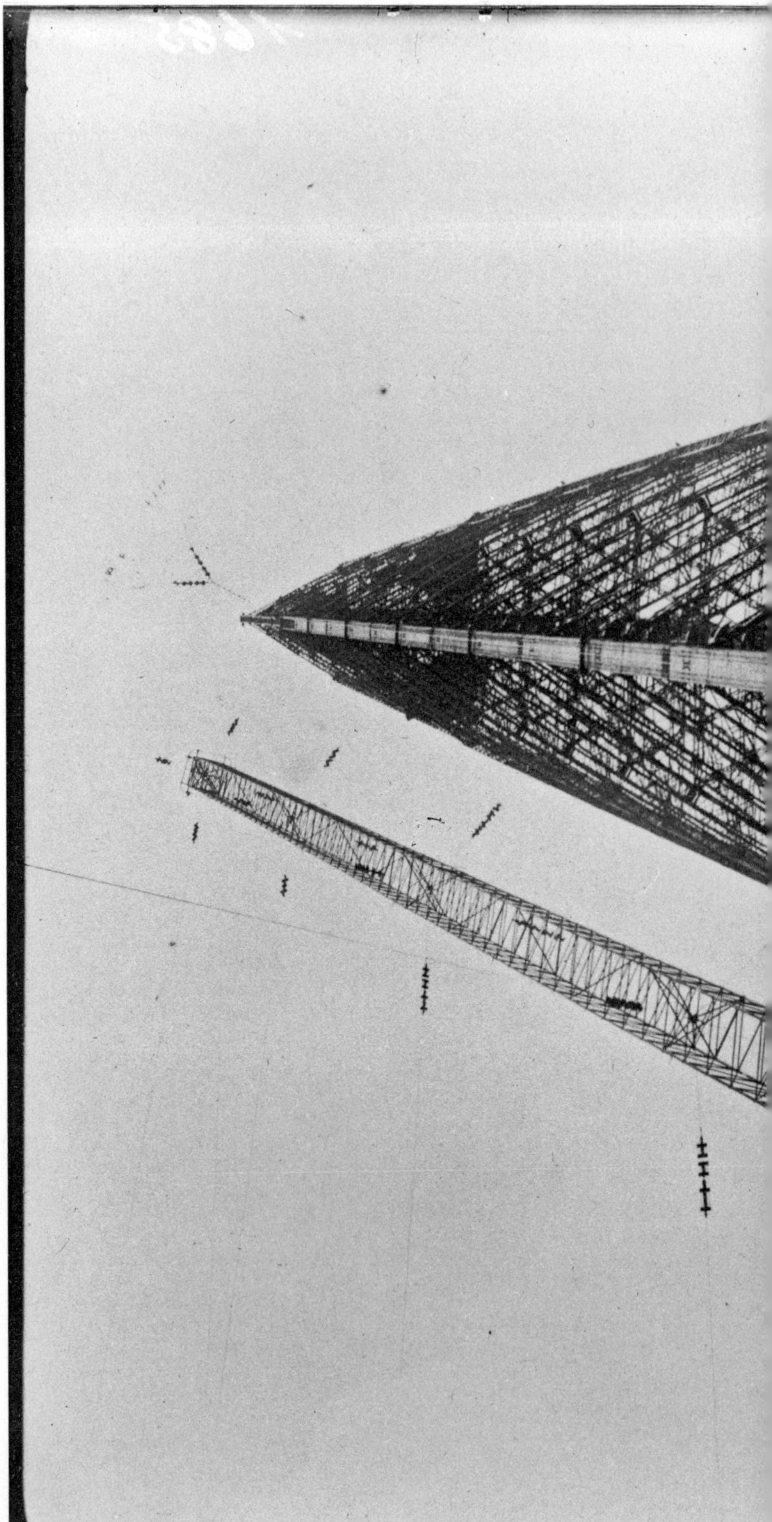